AF564934

CROP BIOTECHNOLOGY

CROP BIOTECHNOLOGY

By

Dr. P.R. Yadav
Lecturer
Department of Zoology
D.A.V. College
Muzaffarnagar (U.P.)

&

Dr. Rajiv Tyagi
Department of Zoology
M.M. College
Modi Nagar (U.P.)

DISCOVERY PUBLISHING HOUSE
NEW DELHI-110002

First Published-2006

ISBN 81-8356-082-2

Published by

DISCOVERY PUBLISHING HOUSE
4831/24, Ansari Road, Prahlad Street,
Darya Ganj, New Delhi-110002 (India)
Phone: 23279245 • Fax: 91-11-23253475
E-mail:dphtemp@indiatimes.com

Printed at
Arora Offset Press
Laxmi Nagar, Delhi–92

Preface

The present title "Crop Biotechnology" has been written for undergraduate, post-graduate, researchers, experts and novices alike to provide a basic orientation in the regulatory process for crop biotechnology. Regulatory and environmental considerations are of the utmost importance in the safety evaluation of either biotherapeutic agents or genetically engineered food crops. The aim of this present title is to provide enough information and examples to give the readers a sound knowledge of plant biotechnology in all its various quises, but particularly those related to genetic manipulation of crop plants. The present title also provides the corner stone of the world's diet, and it is here that the potential benefits and drawbacks of genetic modifications are being most passionately debated. Genetically modified organisms can also be used to increase the nutritional value and taste of crops.

To make the work more comprehensive and informative, the author has consulted many authoritative books, research journals, abstracts, monographs etc. He is grateful to all those great scholars whose work are cited or substantially reproduced.

There can be no claim to originality except in the manner of treatment and much of the information has been obtained from the books and scientific journals available in the different libraries.

The author expresses his thanks to his friends and colleagues whose continue inspirations have initiated him to bring out this book.

The author is painfully aware of the shortcomings, errors and misprints that have crept in, and shall be greateful to receive suggestion for improvement of the next edition from all the readers.

The author expresses his gratitude to Mr. Wasan and staff of M/s Discovery Publishing House for their whole hearted co-operation in the publication of this book.

Author

CONTENTS

1

INTRODUCTION

Plant-derived substances (so called 'secondary products' or 'secondary metabolites') have been used by Man for years and include pharmaceuticals, dyes, colours, perfumes, and insecticides. The use of plants or plant extracts to cure disease or alleviate illness has been known for years, long before the development of modern drugs and antibiotics. Plant-derived compounds still provide a substantial proportion of all the prescribed drugs used today.

Table 1.1. Examples of plant products used in industry

Product	*Plant*	*Application*
Codeine	*Papaver somniferum* (opium poppy)	Analgesic (pain-killer), sedative drug
Quinine	*Cinchona ledgeriana*	Anti-malarial drug, bittering agent
Digoxin	*Digitalis lanata* (foxglove)	Heart drug
Atropine	*Atropa belladona* (deadly nightshade)	Anticholingic drug, suppresses para-sympathetic nerves
Vincristine	*Catharanthus roseus* (periwinkle)	Anti-leukaemic drug
Crocin (saffron)	*Crocus sativus*	Flavour and colour for food
Vanillin	*Vanilla fragrans* (vanilla bean)	Flavour for food
Jasmine oil	*Jasminium sp.*	Perfume

It is clear that to be commercially viable, a new and untried process must aim for a product which is of high value and required in relatively small quantities. As can be seen, all are single compounds of high value used in the pharmaceutical industry. The advantage of the *in vitro* based production of such compounds were seen as:

1. Independence from environmental factors such as climate, pests and diseases;
2. A defined production system, eliminating seasonal variations;
3. Consistent product quality and quantity;
4. Reduced land use;
5. Freedom from political perturbations of supply.

Such a system for the production of, say, a pharmaceutical which was either expensive or potentially in short supply had obvious commercial value. However, this idea was oversold by people who had, perhaps, little idea of the problems inherent in the large-scale cultivation of plant cells *in vitro* and who were over-optimistic. Great encouragement was given to the development of plant cell culture processes when, in 1983, Mitsui Petrochemical Co. announced the commercial production of 'shikonin' in Japan. Shikonin is a traditional medicine in Japan. It is extracted from the root of *Lithospermum erythrorhizon* which is usually grown in Korea. It commanded a price of about \$4500 kg^{-1} and so was a suitable candidate for commercial production. The process was the first commercial one to be developed. Note the two main stages: MG5 medium supports the growth of cells, M9 supports the production of shikonin. Shikonin synthesis is an example in which secondary product synthesis was dissociated from growth. Since the shikonin process was introduced, several potential processes

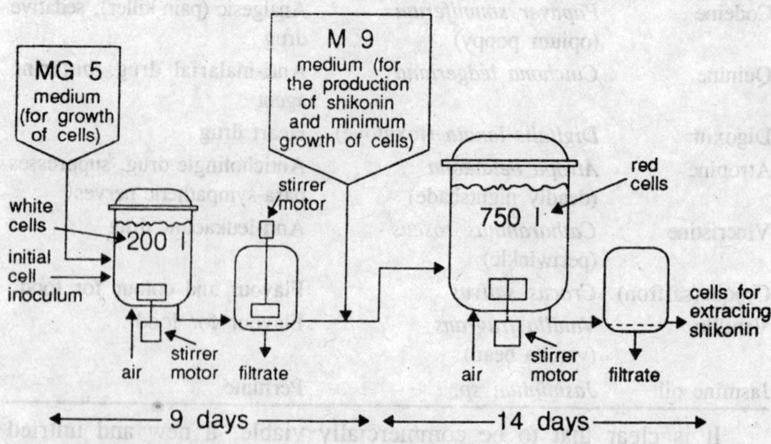

Fig. 1.1. The Mitsui process for the production of shikonin by cultures of Lithospermum erythrorhizon.

or systems have been under development including systems to produce ginseng and berberine.

While the production of pharmaceuticals using plant cell cultures was predominant, other classes of compounds have been suggested and these all fall into the following groups, in terms of production:

1. As a new route for synthesis for established products;
2. As a route for synthesis for products from plants difficult to grow, or in short supply;
3. As a source of novel chemicals;
4. As biotransformation systems.

Much of the research concentrated on established products such as there was already a market and price structure for these. It was hoped that these factors would bring the process to a commercial scale more rapidly. In retrospect, it is perhaps the second area where plant cell culture may see further commercialization. Here we will particularly focus on the problems associated with the production of compounds by plants that are difficult to grow or in short supply.

Products from Plants Difficult to Grow or in Short Supply

One of the first examples of a product from plants in short supply was 'diosgenin' which is normally extracted from the Mexican yam and used in steroid synthesis. The supply failed to keep up with the demand and therefore plant cell culture was investigated as an alternative source. Cultures with concentrations of 2% dry weight have been developed, but this was insufficient for commercialization. More recent examples have been the compounds 'castanospermine' and 'taxol'. The Royal Botanic Gardens, at Kew in England, surveyed plants for extracts which inhibit glycosidases. These enzymes were found to inhibit the replication of the AIDS virus, which has a highly glycosylated coat protein. One of these compounds was castanospermine extracted from Castanospermum australe, the Moreton Bay chestnut. However, the trees are slow-growing and their supply is limited. Examine the structure of this compound but do not try to remember it. Clearly castanospermum is a complex molecule.

Chemical synthesis is not a viable commercial option, so plant cell cultures formed the only realistic alternative source of the compound. The second compound, taxol is extracted from the bark of the Pacific yew, *Taxus brevifolia*. Clinical trials with taxol have shown promising results in the treatment of ovarian, breast and other cancers. The Pacific yew is a slow-growing evergreen and the taxol concentrations are only 0.01-0.03% (dry weight) in the bark. Each

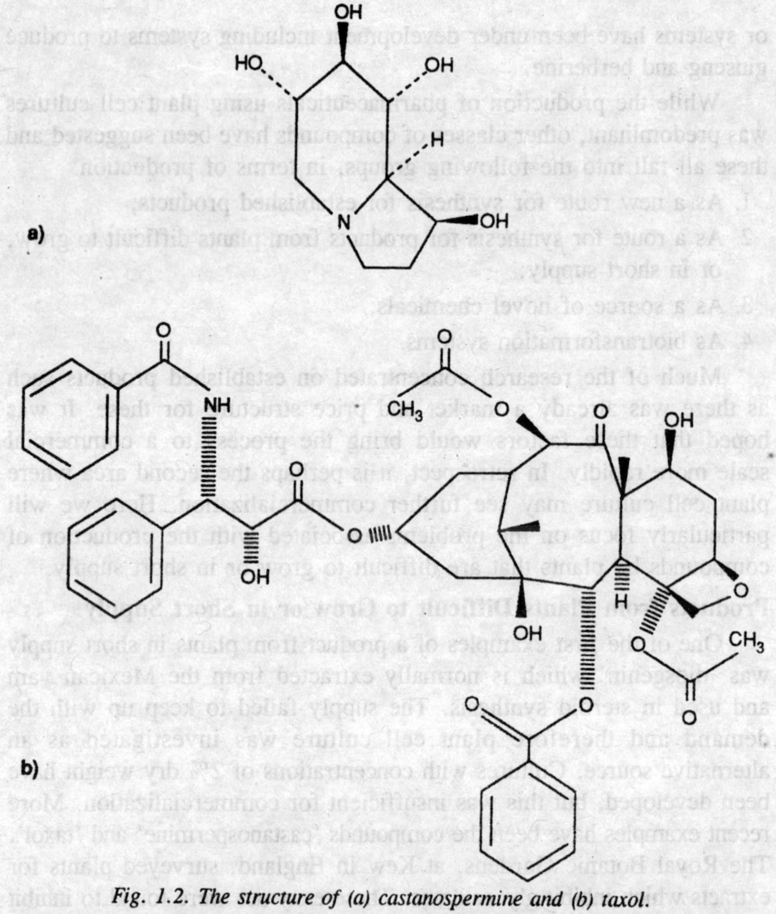

a)

b)

Fig. 1.2. The structure of (a) castanospermine and (b) taxol.

clinical case requires between 2.5-4 g of taxol—equivalent to about 30 kg of bark. As is the case for the castanospermine, chemical synthesis is not possible and the supply of yew bark will not sustain demand. The 'environmental lobby' is also against the felling of large quantities of yew. In this case, plant cell culture is the only alternative supply and the American National Cancer Institute has provided a large grant for its development. Castanospermine, in contrast, has proved to be too toxic for human use but work still continues on similar molecules.

Source of Novel Chemicals

Plant cell cultures have been shown in a number of cases to produce compounds not found in the original plant. The pharmaceutical potential of these compounds is under investigation but no definite results have been obtained yet.

Biotransformations

One use of plant cell culture is in biotransformation, a process where functional groups of organic compounds are modified. Biotransformations using microbial cells have been used for some time in processes such as steroid synthesis. Plant cell cultures, because of their slow growth rate and other difficulties, cannot compete directly with micro-organisms, but plants are diverse and contain a large number of unique pathways and enzymes. Some of these unique reactions have industrial potential. The best known biotransformation using plant cells and that which came closest to a commercial process is the conversion of β-methyl-digitoxin to β-methyl-digoxin using a culture of *Digitalis lanata*. It is the β-methyl-digoxin which is of value as a cardiac drug but both the compounds described above are extracted from *D. lanata* plants.

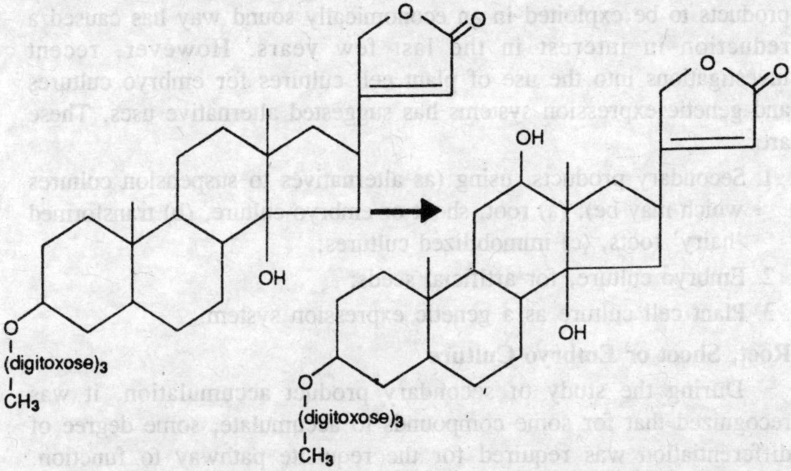

Fig. 1.3. The biotransformation of β-methyldigitoxin to β-methyl-digoxin.

A process was investigated whereby *D. lanata* culture was grown in a 30 l airlift bioreactor before being transferred to a 200 l airlift bioreactor for the biotransformation process. The cells were retained in the 200 l airlift bioreactor for 15-20 days and gave high conversion rates, but failed to compete on economic grounds with a recently developed chemical conversion process. However, the ability of plant cell cultures to transform a wide range of compounds should not be ignored as only a few systems have been investigated in any great detail.

Table 1.2. Some examples of biotransformations carried out by plant cells

Class of compound	*Culture*	*Substence*	*Product*
Terpenoid	*Nicotiana tabacum*	Linalool	8-hydroxy linalool
	Cannabis sativa	Geraniol	Citral, nerol
	Mentha spp	Pulegone	Isomenthone
	Digitalis purpurea	Digitoxin	Purpurea glycoside β
	Panax ginseng	Panaxatriol	Panaxtriol 3-β
Alkaloid	*Papaver somniferum*	(–) *codeinone*	(–) *codeine*
Acid	*Aconitum japonicum*	*Phenylacetate*	β-glucoside

Other Options for Plant Cell Culture

Failure of suspension cultures to accumulate sufficient secondary products to be exploited in an economically sound way has caused a reduction in interest in the last few years. However, recent investigations into the use of plant cell cultures for embryo cultures and genetic expression systems has suggested alternative uses. These are:

1. Secondary products, using (as alternatives to suspension cultures which may be): (a) root, shoot or embryo culture, (b) transformed 'hairy' roots, (c) immobilized cultures;
2. Embryo culture, for artificial seeds;
3. Plant cell culture as a genetic expression system.

Root, Shoot or Embryo Culture

During the study of secondary product accumulation, it was recognized that for some compounds to accumulate, some degree of differentiation was required for the requisite pathway to function. Specific cells had to be developed for their accumulation and storage. Therefore root, shoot and embryo cultures have been used for secondary product accumulation. It has been shown in cultures of *Pelargonium* that the development of glandular hairs was required for essential oil accumulation.

In other cases, accumulation sites have been provided by the addition of lipophilic compounds such as Miglyol (a mixture of triglycerides containing C_{20} and above, fatty acids) or ion-exchange resins to the medium. The addition of Miglyol to cultures of *Thuja*, a plant which had been in culture for some 17 years without product accumulation, initiated the accumulation of monoterpenoids.

Table 1.3. Some examples of secondary products accumulated by root and shoot cultures

Culture	*Secondary Product Accumulated*
Root cultures	
Atropa belladonna	Hyoscyamine, scopolamine (tropane alkaloids)
Catharanthus roseus	Ajmalicine (indole alkaloid)
Papaver bracteatum	Thebane (isoquinoline alkaloid)
Papaver somniferum	Codeine (isoquinoline alkaloid)
Digitalis lanata	Digoxin (steroid)
Shoot cultures	
Rauwolfia serpentina	Ajmalicine (indole alkaloid)
Digitalis purpurea	Digitoxin (isoprenoid)
Pelargonium fragrans	Pinene (monoterpene)

You should have recalled that many secondary products are only synthesized by particular differentiated (specialized) cells. Differentiation is a process by which selected genes are switched on giving rise to cells with particular properties. In some cases the specific genes that are expressed include those which code for enzymes involved in secondary product synthesis.

Transformed or 'Hairy Roots'

Agrobacterium rhizogenes, a Gram-negative bacterium, is the causative agent of 'hairy root' disease; the site of infection produces roots profusely. These induced roots produce specific metabolites, opines, which are secreted into the soil where they are used by the free-living *Agrobacteria*. The induction of hairy roots is due to the transfer of a part of a plasmid (Ri plasmid) to the plant where it is integrated into the genome. These transformed roots can be removed from the plant or explant and cultured in simple PGR-free media. The cultures can be freed from the bacteria by antibiotic treatment and are characterized by:

1. High degrees of genetic and biochemical stability;
2. Rapid growth;
3. Accumulation of root-derived secondary products.

Immobilized Plant Cells

Cells and small aggregates of cells can be grown in bioreactors either as 'free cells' in suspension, or they can be 'immobilized'—i.e. fixed in, or to, some inert support. Immobilization has been applied to microbial and animal cells for many years and relatively recently this technique has also been applied to plant cells. Three main methods are used to immobilize cells; adsorption onto a support, covalent attachment onto a support and entrapment in a range of polymers or within semi-permeable membranes.

The processes of passive entrapment in polyurethane foam or in alginate gels have been used most widely because these methods are gentle and cause least stress to the cells and small aggregates. Immobilization has a number of advantages in comparison with suspension cultures:

1. It is possible to use higher cell densities;
2. Cells remain active for longer periods;
3. Cells are protected from shear stresses;
4. Separation of cells from medium allows for a continuous process to be developed;
5. Cell products or inhibitors can be removed easily.

We might also include the fact that the increased cell-to-cell contact might enhance secondary product production.

The disadvantages are:

1. The system must operate under zero or low growth conditions;
2. Product release from cells can present operational problems.

In general, secondary products are retained in the vacuole of plant cells. Changes in medium pH, ionic concentration, or the addition of permeabilizing agents such as chloroform or DMSO (dimethyl sulphoxide), have all been used to release cell secondary products, but product release is often followed by loss of cell viability. Physical methods such as electroporation and ultrasound have also been tried with mixed success.

There are insufficient examples of comparisons with suspension culture to say which is the best method, but there has been industrial interest, for example, from a major flavour company in the UK for the production of capsaicin from polyurethane foam-immobilized *Capsicum frutescens* (green pepper). The critical feature of successful immobilization is the release of the product into the medium without harming the cells. Many of the conditions which favour secondary product formation involve some form of physiological stress and it may well be that this also occurs as a result of immobilization.

Embryo Cultures

Somatic embryo formation in liquid media can not only be used for secondary product accumulation but may also be a part of a micropropagation system. Micropropagation is expensive because of the extensive handling and therefore the product cost. It should help towards the development of the 'artificial seed' when the embryo would be encapsulated, which probably would also allow for fluid drilling of embryos.

Genetic Expression Systems

Secondary product expression systems, using the plasmids from *A. tumifaciens* as plant cloning vectors, have been developed. The gene for an antibody has been expressed in tobacco plants and an antibiotic resistance gene in suspension cells. A plant suspension culture of *Nicotiana tabacum* has been genetically engineered to produce the enzyme chloramphenicol acetyltransferase (CAT). We will not go into details of the genetic engineering of plant cells here, but we will examine several advantages, that might be derived from using plant cell suspensions for production of foreign proteins. These advantages are:

1. The protein products from genetically engineered plant cells may be more function and potent as pharmaceuticals than the products from the same genes transferred to microbial cells. This is because plant cells are capable of carrying out post-translational modification of proteins (e.g. glycosylation). Prokaryotes do not carry out such processes;
2. In comparison plant culture media is inexpensive compared to media used for the cultivation of mammalian cells. Thus commercially it may be more attractive to use plant cells rather than animal cells to produce particular proteins;
3. Suspension of plant cells can reach a high density (up to 50 g dry weight l^{-1}). This is much higher than can be achieved with many animal and microbial cell systems.

Thus, plant cell culture may offer an alternative expression system for genetically engineered proteins.

Main Requirements of Bioreactors for the Growth of Plant Cells or Tissue

In this section we will examine the use of bioreactors for the large-scale cultivation of plant cells. We will focus on the issues which arise specifically from cultivating plant cells and not on the principles of process technology. We will remind you of the special requirements of plant cell suspension cultures before exploring the requirements of the more organized cultures such as hairy roots. Plant cell suspensions have specific properties which affect their growth in bioreactors. Plant cells are, in general, large (length up to 200 μm). They have a relatively rigid cell wall and a culture will contain a wide range of cell shapes and sizes.

Unlike many micro-organisms, plant cells in suspension culture occur as groups or aggregates. The aggregate structure has also been implicated in secondary product accumulation, as it may act as a diffusional barrier and so form micro-environments within the aggregate, which nevertheless stimulate secondary product synthesis. These aggregates can be up to 2 mm in diameter containing many thousands of cells. Whether these aggregates arise due to non-separation of cells after cell division or by cell aggregation is unknown, but they are loose structures whose average size and size distribution can alter with culture conditions. A consequence of these characteristics is that the culture will settle out rapidly once bioreactor mixing is stopped and, therefore, sampling will be difficult.

The large size and rigid cell wall have long been regarded as the reasons for plant cells' apparent sensitivity to shear. Shear is the force that arises from the movement of a fluid layer relative to that of an adjacent layer. Such velocity profiles occur in the vicinity of a bioreactor impeller and cells exposed to such velocity profiles can suffer damage. Many of the failed attempts to grow plant cells in bioreactors in the early days were put down to the effects of shear stress. The growth rate of plant cells is slow, with doubling times of 2-3 days or more, as a consequence their oxygen requirements are also low. The effect on bioreactor use is that the culture 'runs' will be of 2-3 weeks in duration: this will mean that much attention will need to be paid to maintaining asepsis, as any microbial contaminant would soon overtake the growth of the plant cells.

It will also mean that the number of runs per year may be limited to as few as one per month, which will give a low overall productivity for any single bioreactor. This feature, along with the large volume of inoculum required, (below a certain concentration usually about 5-10%, no growth will occur) means that the scale-up steps will have to be smaller (1→100 fold) compared to those needed with micro-organisms (1→100) and therefore the time taken to reach production volume will be longer. Coupled with the fact that, for many plant species, accumulation of the product only starts after cell growth had ceased, explaining why often very long bioreactor runs are needed. One minor advantage of the slow growth is the low aeration requirement (1-10 mmol O_2 l^{-1} h^{-1})- 1-10% compared to that required by the culture of micro-organisms.

The supply of oxygen is probably not, therefore, the most critical feature of bioreactor culture. This allows the use or development of bioreactors with somewhat lower aeration potential (K_La) in which we can therefore concentrate on other requirements such as mixing. In fact, high aeration rates have been shown to reduce growth, probably by stripping out carbon dioxide and other 'beneficial' gases or volatiles from the culture. The main disadvantages of growing plant cell suspension culture relate to this slow growth rate, (fewer runs per year therefore less yield, greater risk of microbial contamination) and low yield. However the slow growth rate is reflected in a low rate of oxygen consumption. Thus oxygen supply is not often critical. (Note that with many microbial systems, oxygen supply is often rate limiting).

Development of an Industrial Process

A commercial process requires a number of criteria to be met in order to produce a consistent product at an economic price. These are:

1. Rapid growth of cells in large volumes;
2. High yield of product;
3. Stable yield with time and/ or scale-up;
4. Downstream extraction and purification of product with minimum losses.

The first two combine to constitute 'productivity', measured as product per unit volume per day or hour ($g\ l^{-1}\ d^{-1}$). Productivity can also be affected by the process format, for example whether it is 'continuous', or 'batch'. As plant cells grow slowly, in order to increase productivity, both the yields and biomass levels need to be high. The yield of some secondary products is high but often those of commercial interest remain stubbornly low. Higher 'biomass loadings'—i.e. increasing the biomass per volume of bioreactor is an alternative strategy to increase productivity.

The water content of plant cells is around 80-90% so that 1 litre of cells would represent 100-200 $g\ l^{-1}$ dry weight. As plant cells are uneven in shape and will need to be mixed, the practical maximum level of biomass should be between 40-60 $g\ l^{-1}$ dry weight. Plant cell cultures are normally grown using 2 or 3% w/v (20-30 $g\ l^{-1}$) sucrose which yields biomass at 10-15 $g\ l^{-1}$ dry weight of cells. Increasing the sugar content of the medium up to 10% (100 $g\ l^{-1}$) has given biomass yields of up to 35 $g\ l^{-1}$ dry weights. However above 8%, growth is often reduced and at these high concentrations of sucrose it takes longer to reach maximum biomass, so reducing overall productivity.

Two strategies have been used to overcome these problems. The first is termed 'fed-batch', where sucrose is added in aliquots as growth proceeds, rather than all at once at the beginning of growth. The results have been somewhat disappointing; the growth rate and final biomass were not improved. The second method was to continuously replace the medium while retaining the cells, a sort of perfusion culture. This gave values of 78 g l^{-1} dry weight for cultures, but as it has only been tried using a limited number of cultures, its full application is unknown.

Table 1.4. Example of the high yields of secondary products achieved with plant cell cultures

Product	*Species*	*Yield (%dry weight)*
Rosmarinic acid	*Coleus blumei*	21.4
Anthraquinone	*Morinda citrifolia*	18.0
Benzylisoquinolines	*Coptis japonica*	15.0
Acetoside	*Syringa vulgaris*	15.0
Shikonin	*Lithospermum erythrorhizon*	12.4
Berberine alkaloids	*Berberis wilsonae*	10.0
Shikimic acid	*Galium mollugo*	10.0
Diosgenin	*Dioscorea deltoidea*	7.8
Nicotine	*Nicotiana tabacum*	5.0
Serpentine	*Catharanthus roseus*	2.2
Ajmalicine	*Catharanthus roseus*	1.8

Process Format

Another influence on productivity is the 'process format', or how the cells are grown in the bioreactors. The normal culture method is 'batch', where the bioreactor is inoculated with cells and these are harvested at the end of the culture period. The bioreactor is then cleaned, sterilized and filled with sterile medium and the process is repeated. The time taken for cleaning etc. can be considerable but can be eliminated by the use of continuous culture. 'Continuous' culture is achieved by adding a constant supply of fresh medium to the bioreactor and the simultaneous removal of an equivalent volume of culture. Continuous culture of plant cells is difficult because of the presence of aggregates which give a non-homogeneous culture and make sampling difficult. The adhesion of the cells to the walls of the bioreactor also gives problems. The slow growth of the culture means that the supply of fresh medium has to be at a slow rate which, in practice, is difficult. One method of avoiding continuous culture is 'draw-fill' or 'semi-continuous' culture where, at the end of the culture period, 90% of the culture is removed and the remaining 10% topped up with fresh, sterile medium. This avoids cleaning and sterilizing the vessel between runs.

Batch, continuous or draw-fill cultures are suitable for production of secondary products that are 'growth related'. However, secondary products are generally non-growth related and accumulate only after growth has ceased. To improve or stimulate secondary product yield, the medium or culture conditions are often changed, thus continuous culture or draw-fill are not suitable. The non-growth related accumulation requires a two-stage process, like that used for shikonin accumulation. Such a two-stage process can be organized in a number of ways using batch, draw-fill or continuous culture for the first stage with a possibility of any of the three for the second stage.

The Mitsui process is clearly a two-stage system. The cells are grown in the 200 l vessel and then transferred to the 750 l vessel for the production of shikonin. There is no direct evidence that it is operated as a batch process. The clue to this is at the bottom of the Figure where is shows that the growth stage takes 9 days and the production phase takes 14 days. It could, however, on the information given be operating as a draw-fill system.

Problems of High Biomass Levels

The high biomass loading ($\sim$40-60 g l^{-1}) proposed means that the resulting culture may be viscous and mixing in the bioreactor will be a problem. The viscosity of a suspension culture is measured under conditions of laminar flow using a viscometer consisting of a cone and plate, or a cup and bob. These methods however present difficulties because in a cone and plate viscometer the particles (cells) get trapped; while in the cup and bob viscometer they settle out. Thus it is difficult to measure the true viscosity of plant cell suspensions (a similar problem has also been found with pelleted fungal cultures). However, the viscosity of the fungal cultures has been measured by using a turbine or anchor impeller in place of the bob. This will give turbulent conditions which will stop phase separation but will, of course, not yield absolute viscosity values because of non-laminar flow conditions.

Using this method, plant cell cultures of up to 40 g l^{-1} had apparent viscosities of only up to 20 mPa, which is low compared with $\sim$1000 mPa found for some fungal cultures. Therefore, viscosity may not in the end be a real problem at high biomass loadings. The other problem associated with high biomass loading, that of mixing, does present

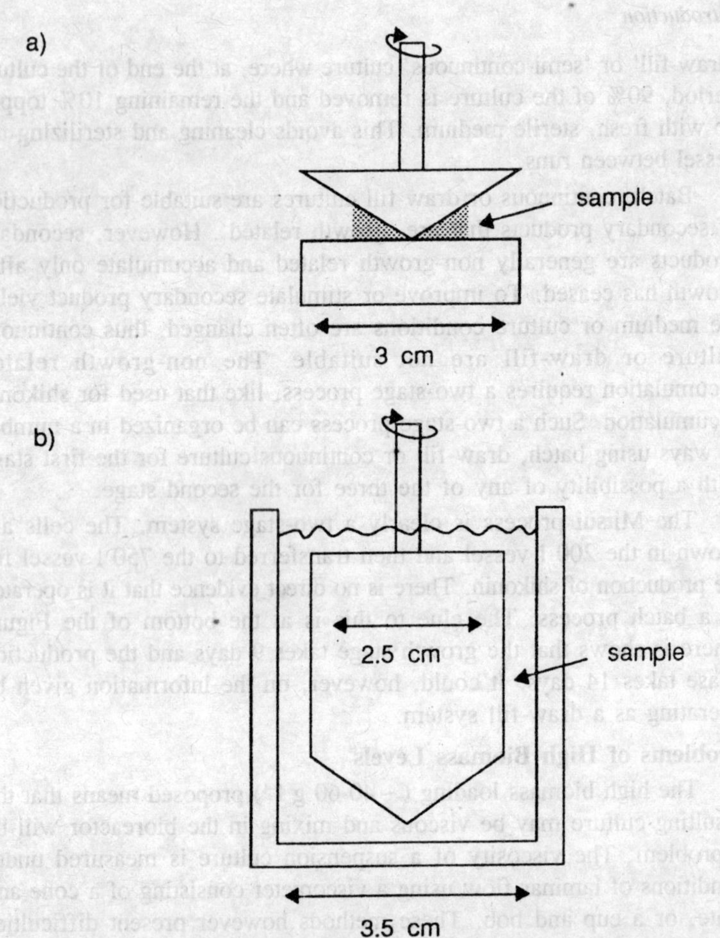

Fig. 1.4. Two types of viscometer: (a) cone and plate; (b) cup and bob.

problems. The solution, for microbial cultures, would be to increase the impeller speed and power input, but plant cells are presumed to be sensitive to shear and this would increase shear greatly. It has emerged recently that some plant cells are considerably more tolerant to shear stress than was at first thought. Exposure of a number of plant cell cultures to 1000 rpm using a turbine impeller for up to 5 hours did not kill the cells.

A few cultures have also been shown to grow (long-term exposure) at 100 rpm. The shear-tolerance is a dynamic characteristic which appears to be associated with fine cultures (low mean aggregate size) with good growth rates. It means that the traditional stirred-tank bioreactor can be considered for the culture at high biomass loading using higher impeller speeds than had previously been considered.

Scale-up

The development of a commercial process usually proceeds from the laboratory scale via a pilot plant to a full scale plant. The process is generally optimized at the laboratory scale and the experimental conditions have had to be translated to the larger scale. While it is possible to maintain conditions such as temperature on scale-up, other physical factors are dependent on size and these will change with the process scale. It is impossible to scale-up for more that one physically based criterion if the geometric similarity of the bioreactor is to be maintained. This is due to the volume: diameter ratio changing on scale-up. In addition, the steps in scale-up for plant cell cultures are much smaller than for micro-organisms. In two cases this has been successfully achieved. A *Catharanthus roseus* culture has been grown up to 5000 l in volume and 75000 l has been achieved for culture of *Echinacea purpurea*.

Bioreactor Design

Bearing in mind the need for mixing rather than aeration for plant cell suspensions and the possible shear stress sensitivity of plant cell culture, a number of different impeller and bioreactor designs has been proposed.

Impeller Designs

Stirred-tank bioreactors equipped with a normal Rushton turbine were used for the early work with plant cells, especially the growth of tobacco cells. The main problem with this type of impeller is that it generates high shear which damage plant cells. To reduce shear stress, these were operated at low speeds—between 20-100 rpm. Since then, a number of impeller designs has been tried including anchor and angled turbines and spiral designs. The angled turbine was found to be particularly effective with *Panax ginseng* and the spiral with *Coleus* cultures. Angled impellers have also been shown to be more effective at higher impeller speeds (1000 rpm) with *Catharanthus roseus* cultures. A cell-lift impeller, originally designed for animal cells, has been shown to work well with *Glycine* max and *Parias elliotti* cultures. A large, flat-bladed impeller has also been proposed as a solution to mixing problems. All these designs have been proposed as solutions to good mixing in low shear stress conditions, but no single design has proved to be universally applicable.

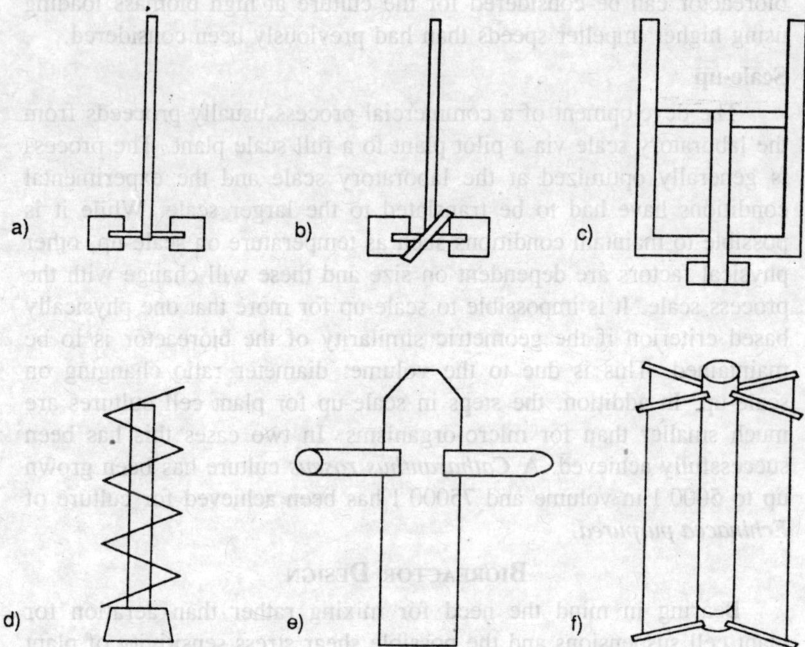

Fig. 1.5. Various forms of bioreactor impellers: (a) Rushton turbine; (b) Angled vanc; (c) Anchor; (d) Spiral; (e) Cell-lift; (f) Large vane.

Alternative Bioreactor Designs

As stated previously, stirred-tank bioreactors were used initially for plant cell cultures, but with the introduction of the airlift bioreactor in the 1970s the design was adopted for cultured plant cells. The airlift bioreactor was first patented by Lefrancois in 1955 as an air-driven bioreactor which was therefore easier to maintain asepsis and cheaper to run. The airlift also produced lower shear stress conditions and was therefore ideal for the cultivation of both animal and plant cells. It was also shown that the airlift bioreactor gave a better yield of anthraquinones when *Morinda citrifolia* were cultivated than could be obtained with other reactor designs.

Stirred-tank bioreactors have continued to be used and the higher shear stress tolerance of some cultures may see their wider adoption. Stirred-tank bioreactors have featured in the two largest scale-up schemes reported for plant cells; 5000 l for *C. roseus* and 75,000 l for *Echinacea purpurea*. Alternative designs to the stirred-tank or airlift bioreactor have also been reported. *C. roseus* and *L. erythrorhizon* have been grown up to volumes of 1000 l in a rotating drum bioreactor.

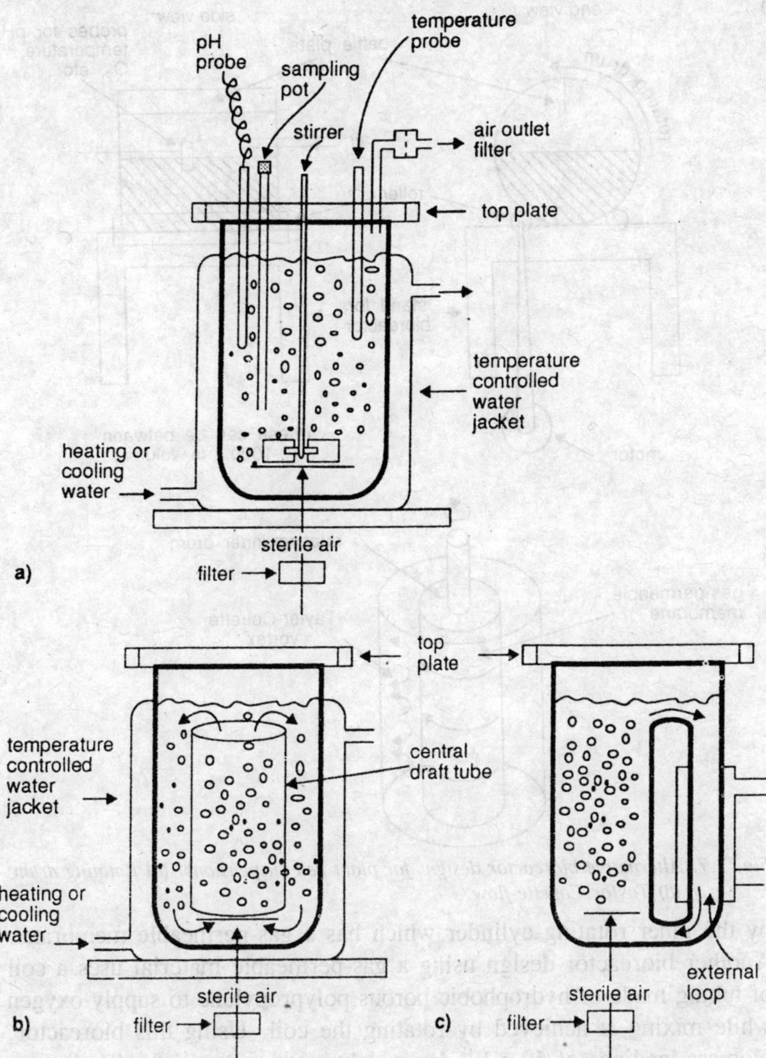

Fig. 1.6. Various bioreactor designs: (a) Stirred-tank (STR); (b) Airlift with internal draft tube; (c) Airlift external loop.

Mixing and aeration are achieved by slowly rotating the drum which is fitted with at least one baffle. This design is said to provide good mixing at low shear levels although no data have been published. A novel approach to low shear mixing has been to use the vortices formed between two cylinders when one is rotated. The flow, known as 'Taylor-Couette flow', is sufficient to mix the culture while oxygen is supplied

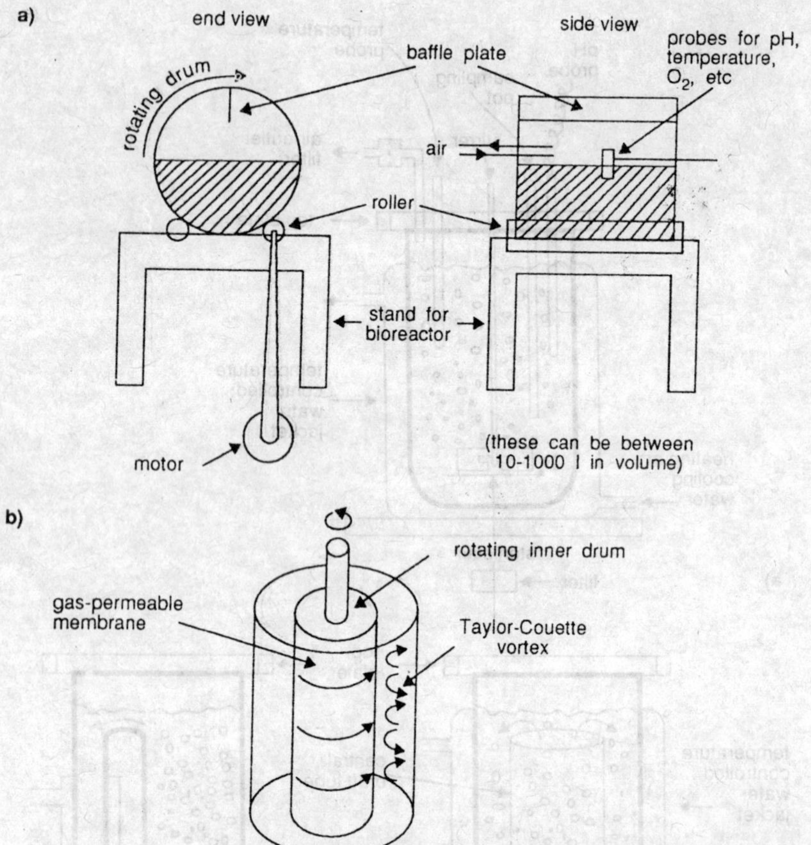

Fig. 1.7. Alternative bioreactor designs for plant cell suspensions: (a) Rotating drum; (b) Taylor Couette flow.

by the inner rotating cylinder which has a gas-permeable membrane. Another bioreactor design using a gas-permeable material uses a coil of tubing made of hydrophobic porous polypropylene to supply oxygen while mixing is achieved by rotating the coil. Using this bioreactor, biomass loadings of 50 g l^{-1} dry weight were reported for *Thalictrum rugosum*.

Bioreactors Designs for Non-suspension Cell Cultures

Immobilized Cells

Depending on the method used to immobilize plant cells, a number of bioreactor types can be used. Immobilization in alginate beads allows their use in a fixed or fluidized bed bioreactor as well as in standard airlift and stirred-tank bioreactors. Polyurethane foam can be used as small cubes in a fixed or fluidized bed bioreactor or it can be used as sheets to replace the draft tube in an airlift bioreactor. Recently, plant cells have been immobilized on the surface of a glass fibre mat or short-fibre polyester material to form a biofilm or surface of immobilized plant cells. These materials have been used in either stirred-tank or airlift bioreactors.

The production of secondary products has been reported to be as good as that found in shake flasks for *C. roseus*. Retaining plant cells behind a semi-permeable membrane has also been practised using either hollow-fibre cartridges or flat-plate membranes. In flat-plate bioreactors, medium is circulated below the membrane and in one case a scraper was provided to maintain an even bed thickness. Here again, it is difficult to compare the results obtained with those from shake flasks as one cell line can differ greatly from another in its response to immobilization. By now you should have been able to make quite a list. The main advantages are that:

1. It is possible to use higher cell densities and thus in principle achieve high productivity per unit volume of bioreactor;
2. Cells may be kept for longer periods and are protected against stress;
3. Cells are easy to separate from the medium which aids product recovery.

Organized Cultures

It is becoming increasing clear that for some secondary products to be formed, a certain degree of differentiation into roots, shoots or embryos is required. The cultivation of roots, shoots or embryos in bioreactors will required gentle mixing as these will certainly be sensitive to shear. Therefore, many of the designs proposed for low shear mixing may be suitable. In addition, other bioreactors have been used for shoot culture. A stirred-tank with large flat-bladed impellers and baffles running at a slow speed 10 rpm) has been used to grow mini potato tubers. The airlift design is clearly suitable and both rectangular and circular designs have been used to grow plantlets and embryos of *Artemesia annua*.

A rotating drum bioreactor has been used to grow shoot cultures and a spin-filter bioreactor has been used for carrot embryos. In the spin-filter bioreactor, a rotating filter mesh mixes the culture and allows the removal of spent medium while retaining the cells. It is effectively an internal biomass filter device and allows us to operate

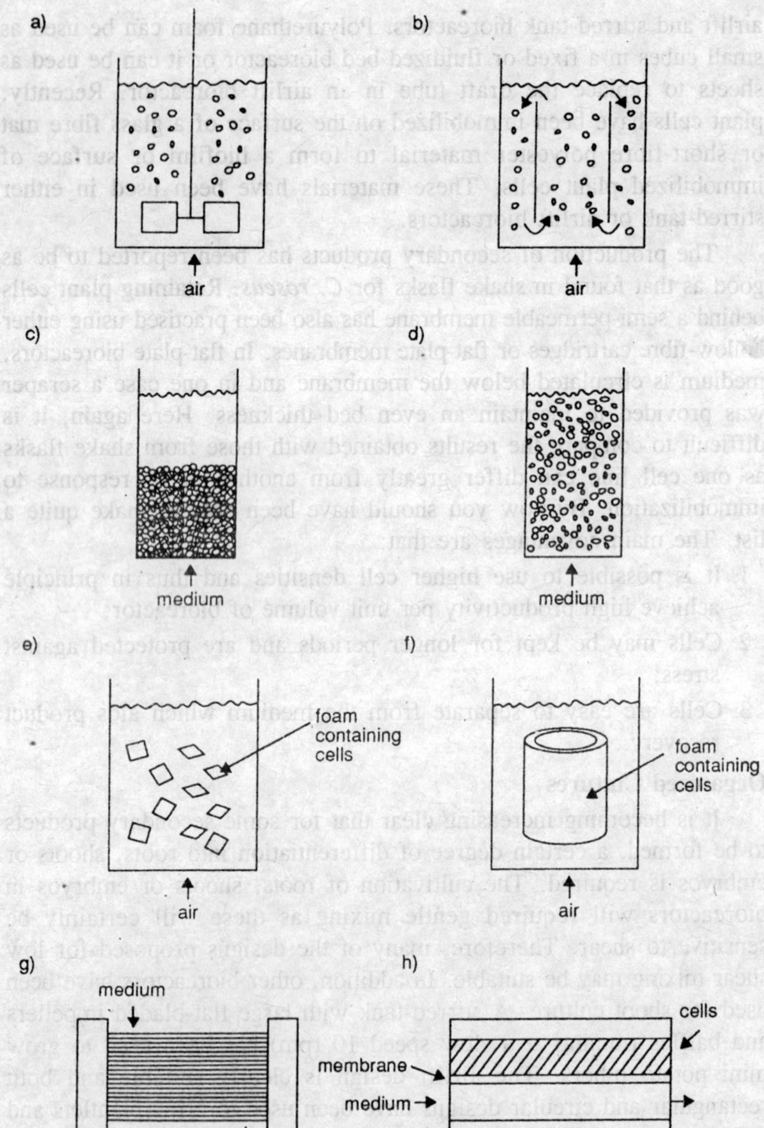

Fig. 1.8. Outline schemes to show a number of bioreactor designs for use with immobilized plant cells. (a) Stirred-tank, cells in beads or foam; (b) Airlift with the same cells; (c) Packed bed; (d) Fluidized bed; (e) Bubble column with the cells in beads or foam; (f) Airlift where foam used to replace the draft tube; (g) Hollow fibre unit; (h) Flat bed reactor.

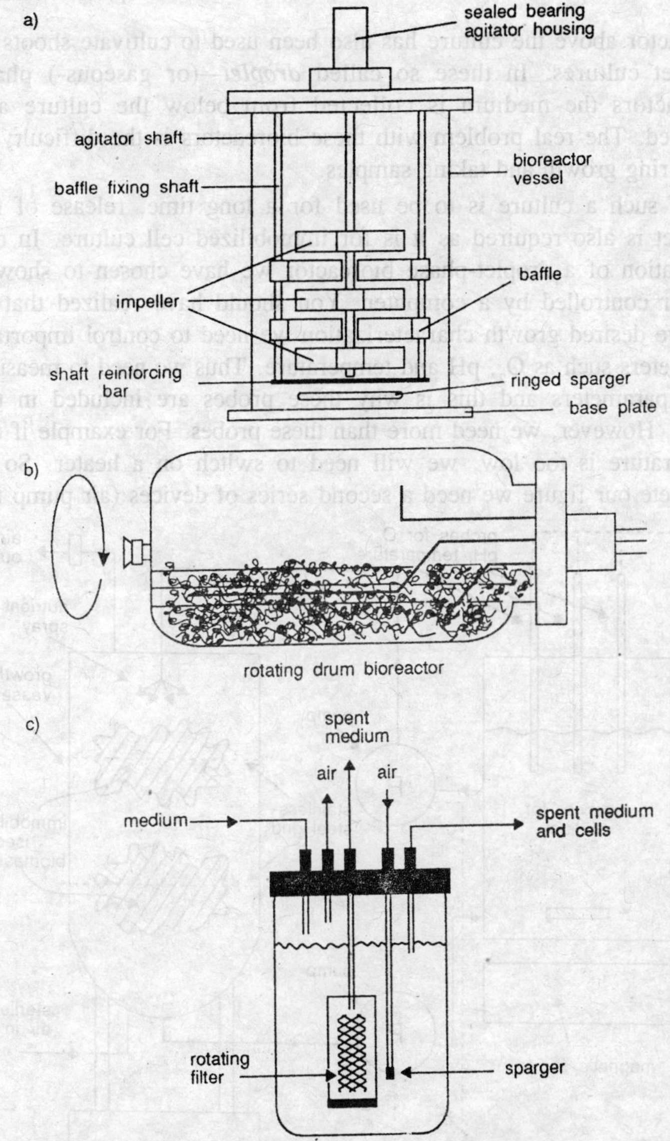

Fig. 1.9. Bioreactor designs used for the cultivation of roots or shoots. (a) A stirred tank with large vanes operated at low speed (10 rpm); (b) A rotating drum bioreactor; (c) A spin-filter bioreactor.

at high biomass concentrations within the bioreactor. A completely different form of bioreactor in which the medium is sprayed into the bioreactor above the culture has also been used to cultivate shoots or plantlet cultures. In these so called *droplet*—(or gaseous-) phase bioreactors the medium is collected from below the culture and recycled. The real problem with these bioreactors is the difficulty in measuring growth and taking samples.

If such a culture is to be used for a long time, release of the product is also required as it is for immobilized cell culture. In our illustration of a droplet-phase bioreactor we have chosen to show a reactor controlled by a computer. You should have realized that to achieve desired growth characterization we need to control important parameters such as O_2, pH and temperature. Thus we need to measure these parameters and this is why these probes are included in the figure. However, we need more than these probes. For example if the temperature is too low, we will need to switch on a heater. So to complete our figure we need a second series of devices (air pump for

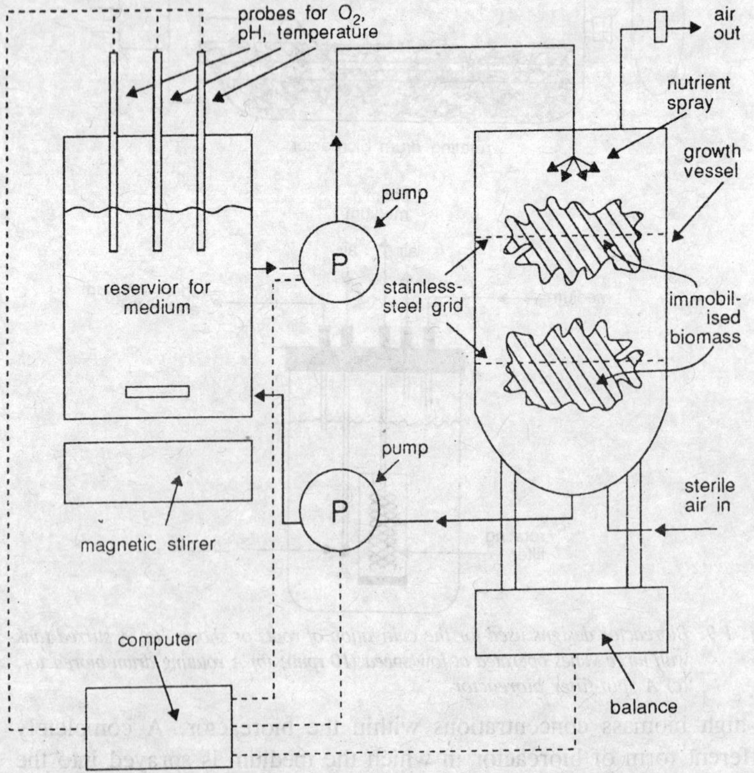

Fig. 1.10. A droplet phase bioreactor used to cultivate organs/plantlets.

oxygen, acid/alkali reservoir and pumps for pH and heater/cooler for temperature) connected to the computer. Note that in the droplet phase bioreactor, these parameters are controlled outside of the reactor itself. The balance gives an indication of the amount of biomass in the bioreactor and this, via the computer, can be used to regulate the rate of spraying media into the reactor. You should however realize that there are many variations on this general design.

Transformed Cultures, 'Hairy Roots'

Hairy roots, like the normal root and shoot cultures, do pose problems different from those of suspension cultures. The hairy roots will also be shear-sensitive, probably more so because of the length of the roots, which form a tangled mass. It has been shown in some cases that too vigorous mixing will cause callus formation rather than root growth. One solution has been to provide a matrix for the roots

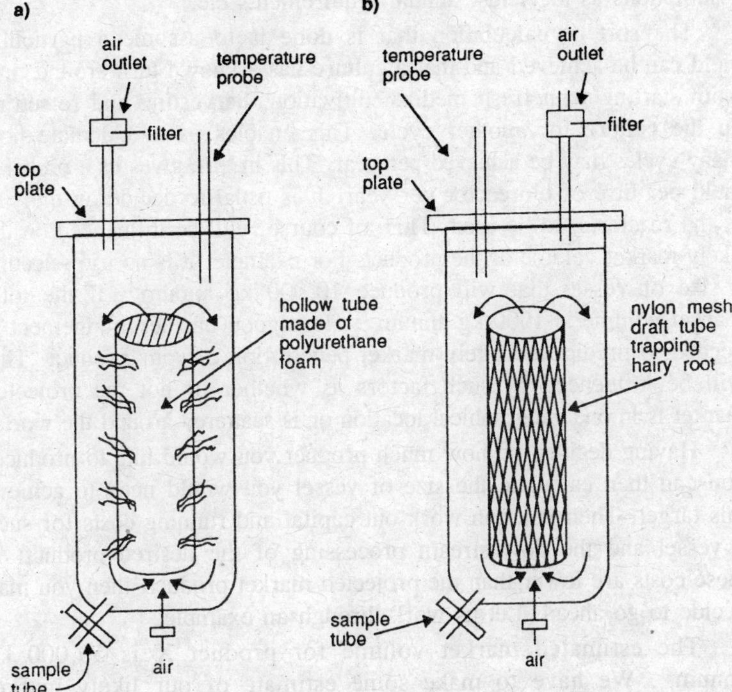

Fig. 1.11. Airlift and bubble column bioreactors used to cultivate 'hairy root' culture. (a) In this bubble column a large column of polyurethane foam is used to anchor the roots; (b) Here the roots are trapped in a nylon mesh which forms the draft tube of the airlift bioreactor.

to grow on and through and then to fit this within a bioreactor. A polypropylene mesh or polyurethane foam have been used in airlift bioreactors. Another design has been to adopt the droplet phase bioreactor format while providing a framework on which the hairy roots can grow. This type of system has been scaled up to a volume of 500 l.

Costing of Plant Cell Culture Processes

In the costing of a process two main factors are normally taken into consideration; the capital cost (the money required to build the plant) and the running costs. Once a process has been developed and the scale-up determined, a plant can be designed and costed. At present, few processes have reached commercial scale, and therefore many of the process parameters, such as biomass yields, are still under investigation. This makes it difficult to outline a process and produce a costing. However, with a limited amount of data, a number of authors have attempted costings but in doing so have had to make a number of assumptions as to yields, annual requirements etc.

The sort of calculation that is done is to assume a particular yield can be achieved and that a culture has a known turn-around time from starting with fresh media, cultivation, harvesting and re-setting up the culture for another cycle. This enables us to calculate how many cycles may be achieved per year. This in turn gives us a probable yield per litre of bioreactor per year. It is usual to decide on the size of the reactor(s) to be used. This, of course, will be influenced on the likely market volume of the product. For example, it is no use selecting a size of vessel that will produce 10,000 kg annum^{-1}, if the total market volume is 1000 kg annum^{-1}. Here good business judgement is needed to predict the likely market penetration of your product. This will be influenced by such factors as whether or not the projected market is in one geographical location or is scattered around the world.

Having decided on how much product you would like to produce, you can then calculate the size of vessel you would need to achieve this target. Then you can work out capital and running costs for such a vessel and the downstream processing of the desired product. If these costs are lower than the projected market product, then you may decide to go ahead. Let us work through an example.

The estimated market volume for product X is 50,000 kg annum^{-1}. We have to make some estimate of our likely market penetration. Let us assume that our best estimate is that we can, if our product was competitively priced, sell about 10% of the total market turnover of X. Thus we will need to produce about 5000 kg annuml^{-1}. Each litre of our culture produces 0.5 g in each 15-day batch culture. If we assume that we can repeat this cycle every 15 days then we might expect to get 366/15= 24 cycles annuml^{-1}. Thus we might expect to produce 0.5 × 24 = 12 g of product for each litre of reactor each year. In practice this yield would be lower. Bioreactors may have quite long down-times. They need to be cleaned. Sometimes they become contaminated and the culture is lost. There may be mechanical breakdown and staff and national holidays add further constraints. Thus the actual yield might be as low as 4.0-10.0 g l^{-1} annum^{-1}. (In some literature down-times as high as 80% of the total time have been reported!). Clearly in practice we must aim for a minimum down-time.

Let us assume that we can produce 5 g l^{-1} annum. Since we wish to produce 5000 kg annum^{-1}, then we would need [(5000 × 10^3/5)] =10^6 l = 10^3 m^3 bioreactor volume. From the data given, the market value of 5000 kg of X is 5000 × 1000$ = 5 × 10^6$. The key question is therefore can we run a 10^3 m^3 bioreactor, process the culture to extract and purify the product, carry out all the quality control measures, package and market the product, allow for depreciation in capital equipment and so on for less than 5 × 10^6$?

The figures we have used in the demonstrations are fairly typical and in the judgement of most commercial organizations unlikely to result in an economically viable operation. The key to bring down costs is to increase productivity. As we have said before if yield could be increased (from say 0.5 → 5.0 g l^{-1} in our example) this would have considerable effects on costs. It would reduce the volume of bioreactors needed tenfold and also we would only need to process 10% of the culture volume.

In practice, the only data which are available are for suspension cultures and therefore no costings can be made, as yet, for the alternative systems. The process success will still depend on two factors —yield and biomass loadings. Yield of the commercially interesting compounds is still low, but there is research being carried out into the biosynthetic pathways and their control and this may help increase yields. High biomass yields are possible and with the use of stirred-tank bioreactors fitted with high flow impeller, mixing should not be a problem. The modification of existing bioreactors for plant cell culture may also reduce the initial capital cost. Therefore, plant cell culture as a way of making chemicals should not be dismissed yet. It will be interesting to see what happens especially in such cases as that of taxol because plant cell culture may be the only source of its supply.

2

Genetic Modification of Crops

For thousands of years agricultural people slowly improved their crops, either unwittingly or sometimes quite consciously, by setting aside part of the harvest as the planting material for the next crop. Crop improvement was greatly accelerated in the 20th century by the rediscovery of the laws of inheritance, first formulated by Gregor Mendel in 1865, and after scientists established that the chromosomes contain the information for specific traits encoded in DNA. The rise of molecular biology, starting in the 1950s with the discovery of DNA's structure, led to a series of further discoveries showing how traits are encoded as proteins in the cell and the plant. This understanding, and new techniques that allow scientists to manipulate DNA in the laboratory, are revolutionizing plant breeding. In the future, increasing yield for traditional crops, improving nutritional or postharvest quality of these crops, adapting them to more stressful environments, domesticating new crops, and converting existing ones to plant factories that produce chemicals for industry will all flow from applying the principles of plant breeding, genetics, molecular biology, and genomics.

"Classic" plant breeders—those who perform mostly field work—will only succeed in making major improvements by exploiting more and more of the tools used in molecular biology. Collaboration between plant breeders, plant pathologists, molecular biologists and geneticists will be essential to expedite the long process of producing new and improved varieties.

Genes are Made of DNA

The 30-year period from the 1930s to the 1960s was a very exciting time for scientific discovery, and must go down in history as the period when one of the most fundamental questions science ever asked was answered: What exactly are genes made of? As you will see in later sections, careful observation of how chromosomes separate during mitosis, and the discovery that sex cells contain only half the number of chromosomes (one set instead of two) that other cells have, led scientists to postulate that Mendel's inherited units, later called *genes*, are organized on the chromosomes. Chromosomes are the filamentous structures in nuclei. Geneticists knew that chromosomes were composed of both protein and DNA (deoxyribonucleic acid).

For many years, biochemists incorrectly believed that genes consisted of protein because they knew proteins were large, complex molecules (they consist of countless sequences of 20 different amino acids) that show great specificity in their activity (mostly enzymatic activity). Biochemists thought that only protein molecules could contain enough information to specify inherited traits. In contrast, they thought DNA was a small and relatively simple molecule because it contains only four nucleotides—surely DNA could not contain enough information to account for the great diversity of life forms on Earth.

Several elegant and decisive experiments conducted with bacteria and viruses by Frederick Griffin, Oswald Avery, Alfred Hershey, and Martha Chase showed that genes consist of DNA and have the necessary information to direct the synthesis of proteins. The researchers showed that mutation of bacterial genes results in the loss of specific enzymatic activities, indicating that genes are responsible for the presence of proteins in the cell. This discovery later led to the "one gene, one enzyme" hypothesis that a distinct gene controls the presence of each enzyme in the cell. Now that biologists were convinced DNA, not protein, was the genetic material, biochemists wanted to answer another fundamental question: What precisely is the structure of DNA? Because they knew this discovery would go down in history, researchers were racing each other to unravel the structure of DNA. With new chemistries developed in the 1940s, Erwin Chargaff determined in detail the base composition of DNA.

Scientists already knew that DNA consists of four different types of nucleotides: two with the purine bases adenine and guanine, and two with the pyrimidine bases thymine and cytosine. The purine bases have a double ring structure, whereas the pyrimidine bases have a

Adenine

Deoxyadenosine monophosphate

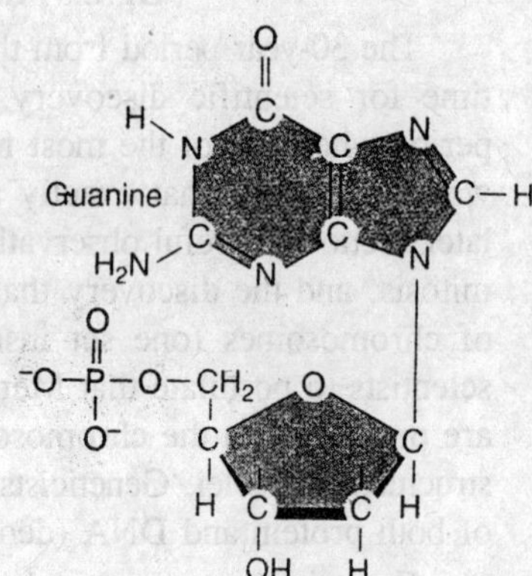

(a) Purine nucleotides

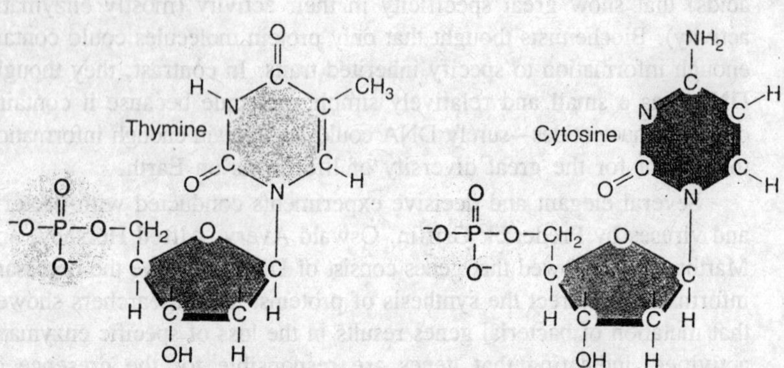

(b) Pyrimidine nucleotides

Fig. 2.1. The building block of DNA.

single ring. These four nucleotides—the building blocks of DNA—consist of a phosphate group, the sugar deoxyribose, and one of the four nitrogen-containing bases (A, C, G, or T). Chargaff showed that in some species, such as *E. coli* and maize, the DNA consists of about 25% of each type of nucleotide, but that in most species, such as humans, this is not the case. Thus DNA shows the variability among species that you might expect of the genetic material of life.

Within a species, however, Chargaff's experiments showed that the amount of A always equals the amount of T, the amount of G always equals the amount of C, and that the percentage of A plus G equals 50% and the percentage of C plus T equals 50%. These data suggested that A is always paired with T and that G is always paired with C. Rosalind Franklin studied the structure of DNA using X-rays and found that DNA has a helical structure.

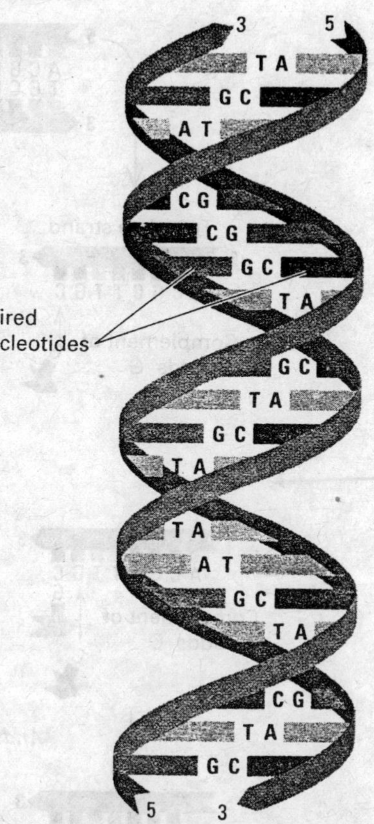

Fig. 2.2. The diagram shows a segment of double-stranded DNA that contains only 18 paired nucleotides.

Using Chargaff's and Franklin's data, and possibly some of Linus Pauling's data, James Watson and Francis Crick constructed a model for the structure of DNA for which they won the Nobel Prize in medicine in 1962. They concluded that DNA is a double-stranded helix with the adenine in one strand always hydrogen-bonded to the thymine in another strand; similarly, cytosine in one strand is always hydrogen-bonded to guanine in the other strand. Together the two strands form the double helix with the sugar-phosphate backbones on the outside and the paired bases on the inside.

The two strands of the DNA molecule are said to be *complementary strands* because if you know the base sequence of one strand, you can deduce the base sequence in the other complementary strand. If one strand contains the sequence ATTGCC, then the other strand must in the same region have the sequence TAACGG, because of the base-pairing rules (A opposite T, and G opposite C). Another requirement of the genetic material is that it must replicate and then transfer to daughter cells. Watson and Crick proposed a mechanism for the copying of DNA based on the specific base-pairing rules, and scientists have now confirmed that DNA replicates by complementary base pairing. DNA copying is called *semi conservative replication* because one of the old strands is present in each new daughter molecule.

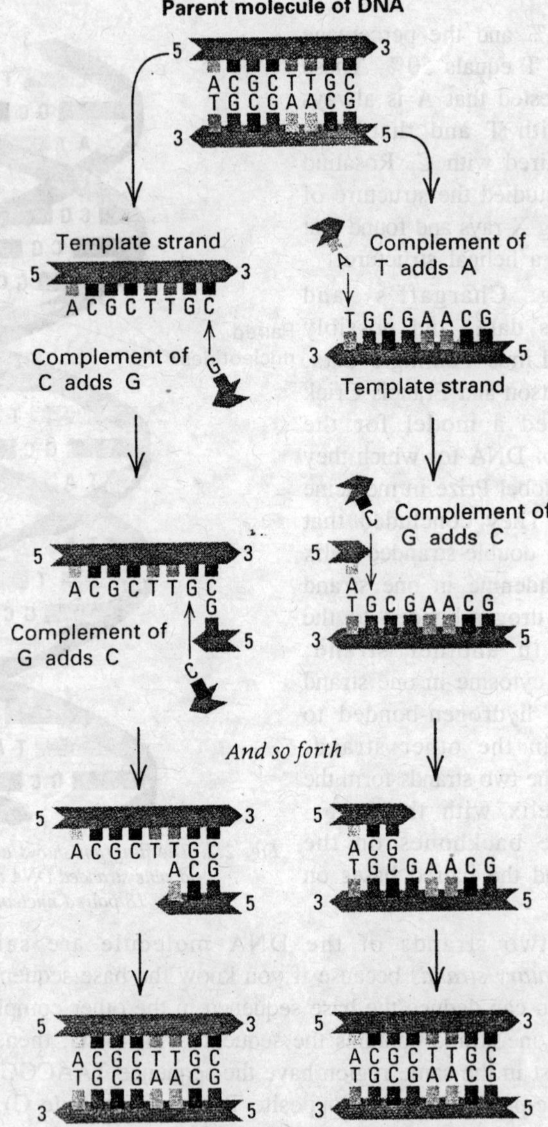

Fig. 2.3. Replication of DNA.

When DNA replicates, the two old strands separate and the complementary strands then synthesize using the existing strands as templates.

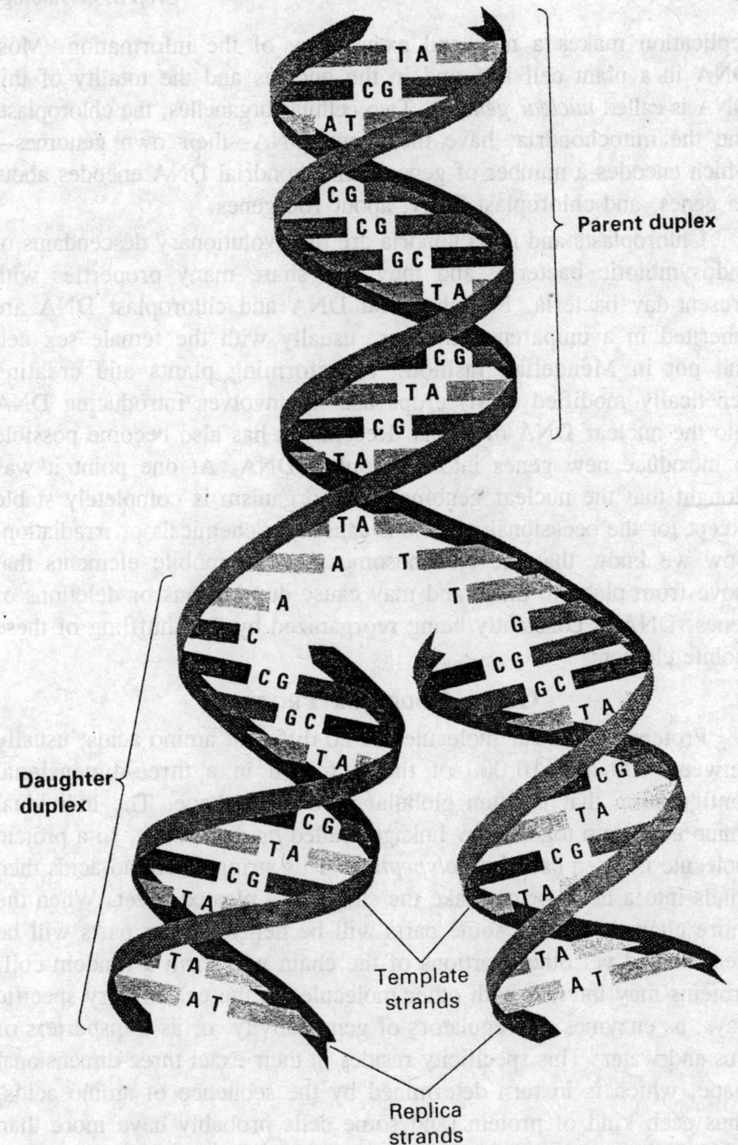

Fig. 2.4. The obligatory pairing of adenines with thymines and guanine with cytosines ensures that the two daughter molecules will be duplicates of the original.

When the two new strands are synthesized, A is always opposite T and C opposite G, so that two identical molecules form. Because the information is contained in the exact sequence of the bases, DNA replication makes a new and exact copy of the information. Most DNA in a plant cell is found in the nucleus and the totality of this DNA is called *nuclear genome*. Two cellular organelles, the chloroplasts and the mitochondria, have their own DNA—their own genomes—which encodes a number of genes. Mitochondrial DNA encodes about 15 genes, and chloroplast DNA, about 100 genes.

Chloroplasts and mitochondria are the evolutionary descendants of endosymbiotic bacteria, and they still share many properties with present-day bacteria. Mitochondrial DNA and chloroplast DNA are inherited in a uniparental manner, usually with the female sex cell and not in Mendelian fashion. Transforming plants and creating genetically modified (GM) crops usually involves introducing DNA into the nuclear DNA of plants. Recently it has also become possible to introduce new genes into chloroplast DNA. At one point it was thought that the nuclear genome of an organism is completely stable except for the occasional mutations caused by chemicals or irradiation. Now we know that the chromosomes contain mobile elements that move from place to place and may cause duplications or deletions of genes. DNA is constantly being reorganized by the shuffling of these mobile elements.

Genes Code for Proteins

Proteins are linear molecules of 20 different amino acids, usually between 100 and 10,000 of them, folded in a three-dimensional configuration that is often globular in overall shape. The individual amino acids join together by linkages called *peptide bonds*, so a protein molecule is often called a *polypeptide*. This string of amino acids then winds into a helix or can take the shape of a pleated sheet. When the entire chain is folded, some parts will be helical, other parts will be pleated, and yet other portions of the chain will form a random coil. Proteins may interact with other molecules in the cell in very specific ways: as enzymes, as regulators of gene activity, or as transporters of ions and water. This specificity resides in their exact three-dimensional shape, which is in turn determined by the sequence of amino acids. Thus each kind of protein (and some cells probably have more than 5,000 different proteins) has a characteristic amino acid sequence.

The information that specifies the order in which the amino acids must assemble when a protein is synthesized, is contained in the DNA, specifically in the sequence of bases attached to the sugar-phosphate backbone. At first sight this coding might seem to be a problem, because DNA has only four different bases and proteins have 20

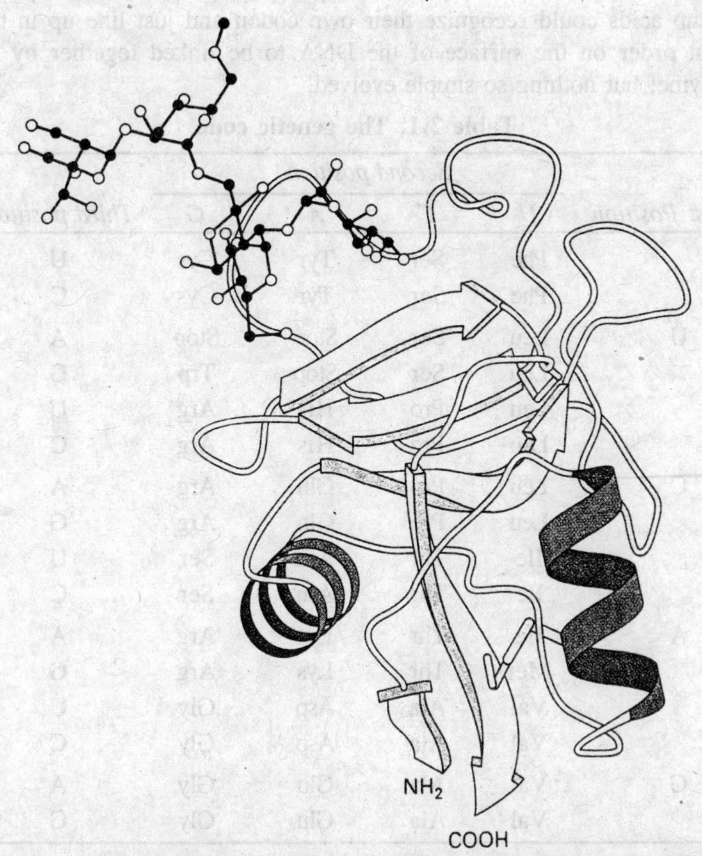

Fig. 2.5: A ribbon diagram showing the ways in which a polypeptide can be folded.

different amino acids. This coding problem is solved by having a sequence of three bases specify one amino acid. When four different nucleotides are arranged in groups of three, 64 different combinations ($4 \times 4 \times 4$) are possible, more than enough to specify 20 amino acids. In the 1960s, Marshall Nirenberg, Heinrich Matthei, and Philip Leader figured out which three-base combination or codon specifies a particular amino acid. The correspondence between a three-base combination and an amino acid is called the *genetic code*. The genetic code is universal: It is the same in all organisms. Most amino acids are represented by more than one codon. Of the 64 different possibilities, 61 codons specify amino acids, and 3 specify "stop" signals marking the end of a protein-coding segment of the DNA. Cells have an elaborate machinery for translating the nucleotide sequences in the

DNA into amino acid sequences in proteins. It would be simple if amino acids could recognize their own codon and just line up in the right order on the surface of the DNA to be linked together by an enzyme, but nothing so simple evolved.

Table 2.1. The genetic code

	Second position				
First Position	*U*	*C*	*A*	*G*	*Third position*
	Phe	Ser	Tyr	Cys	U
	Phe	Ser	Tyr	Cys	C
U	Leu	Ser	Stop	Stop	A
	Leu	Ser	Stop	Trp	G
	Leu	Pro	His	Arg	U
	Leu	Pro	His	Arg	C
C	Leu	Pro	Gln	Arg	A
	Leu	Pro	Gln	Arg	G
	Ile	Thr	Asn	Ser	U
	Ile	Thr	Asn	Ser	C
A	Ile	Thr	Lys	Arg	A
	Met	Thr	Lys	Arg	G
	Val	Ala	Asp	Gly	U
	Val	Ala	Asp	Gly	C
G	Val	Ala	Glu	Gly	A
	Val	Ala	Glu	Gly	G

Rather, cells use a different class of nucleic acids, RNAs (ribonucleic acids), to help translate the information contained in the DNA. RNAs are also strings of nucleotides, and consist of a sugar-phosphate backbone with a base attached to each sugar group, except that the sugar is ribose. Three of the bases in RNA are the same as in DNA (C, A, and G) but instead of thymine (T), RNA uses uracil (U). Gene transcription proceeds via the formation of a messenger RNA (mRNA) molecule.

The mRNA carries the information that will specify the amino acid sequence from the nucleus, where the nucleotide sequence of the DNA contains the message, to the cytoplasm of the cell, where this message is translated and the protein synthesized. Translation of the message involves transfer RNA (tRNA) molecules as well as ribosomes,

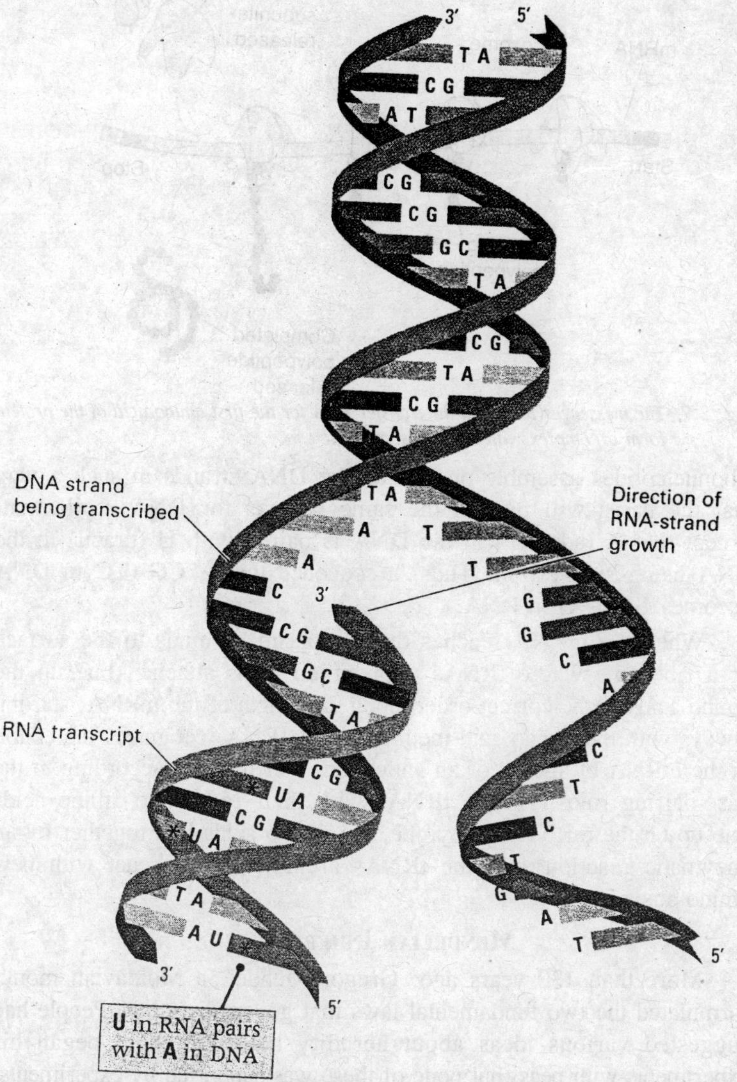

Fig. 2.6. Before protein synthesis can begin, the genetic information (sequence of bases) in the DNA must be copied into an RNA molecule.

specialized structures on whose surface new proteins are assembled. As the amino acid chain lengthens, it begins to fold up, first in a helical configuration, and then in a three-dimensional globular shape. During transcription, the two strands of the DNA separate and

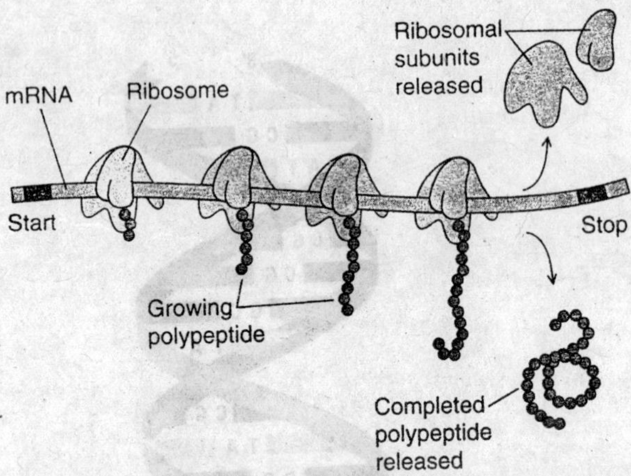

Fig. 2.7. The messenger RNA and the transfer RNA for the first amino acid of the protein form a complex with the ribosome.

ribonucleotides assemble on one of the DNA strands in such a way that the bases will pair in the same way as for DNA replication, except that A (adenine) in the DNA is paired with U (uracil) in the RNA being synthesized. Thus, a sequence that is CGATC in DNA becomes GCUAG in RNA.

When the mRNA reaches the cytoplasm, it binds to the surface of a ribosome where tRNAs with amino acids attached line up the amino acids in the correct order along the length of the mRNA, starting always with the amino acid methionine. A tRNA recognizes the codon in the mRNA by means of an anticodon, which aligns according to the base-pairing rules. As the tRNAs with their respective amino acids line up on the mRNA one by one, the amino acids link together by an enzymatic reaction and, the tRNAs are released to react with new amino acids.

Mendelian Inheritance

More than 130 years ago, Gregor Mendel, a Moldavian monk, formulated the two fundamental laws that govern heredity. People had suggested various ideas about heredity before Mendel began his experiments with peas, but none of these was supported by experiments. When Mendel began his work, most breeders agreed that both sexes contributed to a new individual. They thought that parents of different appearance always produced offspring with an intermediate appearance. Thus, a cross between a plant with red flowers and one with white flowers would lead to only plants with pink flowers. When future generations produced red and white flowers, breeders mistook this to mean that the genetic material was unstable.

Mendel conducted a series of experiments in which he crossed two varieties of pea plants with contrasting characteristics such as white and red flowers or round and wrinkled seeds. Remarkably, the characteristics he used can still be found today, precisely because they all have all-or-none variation. He published his observations in 1865 in a local scientific journal, but his findings went largely unnoticed until other scientists made the same observations 35 years later.

Mendel, who was very familiar with peas, knew that they always breed true (the offspring are exactly like the parent plant and like each other), because peas are self-fertilizers and the pollen fertilizes the pistil (female reproductive organ) even before the flower opens. However, by cutting away the anthers (male reproductive organ) and brushing on pollen from another plant, Mendel figured out that peas could be cross-pollinated. Mendel had many types of true-breeding peas growing in his garden, and he wanted to know what would happen if he cross-fertilized lines of his peas with contrasting characteristics. One line of pea always produced smooth, round peas, but another one produced wrinkled peas. When the latter peas dried out at the end of seed maturation, some of the inner tissues collapsed, providing the seed with a wrinkled appearance (phenotype).

When Mendel crossed round and wrinkled peas, he observed that the first generation (the F_1 generation) consisted entirely of round peas. When these round peas were allowed to sprout, grow, and flower, and when the plants were allowed to set seed, most seeds of the F_2 generation (about 75%) were round and a minority (about 25%) were wrinkled. None were in between; none were just a little wrinkled or nearly round. Thus a characteristic that disappeared in the first generation reappeared in the second.

Mendel repeated these experiments using seven other discontinuous characteristics (for example, green and yellow seeds) and confirmed his experiments with round and wrinkled peas. Interestingly, he found that these seven different characteristics were not linked, but were transmitted to the next generation independently. Thus, when he crossed a round yellow pea with a wrinkled green pea, the first generation had only round yellow peas. However, in the second generation he found four types: the two original types and two additional combinations of the characteristics (round green and wrinkled yellow). From such experiments, Mendel drew two important conclusions. First, characteristics or traits transmit to the next generation as discrete units, now called *genes*. Second, an individual must contain two copies of each of these units, and each parent transmits only one copy to the next generation. That is the only way Mendel could account for the disappearance and subsequent reappearance of a characteristic. The implication is that the unit (gene) is always present, but may not be *expressed*, as is the case with the wrinkled or green characteristics in the first generation of crosses of round and wrinkled or green and yellow peas.

Although Charles Darwin and Mendel were contemporaries (Mendel published his work in 1865, fifteen years before Darwin's death), Darwin did not know of Mendel's work. Darwin knew that characteristics could disappear and reappear, but he did not understand their mode of inheritance. In his book, *On the Origin of Species by Means of Natural Selection,* Darwin discussed the notion that a specific trait can be inherited by one child but not by another and that traits of grandparents sometimes appear in the grandchildren although they did not appear in the parents. He did not understand this skipping of a generation, which Mendel's experiments so beautifully explained.

Why are there smooth and wrinkled peas? Biochemists have recently discovered that wrinkled peas lack one of the important enzymes for starch synthesis. During their development, peas import sucrose from the rest of the plant. This sucrose is quickly converted to starch, which makes up 60% of the weight of a mature pea seed. Wrinkled peas lack one of the enzymes for starch synthesis, so when the seeds mature they contain a lot of sucrose and water and much less starch than normal. Then, when the seeds dry out, they wrinkle. The absence of the enzyme is caused by an alteration in the gene that encodes the information for synthesizing this enzyme. A single gene usually controls all-or-none variation, as in the case of the smooth and wrinkled peas.

Many agronomically important characteristics, such as yield or protein content of seeds, show continuous variation, and the inheritance of such traits is controlled by many genes. Such traits are referred to as *multigenic* or *quantitative* traits. Thus, the ability of a plant to take up soil nutrients, photosynthesize, transport photosynthate to the seeds, and withstand drought all affect yield. And each of these characteristics is controlled by many genes, making yield truly a multigene trait.

Most organisms have two copies of every gene in every cell, except in the sex cells (sperm and egg cells), which each only have one copy of each gene. In humans, each cell has 22 pairs of chromosomes and 2 sex chromosomes, for a total of 46. Maize has 10 pairs of chromosomes, and *Haplopappus*, a plant that thrives in dry areas, has only 2 pairs. Because cells have two copies of every chromosome, they also have two copies of every gene, one copy on each of the chromosomes that make up a pair of chromosomes. When these two copies are identical, the organism is said to be *homozygous* for that gene. If one of the two gene copies has mutated, they will be different, or *heterozygous*. These different forms of the gene are called *alleles*. Plants that normally self-fertilize, such as peas and beans, are homozygous for a very large number of genes, whereas plants that normally outcross, such as maize, may be heterozygous for most of their genes.

Inheritance and Cell Division

As noted earlier, genes are located on filamentous structures, called *chromosomes*, in the cell nucleus. Most complex organisms such as flowering plants, insects or mammals have in their cells some 20,000-60,000 genes but only a small number of chromosomes. Thus thousands of genes are linearly arranged on each chromosome. The chromosomes playa vital role in passing on genes, and therefore phenotypic characteristics, from a mother cell to the daughter cells during cell division, and from the parents to the offspring during reproduction. Cell division, which in plants occurs in meristems, is preceded by mitosis, a process of chromosome duplication, and the subsequent separation of the two sets of chromosomes. Thus for a short while a cell contains four copies of every gene, but as the new chromosomes separate, each daughter cell again contains two copies of every gene.

During the formation of egg cells and sperm cells, a different type of cell division occurs. A single cell undergoes one round of chromosome replication followed by two rounds of chromosome separation and cell division. Four cells are formed in this process, which is called *meiosis*, and each cell ends up with a single copy of each gene. If the organism is heterozygous for that gene, then half the sex cells will have one allele of the gene and the other half will have the other allele. When two sex cells fuse during fertilization to start a new organism, the cells of this new organism will again contain two copies of every gene.

A plant homozygous for a given gene will produce sex cells, all of which have identical forms of that gene. That is, in terms of that gene, it produces only one type of reproductive cell. However, a plant that is heterozygous for a certain gene, with (for example) one normal and one mutant copy, will produce two types of reproductive cells. Half the sex cells will carry the normal gene, half will carry the mutant gene. The fact that many plants are normally heterozygous for many genes creates many possibilities for variation in the offspring, because random assortments of genes are brought together when the sperm cell fertilizes the egg cell. Let us consider a plant with a haploid chromosome number of 2. Its normal diploid cells will have two pairs of chromosomes in each cell. Such a cell is shown with two heterozygous genes. T and t are alleles of the same gene and are located on the long chromosome (note that T came from the male parent and t from the female). A and a are alleles of a different gene and are located on the short chromosome. Before mitosis begins, the cell has four chromosomes in its nucleus: a pair of each of two types. Mitosis involves three steps:

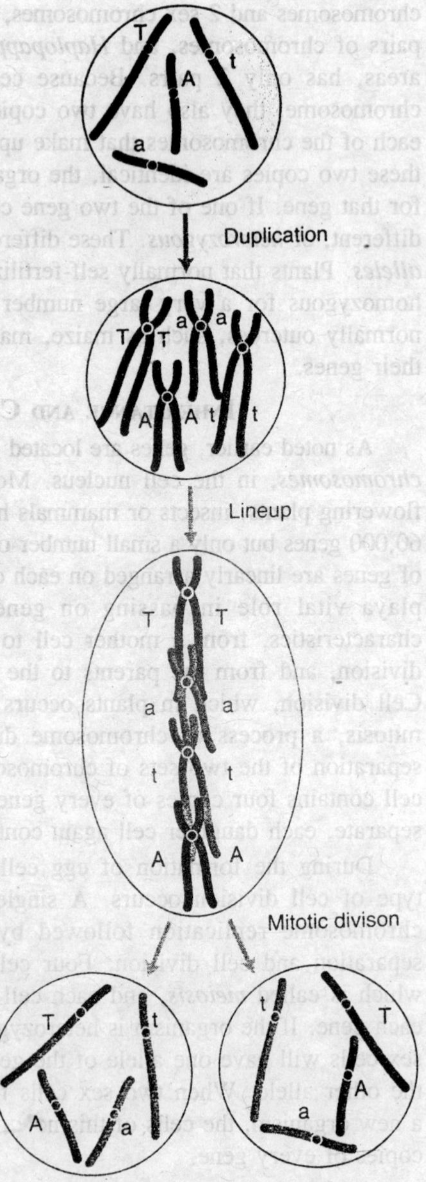

Fig. 2.8. The mitotic method of cell reproduction occurs in plant meristems.

1. *Duplication* of the chromosomes' genetic material (DNA); this actually occurs during the time that the cell nucleus is visible and chromosomes cannot be seen as distinct structures.

2. *Lining up* of the duplicated chromosomes, usually at the center of the cell.
3. *Separation* of each member of the duplicated chromosomes from its partner, such that each of the two new cells gets one copy of the pair.

Because the two members of the duplicated chromosome are genetically identical, the two new cells that are formed contain the same genes (T and t, A and a) as the cell that produced them.

The earliest part of meiosis is identical to mitosis: The chromosomes are *duplicated* and each one becomes a double strand. The chromosomes *line up* in the center of the cell; however, in this case, they line up as homologous pairs. Now, *separation* occurs such that the duplicated chromosomes separate from each other, one duplicated member going into each of two new cells. The resulting cells clearly have only half the number of chromosomes (in this case, 2) as the originating cell (4). These two chromosomes are double, so that in meiosis II, the two members of the doublet separate, just like in mitosis. Now there are four cells formed, each with a haploid chromosome number (2).

At the lineup and separation stages of meiosis I, the only requirement is that each new cell gets one of the two members of the homologous pair. So it is equally probable that one of the new cells will get the "A" and "T" chromosomes and the other "a" and "t" as it is for them to get the "A" along with "t" and "a" with "T." So in our simple case, there are four possible chromosome combinations in the gametes: AT, at, At, and aT.

Few plants have just two pairs of chromosomes (diploid number of 2). For example, corn has 10. As we have just seen, the number of possible gametes formed with two chromosome pairs, just with regard to paternally and maternally derived chromosomes, is 4 (2^2). For 10 chromosomes, the number is 1,024 (2^{10}). If we consider that there are often allelic differences between the two chromosomes (see our example, T and t, and A and a), there are 1,024 genetically different gametes possible. These combinations mean that in sexual reproduction, the offspring will not be genetically identical to the parents. This provides much genetic variation, the raw material of evolution by natural selection. It also provides a challenge to plant breeders, who seek to minimize variation in offspring. Additional genetic variation is created when chromosomes exchange portions during crossing over in meiosis.

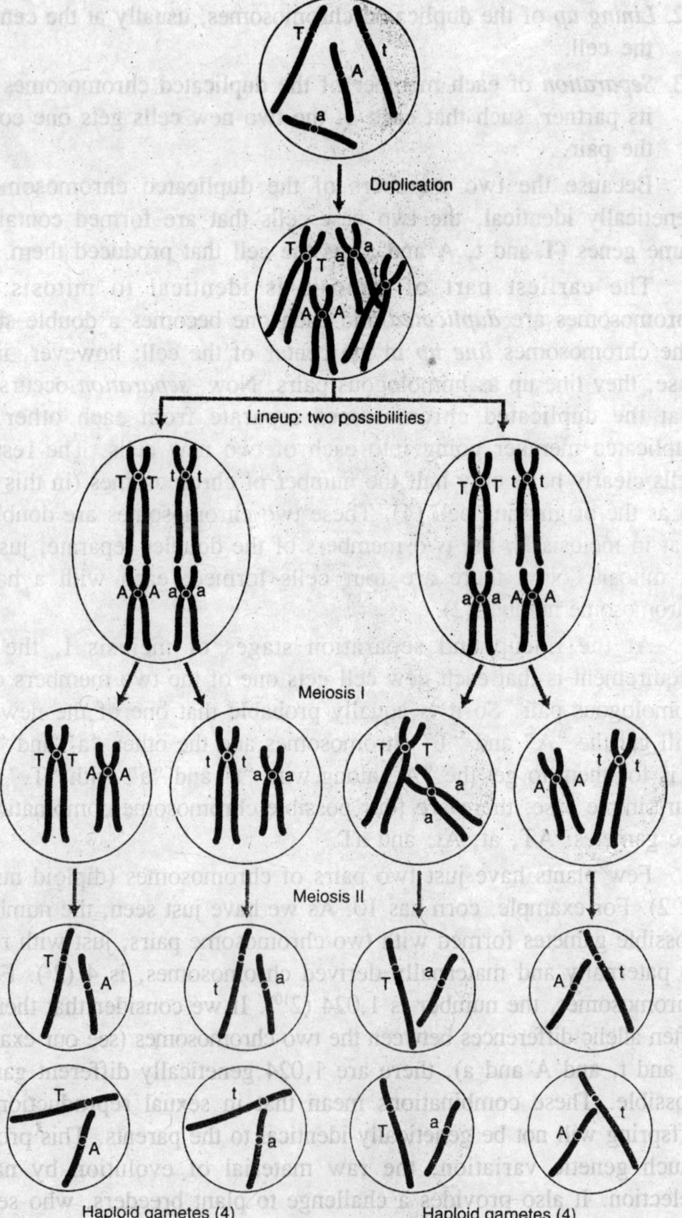

Fig. 2.9. The meiotic method of cell reproduction results in the formation of sex cells (pollen or egg cells).

Multipartite Nature of Gene

Orderly development of an organism requires that the correct proteins be synthesized at the correct time and place during development, and in response to the proper stimuli. For example, the enzymes that cause softening of cell walls in the ripening fruit are only synthesized in the fruit and as a result of an increase of the hormone ethylene. So far, we have described genes only in terms of the information they contain to specify a sequence of amino acids and the structure of a protein. This part of a gene is called the *protein-coding region*. However, a gene is more complex (much longer, really) than just a protein-coding string of nucleotides. First of all, the protein-coding portions, called *exons*, may be interrupted from one to a dozen times by very short to very long stretches of DNA, called *introns*.

Introns are transcribed when RNA is first made in the cell nucleus, but they are later removed when the initial product of transcription is processed into a mature mRNA that is transported to the cytoplasm. These interruptions in the coding sequence are characteristic and different for every gene, and can also be involved in regulating the gene's expression. Second, a DNA segment on each side of the protein-coding region specifies when and where the gene is to be activated during development, to which stimuli (hormonal or environmental) the gene should respond, in which cells it should be active, how much mRNA is to be made, and when the gene is to be turned "on" and "off." These *regulatory regions* are, of course, of great importance for the orderly development of a cell and an organism. Thus, genes that encode enzymes for the synthesis of chlorophyll are "on" in the leaves, but "off" in the roots and the flowers.

Furthermore, they are "off" in the dark and "on" in the light. When a root finds itself in soil that is rich in nitrate, genes are turned "on" that encode proteins needed to take up nitrate, its transformation into ammonia, and the use of this ammonia in amino acid biosynthesis. When the spore of a plant pathogen germinates on the surface of a leaf and tries to penetrate into the leaf cells, defense genes are rapidly turned on in the cells of resistant plant varieties. These defense genes encode enzymes that synthesize toxic compounds that will kill the invader. When light strikes a seedling that has been growing in the dark underneath the soil surface, hundred of genes that were completely off or barely on are turned on and the cells start making hundreds of new proteins that let the chloroplasts develop, and photosynthesis and autotrophic growth begin. These are only a few examples of the gene activity regulation in response to specific stimuli. Regulating gene activity is the responsibility of proteins called *transcription factors* that bind to the regulatory regions of genes, and these transcription factors are themselves the products of genes. Thus, turning genes "on" and "off" is not a simple matter and may involve a regulatory cascade in which the product of gene A activates gene B, whose product activates gene C, whose product activates gene D by binding to its regulatory region.

Plant breeding and genetic engineering both involve transfer of genes from one organism to another. In plant breeding, the breeder selects for an ultimate outcome or a phenotype, thus ensuring that the whole gene regulation pathway will operate correctly. Let us take the example of transferring resistance to a specific pathogen from a wild relative of wheat to a cultivated wheat variety. A resistance gene may encode a protein that lets the plant detect the invader quickly so the plant can turn on its defenses. The gene that encodes this detector protein responds to a stimulus from the pathogen—perhaps a chemical or metabolite made by the pathogen—and this response involves several other genes. Thus when the breeder transfers resistance to that specific pathogen from a wild variety of wheat to a domesticated wheat by selecting for pathogen resistance in the field, the entire regulatory cascade must work correctly.

If the new combination carries the resistance gene but is not correctly expressed, then the new variety would not resist the pathogen. Genetic engineers transfer one gene at a time and need to understand how the gene they transfer is regulated. First, they need to transfer not only the protein-coding part of a gene but also its regulatory region. Or they can equip the gene with a new regulatory region that ensures correct expression. For example, regulatory regions of bacterial genes generally do not work in plants because the gene transcription machinery of bacteria differs from that of plants. Similarly, a regulatory region of the gene from a monocotyledon (maize) may not work in a dicotyledon (bean). Thus, a full understanding of the regulation of gene expression is of great interest not only to plant biologists who want to know how plant development is regulated, but also to genetic engineers who transfer genes between organisms.

Restriction Enzymes

Since the 1960s, molecular geneticists have made tremendous advances that have led to the development of exciting new areas of science in DNA technology and plant biotechnology. Biotechnology is the use and manipulation of living organisms, or substances obtained from these organisms, to make products of value to humanity. Although *biotechnology* is a relatively new term, this idea is not new—people have bred plants and animals to express particular traits since the beginning of civilization. However, biotechnology based on manipulating DNA outside of the living cell is a newer and much more powerful application of the technology. These new techniques let scientists isolate genes from one organism and then insert them into another, to produce an organism with a new trait. Before these developments, plant breeders were largely limited to genetic exchanges within a new species and between closely related species.

Now scientists can transfer genes (genetic engineering) and therefore inherited traits between very different organisms. This ability to transfer genes among humans, plants, and bacteria has revolutionized biotechnology. Progress in genetic engineering depends on a new technology that lets scientists isolate, identify, and clone (produce multiple copies of) genes. This recombinant DNA technology, called *gene cloning*, was developed in the 1980s, but originated with the discovery of plasmids and restriction enzymes in bacteria in the 1970s. Scientists make recombinant DNA (rDNA) by joining, or "recombining," DNA segments from different sources: plant and bacterial DNA, or plant and animal DNA. Because DNA strands from all organisms have the same chemical structure, they can be cut into segments and the segments linked together again in new and different ways. Cutting the DNA is done with special enzymes called *restriction enzymes*, which occur in bacteria as part of a natural defense system against invading viruses.

Restriction enzymes are highly specific to the nucleotide sequence, and each enzyme only cuts at one specific short base sequence in the DNA. For example, the enzyme *EcoR1,* found in the intestinal bacterium *Escherichia coli*, cuts only at GAATTC (the complementary strand, CTTAAG in this case, is assumed), and such a sequence occurs on the average only once every 4,000 nucleotides of DNA. An interesting feature of most restriction enzymes is that they make a staggered cut across the two strands of DNA, producing what molecular biologists call "sticky ends." Another enzyme, DNA ligase, can rejoin (or ligate) those sticky ends together again and this can happen whether the DNA is from the same organism or from different organisms. Such manipulations of DNA in the laboratory are often called *gene splicing*. Scientists carry out the job of cloning—of producing multiple copies

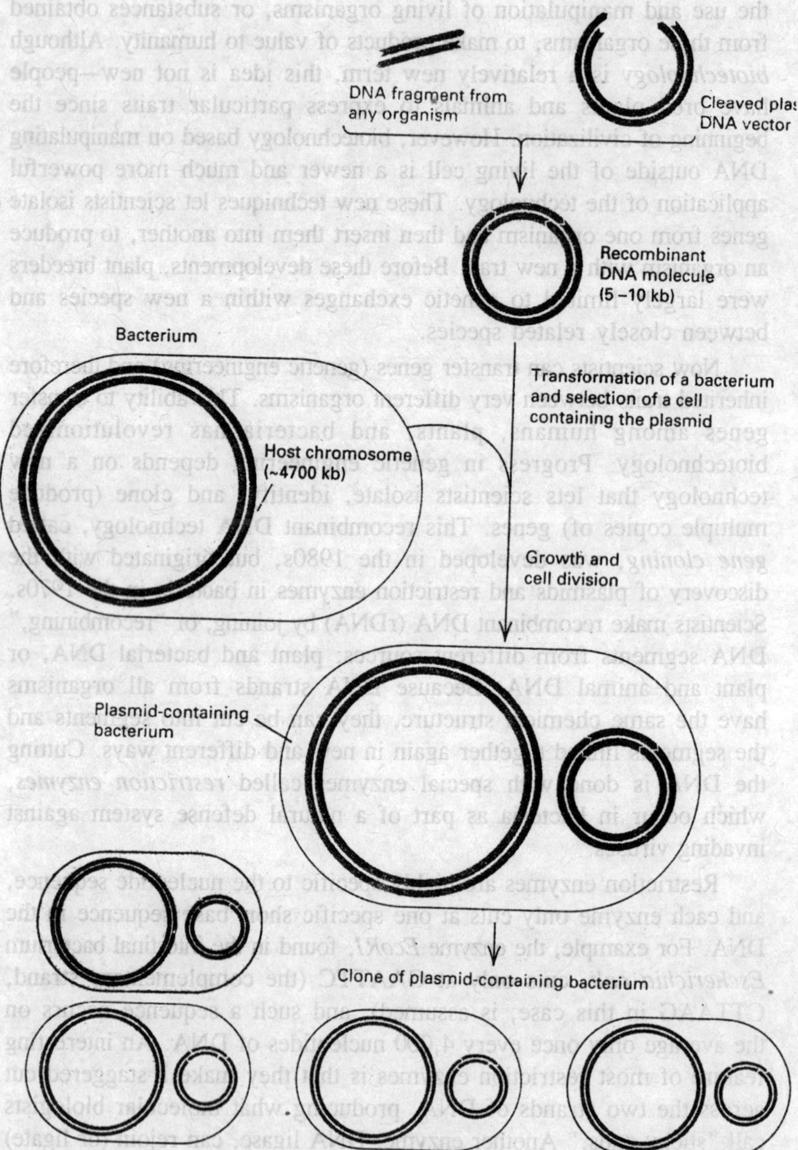

Fig. 2.10. An example of cloning.

of a gene or DNA segment—with the help of what is called a *vector*. The most common vectors are plasmids, and these are the delivery vehicles scientists use to introduce recombinant DNA into a host cell.

Plasmids are (small) double-stranded circular DNA molecules that replicate in bacteria. An important property of plasmids is that they can sometimes be transferred from one organism to another. When a foreign gene is introduced into a plasmid, the plasmid can still replicate. By using a restriction enzyme, a scientist can cut the circular plasmid open. If DNA from another organism has been cut with the same restriction enzyme so that the same sticky ends were created, DNA ligase can be used to reform circular molecules that contain a segment of foreign DNA spliced into the site where the plasmid was cut. The new plasmid is transferred to bacteria that have been specially treated for this purpose; when the bacteria are allowed to multiply, the plasmids are copied within the bacteria. Millions of copies of the plasmid and the inserted gene can be reisolated from these bacterial cells.

Regeneration Technology

When small pieces of plant tissue are put in sterile culture on a solid nutrient medium, the cells proliferate and a callus forms, and if auxin and cytokinin are present in the correct amounts, shoots will form. Shoots normally arise from small groups of rapidly dividing cells within the callus. This discovery led scientists to ask a fundamental question: Can a whole plant be grown from a single cell, given the correct nutritional and hormonal environment? Many years ago, German, Japanese, and U.S scientists discovered independently that this is indeed the case; and this discovery let them draw important conclusions about genes and how they function.

Making an entire plant depends on the correct expression of at least 10,000 genes, perhaps more. When a living plant cell is isolated from a mature tissue, it can be induced to start dividing again and all the genes necessary to make an entire organism can be induced to function again in the correct sequence. This ability of a single mature plant cell to give rise to an entire organism is called *totipotency*. Because plant cells are interconnected by their cell walls, in a kind of honeycomb, it is not so easy to isolate single cells. It is much easier to digest the cell walls of a small piece of plant tissue with enzymes and then isolate the protoplasts (naked cells without walls). The protoplasts are quite fragile, but they too can be cultured and regenerated into entire plants.

Regenerating whole plants from single cells (whether a protoplast or a cell that is still part of a tissue) is an important aspect of genetic engineering. Genetic engineering involves introducing a gene, usually from a different plant or an unrelated organism such as a bacterium, into a plant. If you can get the gene incorporated into the genome of just one cell and then encourage that cell to divide and form a whole organism, then you can get that gene into all the cells of that new plant. Two methods of introducing a new gene are commonly used, *Agrobacterium*-mediated gene transfer or a gene gun. With either method only a few cells are transformed; most cells remain untransformed. The latter must be prevented from growing into whole plants, otherwise you won't know which new plants are transformed and which are not.

To kill all the untransformed cells, scientists use a herbicide or an antibiotic in the culture medium, and let the transformed cells survive by linking them to a second gene that inactivates the herbicide or antibiotic as soon as it enters the cell. This gene is called the *selectable marker*. Genetic engineers usually introduce two genes at the same time: the gene that encodes the novel trait they wish to introduce—also called the *gene of interest* and the selectable marker gene. Molecular manipulations later eliminate the molecular marker gene. However, during the tissue culture phase the selectable marker gene lets the cells synthesize the enzyme that breaks down the antibiotic or herbicide, so only transformed cells survive and generate transformed plants. Plants produced with these techniques are usually called *genetically modified* (*GM*) or *genetically enhanced* (*GE*) *plants*. We use this terminology here for consistency, recognizing that plants improved by classical breeding are also genetically modified or enhanced.

Plant Transformation

The goal of GM technology is to isolate one or more specific genes and introduce these into plants. GM crops (or foods) are crops produced by using molecular techniques to introduce new genes, followed by several years of plant breeding. For many plant species, one or a few genes can be introduced via the natural gene transfer system of the pathogenic soil bacterium *Agrobacterium tumefaciens*, which causes tumors (called *crown galls*) in many plants. When these bacteria infect a wound site, usually on the stem close to the ground, the infection disturbs the normal healing process. Instead of making a protective tissue that covers the wound, cells proliferate into cancerous growth. Cells from this tumor can be grown in tissue culture, and unlike normal plant cells, they continuously proliferate even when hormones are absent from the culture medium. Molecular biologists in Belgium, the Netherlands, and the United States found that after the bacteria

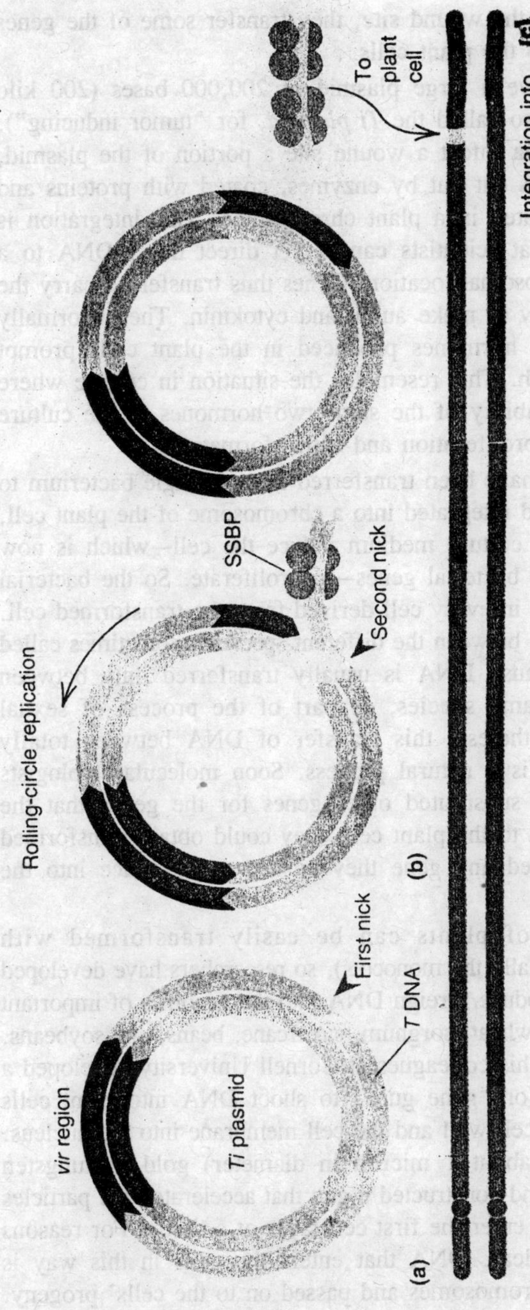

Fig. 2.11 Genetic transformation of a plant genome by T-DNA from the Ti plasmid. (a) A nick forms at one end of the T-DNA after the bacteria have attached to a wound site on the plant. (b) Replication elongates one end and displaces the other end where the first nick was formed. The single-stranded binding protein (SSBP) that is coded by one of the vir genes then stabilizes this displaced piece of single-stranded DNA. A second nick terminates replication. (c) The single-stranded DNA that is coated with the SSBP is transferred into a plant cell and integrates into the plant genome.

attach themselves to the wound site, they transfer some of the genes on the Ti plasmid to the plant cells.

The bacteria have a large plasmid of 200,000 bases (200 kilo bases or 200 kb) (also called the *Ti plasmid*, for "tumor inducing"), and when the bacteria infect a wound site a portion of the plasmid, called the T-DNA, is cut out by enzymes, coated with proteins and then becomes integrated in a plant chromosome. This integration is random, meaning that scientists cannot yet direct the T-DNA to a specific plant chromosomal location. Genes thus transferred carry the information necessary to make auxin and cytokinin. The abnormally high levels of these hormones produced in the plant cells prompt cancerous cell growth. This resembles the situation in culture where the continuous availability of the same two hormones in the culture medium causes cell proliferation and callus formation.

After the genes have been transferred from a single bacterium to a single plant cell and integrated into a chromosome of the plant cell, the hormones in the culture medium induce the cell—which is now transformed with the bacterial genes—to proliferate. So the bacterial genes will be present in every cell derived from the transformed cell. This transfer of DNA between the different species is sometimes called "promiscuous," because DNA is usually transferred only between individuals of the same species, as part of the process of sexual reproduction. Nevertheless, this transfer of DNA between totally unrelated organisms is a natural process. Soon molecular biologists realized that if they substituted other genes for the genes that the bacterium transferred to the plant cell, they could obtain transformed plant cells that carried any gene they wished to introduce into the plant.

Not all types of plants can be easily transformed with *Agrobacterium* (especially the monocots), so researchers have developed other methods to introduce foreign DNA into the genomes of important crops such as maize, wheat, sorghum, sugarcane, beans, and soybeans. John C. Sanford and his colleagues at Cornell University developed a microprojectile gun (or "gene gun") to shoot DNA into plant cells directly, through the cell wall and the cell membrane into the nucleus. They coated small (about 1 micron in diameter) gold or tungsten particles with DNA and constructed a gun that accelerated the particles with enough speed to enter the first cell layer of a tissue. For reasons not yet completely clear, DNA that enters the cells in this way is integrated into the chromosomes and passed on to the cells' progeny.

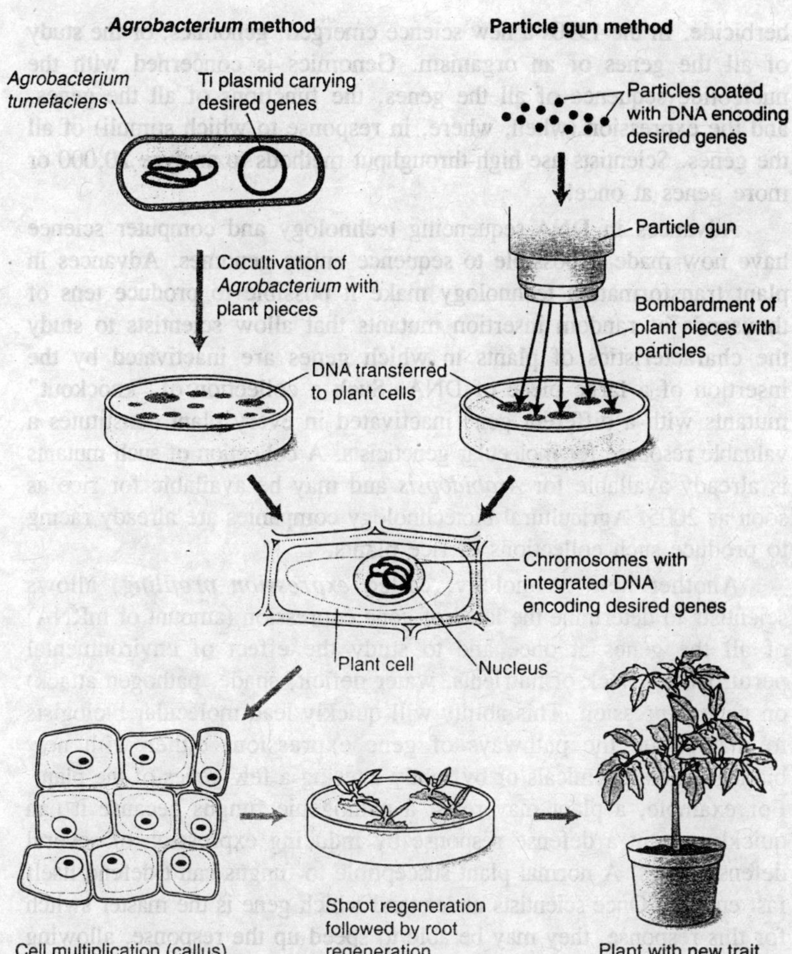

Fig. 2.12. Schematic representation of two different ways to create transgenic plants.

By using tissue culture techniques, as described earlier, researchers can generate a whole new plant from this transformed cell.

Crop Improvement

Gene technologies and tissue culture techniques have dramatically increased the possibilities for crop improvement. It is no longer necessary to rely only on the genes of close relatives of a particular crop because biologists can use genes from any source and can introduce a particularly useful gene into many crops. Using the same bacterial gene, for example, they can make 10 different crops tolerate the same herbicide. In the 1990s a new science emerged: genomics, or the study of all the genes of an organism. Genomics is concerned with the nucleotide sequence of all the genes, the functions of all the genes, and the expression (when, where, in response to which stimuli) of all the genes. Scientists use high-throughput methods to analyze 20,000 or more genes at once!

Advances in DNA-sequencing technology and computer science have now made it possible to sequence entire genomes. Advances in plant transformation technology make it possible to produce tens of thousands of random insertion mutants that allow scientists to study the characteristics of plants in which genes are inactivated by the insertion of a large piece of DNA. Such a collection of "knockout" mutants with a different gene inactivated in every plant constitutes a valuable resource for molecular geneticists. A collection of such mutants is already available for *Arabidopsis* and may be available for rice as soon as 2005. Agricultural biotechnology companies are already racing to produce such collections of rice plants.

Another new technology, called *expression profiling*, allows scientists to determine the level of gene expression (amount of mRNA) of all the genes at once and to study the effect of environmental perturbations (lack of nutrients, water deficit, shade, pathogen attack) on gene expression. This ability will quickly lead molecular biologists to manipulate the pathways of gene expression, either with new biodegradable chemicals or by subtly altering a few genes of the plant. For example, a plant may resist a pathogenic fungus because it can quickly mount a defense response by inducing expression of several defense genes. A normal plant susceptible to fungus can't defend itself fast enough. Once scientists understand which gene is the master switch for this response, they may be able to speed up the response, allowing the plant to mount an effective defense. This type of genetic engineering will require subtle changes in genes and will resemble the normal process of evolution more closely than do current genetic technologies that introduce completely new genes.

3

Crop Improvement

Plant growth requires not only carbon dioxide and oxygen from the air but also water and mineral nutrients from the soil. Soil has been called the "placenta of life," because it supplies essential nutrients to all land plants, and the plants in turn feed all the terrestrial ecosystems. Throughout history, humanity's standard of living has depended on the fertility and productivity of the soil. The Fertile Crescent of the Middle East, one of the areas in the world where agriculture originated, is considerably less productive today than it was 10,000 years ago. This decrease in productivity is caused, in part, by changes in weather patterns—possibly as a result of deforestation of the Mediterranean basin—and, in part, by failure of the inhabitants to maintain soil productivity. Soil erosion and salinization are accelerated by poor agronomic practices. Mismanagement and neglect of soil can ruin the arable land, which is a fragile and precious resource. The Harappan civilization in western India, Mesopotamia in Asia Minor, and the Mayan culture in Central America all collapsed partly because of soil degradation. Maintaining productive soils should be one of society's important goals.

A Vital Resource

The formation of soil is a long and complex process involving breakdown of the parent rock into small mineral particles, chemical modification of these particles, and finally continuous addition and decomposition of organic residues from plants, animals, and microorganisms. Solid rock is continually broken down into small particles, a process termed *weathering*. Heating and cooling of rocks, freezing and thawing of water that seeps into cracks, running water, the scouring action of winds carrying small particles, glaciers that creep over the rocks, grinding them into small particles, all contribute to the weathering process. Physical forces break up rocks into smaller particles, and chemical forces can change their chemical properties.

The more a rock is fractured by physical weathering, the faster chemical weathering will occur. Earth's crust contains more than 90 different elements that exist in certain combinations called *minerals*. These minerals usually form small crystalline grains that become cemented together to form rocks. Many minerals are specific combinations of silicon, aluminum, iron, and oxygen, because these four elements are by far the most abundant in Earth's crust. Once the *parent rock* has been broken down into the individual *mineral grains*, these grains are differentially affected by temperature, rainfall, and vegetation.

Table 3.1. Some primary minerals

Minerals group	*Typical composition*
Mica	$KH_2Al_3(SiO_4)_3$
Feldspar	$KalSi_3O_8$
Quartz	SiO_2
Iron oxide	Fe_2O_3
Carbonate	$CaCO_3$

Some, such as quartz grains, are not changed, but are slowly broken into even smaller particles, whereas others, such as feldspar or mica, are actually modified. Feldspar reacts with the carbonic acid in rainwater and forms soluble potassium carbonate and kaolinite, a clay mineral. Mica is even further degraded. It falls apart into potassium carbonate, oxides of iron and aluminum, and clay. This complete dissolving of certain components of the mineral particles is an important aspect of weathering. Indeed, the minerals must be dissolved in the soil solution before plants can take them up. Once the minerals have been dissolved, they can remain in solution, bind to the outside surface of the soil particles, or react with other dissolved minerals to form insoluble compounds. Soil formation requires the accumulation of soil particles.

Both water and wind can carry soil particles from their site of formation to other regions. For example, some of the soils that are the most agriculturally important today formed 10,000 years ago, when winds deposited enormous amounts of clay and silt particles in certain areas. But just as these factors help to form soils, so too they can remove them, in the process of *erosion*. Plants play an important role in preventing erosion, because their roots hold the soil together—a lesson painfully learned in the United States, in Oklahoma and Texas, during the 1930s "dust bowls." Whenever the soil is not covered by plants, winds can whip up huge dust storms, carrying away the soil and depositing it elsewhere. A layer of soil particles may form rapidly or slowly, depending on the nature of the parent rock, weather, vegetation, and topography. Level terrain with forest vegetation, with parent rock that is easily broken down, and with a warm, humid climate, all favour rapid soil formation.

High rainfall and high temperatures promote rapid chemical weathering, because the downward movement of water carries dissolved mineral components to lower layers of the soil and also because chemical reactions go faster at higher temperatures. The percolating water also leaches acids out of the decaying plant residues. Decaying

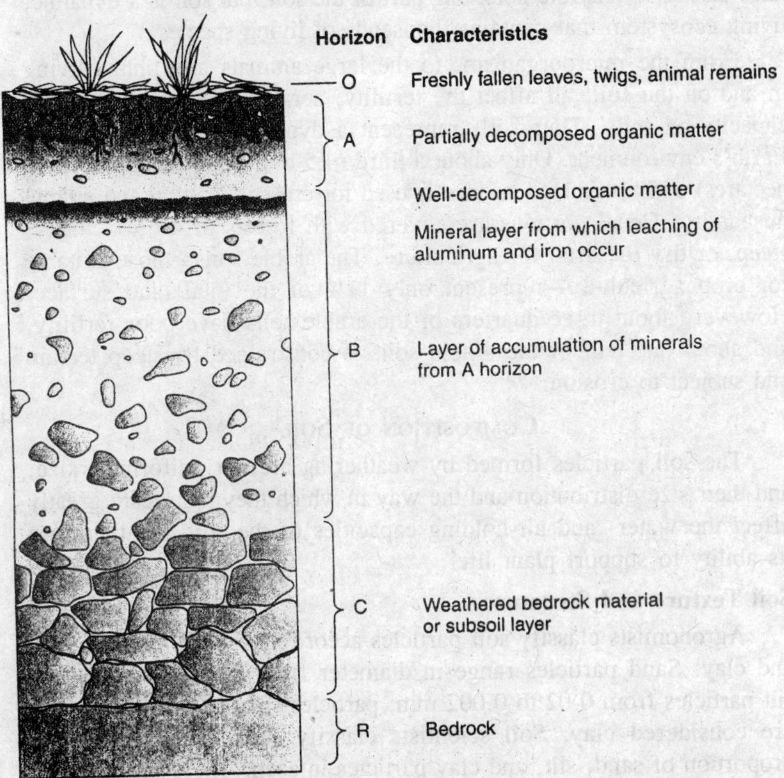

Fig. 3.1. A soil profile, showing the different soil horizons.

needles of conifers produce more acid than leaf litter of deciduous forests, which in turn are more acidic than the decay products of prairie grasses. Thus, the type of vegetation affects chemical weathering. When rain percolates into the soil, there is a slow downward movement of dissolved minerals, bits of organic matter, and the smallest mineral particles.

Once formed, a mature soil usually shows at least three distinct layers or horizons. The A horizon, or topsoil, is the layer richest in the decaying organic matter and dissolved minerals that sustain plant growth. On the average, it is no more than 20 to 30 cm thick. Below the topsoil is the B horizon, or subsoil. Minerals leached out of the A horizon accumulate here. Most of a plant's roots are in the topsoil, but good farming practice often involves breaking up the subsoil to let plants find additional water and nutrients. The third layer, or C horizon, consists of parent rock in the process of being broken up. So far we have discussed only the nonliving part of the soil, but soil is a dynamic, living ecosystem that contains thousands of living species.

From the microorganisms to the large animals and plants living in and on the soil, all affect the fertility, aeration, and water-holding capacity of soils. Thus soils represent a dynamic and vital niche in Earth's environment. Only about a third (4.5 billion of the 13.2 billion hectares) of Earth's land surface is used for crop cultivation and animal husbandry. Equal portions are covered with forests or are too rocky, steep, or dry (deserts) for agriculture. The arable soils—those suitable for crop agriculture—represent only 11% of the total land surface. However, about three quarters of the arable soils have poor fertility, and about one half of the arable soils is constrained by steep terrain and subject to erosion.

Composition of Soil

The soil particles formed by weathering are not uniform in size, and their size distribution and the way in which they aggregate greatly affect the water- and air-holding capacities of the soil and therefore its ability to support plant life.

Soil Texture and Structure

Agronomists classify soil particles according to size as sand, silt, and clay. Sand particles range in diameter from 2 to 0.02 mm, and silt particles from 0.02 to 0.002 mm; particles smaller than 0.002 mm are considered clay. Soil scientists classify soils according to the proportion of sand, silt, and clay particles in each. These proportions, generally called the *soil texture*, are determined by parent material and by extent of weathering. Sandy soils are often formed on sandstone, whereas limestone gives rise to loam soils, and shale results in clay-rich soils. Both water and nutrients are *adsorbed* (bound) on the surface of the soil particles.

Table 3.2. Composition of three typical soils classified according to size of mineral particles

	Types of Particle		
Soil type	*Sand*	*Slit*	*Clay*
Sandy loam	85%	6%	9%
Loam	59%	21%	20%
Clay	10%	22%	68%

Obviously, the more finely divided the particles, the greater their surface area per unit mass, and the greater their capacity to bind water and nutrients. Clay particles have often weathered so much that they have become porous and thus present an even greater binding surface. Sandy soils, having relatively more large particles, do not bind much water or plant nutrients in comparison to clay soils. The individual soil particles generally clump together to form aggregates of varying size and shape. This is especially common with silt and clay particles. The "glue" that sticks the particles together is a combination of lime, iron hydroxides, and humus and other decaying organic matter.

The degree of aggregation of the particles is generally referred to as the *soil structure*. Although soil structure is difficult to define, you can readily recognize it from the size and shape of the lumps formed when a soil is crumbled. Another important property of an agricultural soil is its tilth. When a soil is tilled (plowed, leveled, raked, or rolled), the original structure is partially disturbed and new air spaces are created. The new pores may occupy up to half the soil's volume. The tilth of a soil determines how well water percolates into it and whether seedlings can easily push through the surface layer. With time and as a result of gravity and rain, the tilth will change and become less favourable. In soils that are never tilled—as in no-till agriculture—the continuous channels created by earthworms and roots that permit water percolation and gas exchange are very important. Tilling disturbs these natural continuous channels, but creates new channels for water percolation and gas exchange.

Air and Water

The size of the individual pores is related to the size of the particles or aggregates. Small pores normally occur between small particles, whereas larger pores exist between large particles or aggregates. The small pores, called *capillaries*, are usually filled with water, whereas the larger ones are filled with air. A good, fertile soil should have half the pore space filled with water and the other half with air. Such a distribution provides a good balance among aeration, water percolation, and water storage capacity. The importance of air for plant roots and most other soil organisms is commonly underestimated. Plant roots need energy to grow and to take up minerals, and they obtain this energy by respiring sugars made in the leaves. This respiration requires oxygen and causes CO_2 to be given off.

To maintain the proper balance of oxygen and CO_2 in the soil, gases must move continuously in and out of the soil. Oxygen from the atmosphere must move into the soil, and CO_2 must escape or it will build up to toxic levels. This gas movement requires the presence of continuous air channels. Such channels are made by small burrowing animals, such as worms, or are formed when dead plant roots decay. Gas exchange between the atmosphere and the soil readily occurs when the soil has good structure and good tilth. What happens to dry soil during a prolonged slow rainfall or irrigation? As soon as the water touches the soil particles and organic matter, it binds to them.

A soil thus saturated with water is said to be at field capacity: The small pores are filled with water, and the large pores are filled with air. If still more water is added, it will fill not only the small

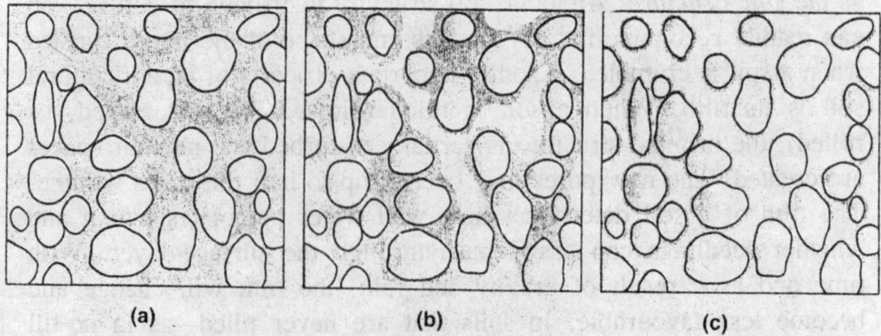

Fig. 3.2. Distribution of water and air in a structured soil. In (a) the soil is waterlogged and all space are filled with water. In (b) the soil is at field capacity; there is air in the air spaces and water in the capillary space. In (c) the soil is beginning to dry out; most of the capillary water is gone and only a thin layer of water surrounds each particle.

pores but also the large ones, expelling the air. The soil is now waterlogged. A waterlogged soil literally suffocates most plants and soil organisms because they die from lack of oxygen. Some plants, such as rice, are specially adapted to thrive in waterlogged soils. The texture, structure, and tilth of a soil together determine how well it can store water and permit water percolation. A clay soil, with its preponderance of small particles, has many small pores and a large water storage capacity.

There may be so few large pores, however, that water cannot percolate downward. Rainwater then gathers on the surface of the soil and eventually runs off, often taking topsoil with it. Adding organic matter to such soils causing aggregates to form, creating larger pores and allowing better percolation. Maintaining plant cover on the land thus helps prevent erosion in three different ways: (1) The roots of the plants stabilize the topsoil, keeping it from being carried away; (2) decaying plants also supply organic matter necessary for aggregate formation, allowing more rapid percolation; and (3) root penetration of the soil offers ways for water to enter after these roots die and decay. The decomposition of organic matter by microorganisms, called *mineralization*, produces organic particles that bind together small soil particles into larger aggregates.

The addition of organic matter may increase the water-holding capacity of sandy soils and improve the drainage of heavy, clay soils. Besides being a source of energy for soil microorganisms, organic matter also acts as slow-release fertilizer to provide mineral nutrients for plants. Thus organic matter improves the physical structure of soils, releases acids for chemical weathering, and feeds the soil ecosystem.

Nature of Soil

Because soil acidity influences the physical properties, the availability of certain plant nutrients, and the biological activity of the soil, it greatly affects plant growth. A soil's degree of acidity depends on the concentration of hydrogen ions (H^+) dissolved in the soil water. In a neutral soil, the H^+ concentration is about 1 part per billion parts of water. An acid soil may have a concentration of H^+ that is 100 to 1,000 times higher, whereas an alkaline soil has a lower H^+ concentration. The acidity or alkalinity of a solution is expressed by a single measurement called the *pH*. A pH of 7 is neutral, a pH of 5.0 (100 times more hydrogen ions) is acidic, and a pH of 9 (100 times fewer hydrogen ions) is basic or alkaline. Neither extreme acidity nor extreme alkalinity is suitable for plant growth or for most other soil organisms. Such conditions also upset soil weathering and the availability of nutrients. Although some plants can grow in strongly acidic or alkaline soils, most crop plants grow best in neutral or slightly acidic soils.

Just over a quarter (26%) of the world's arable land is classified as acidic. In the tropics the percentage is even greater (43%). Acidic soils account for 68% of tropical America, 38% of tropical Asia, and 27% of tropical Africa. We have already noted that acidity depends on vegetation. It also varies as a result of the types of fertilizers that are used (long-term use of ammonium or urea fertilizers cause acidification), and as a result of rainfall that is loaded with acids. The pH of rain is normally just below neutral, but pollution from power plants and cars create strong acids (nitrous acid and sulfurous acid) when they dissolve in the rain. When such acid rain falls on the land, it lowers the pH of the soil. The solubility of aluminum ions from the soil minerals increases when soil pH is 5.5 or lower, and can lead to aluminum toxicity in plants.

The direct effects of air pollutants on plants and the indirect effects caused by acid rain and aluminum toxicity are a major cause of decline in forests in large areas of central Europe and North America. The minerals that are dissolved by chemical weathering sometimes interact with each other to form new, insoluble complexes, A soil's acidity plays an important role in this process, and can greatly reduce the availability of certain nutrients to the plants. For example, phosphate, an essential nutrient, can form a variety of insoluble complexes with other ions in the soil solution. If the soil solution is too alkaline, phosphate readily combines with calcium to form insoluble calcium phosphate.

When the soil solution is too acidic, phosphate combines with iron, aluminum, and manganese to form insoluble products. Thus phosphate is most readily available to plants when the soil solution is neutral or slightly acidic. A soil's degree of acidity can be adjusted to make it more nearly neutral by adding lime, Acidic fertilizers, such as ammonium sulfate, can neutralize excess alkalinity with time. The uptake of ammonium stimulates acid excretion from the roots. The most effective method to acidify the soil is to use elemental sulfur. When microorganisms oxidize elemental sulfur, they release H^+ into the soil. Farmers may also use such treatments to optimize the growth of certain crops. For example, potatoes grow best in a somewhat

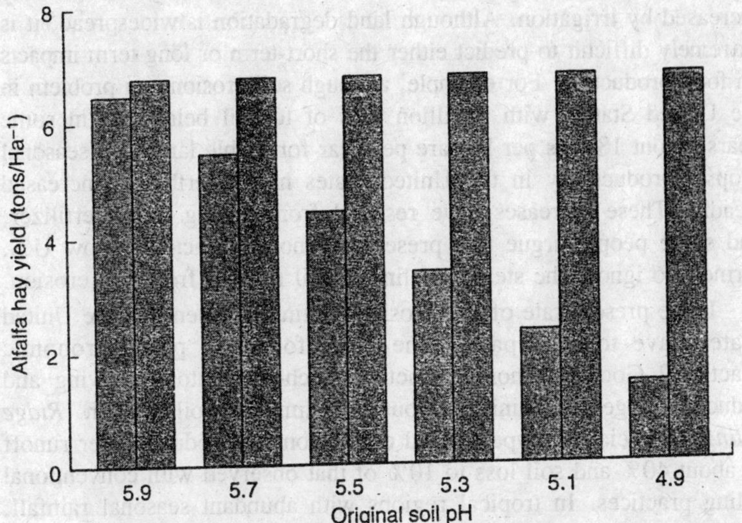

Fig. 3.3. Yields of alfalfa as a function of original soil pH, with or without lime.

acidic soil, whereas alfalfa thrives in a soil that is very slightly alkaline, but grows very poorly on acidic soil.

Soil Degradation

Poor management of natural resources—deforestation and misuse of agricultural—lands has led to extensive soil degradation all over the world. Three quarters of the area degraded by inappropriate agricultural practices, overgrazing, and deforestation are in the developing world. A recent study of all the world's agricultural soils indicates that about 67% of agricultural soils have been and are being degraded by erosion, salinization, compaction, nutrient losses, pollution, and biological deterioration. As much as 40% of the agricultural soils are seriously degraded, reducing world crop productivity by about 16%. Soil degradation is defined as a decline in soil quality that impairs the soil's current or potential capacity to produce crops. It includes physical, chemical, and biological deterioration.

Soil depletion—caused by removal of nutrients by water, or by cropping and removing harvested produce—is a less severe problem because people can remedy it by fertilization, but it is sometimes the beginning of soil degradation. Soil degradation takes many forms: erosion (removal) of the topsoil by wind or water; compacting by traffic; alteration of soil properties by intensive cropping; laterization by deforestation and raised temperature of tropical soils; decline in organic matter and biological activity of soil; and soil salinity or alkalinity increased by irrigation. Although land degradation is widespread, it is extremely difficult to predict either the short-term or long-term impacts on food production. For example, although soil erosion is a problem in the United States, with 5 billion tons of topsoil being lost in some years (about 18 tons per hectare per year for arable land with seasonal crops), productivity in the United States has nevertheless increased steadily These increases have resulted from adding more fertilizer, and some people argue that present agronomic practices allow U.S. farmers to ignore the steady decline in soil fertility from soil erosion.

If the present rate of soil erosion continues, when will the United States have to start paying the price for these poor agronomic practices? Good agronomic practices such as contour plowing and reduced tillage can minimize, but not eliminate soil erosion. *Ridge tilling*, a specialized type of land cultivation, can reduce water runoff to about 40% and soil loss to 10% of that observed with conventional tilling practices. In tropical regions with abundant seasonal rainfall, erosion rates of 50-100 tons per hectare per year are not uncommon and often affect 30-50% of a country's arable land.

Soil erosion generally does not greatly reduce yields on deep soils, especially if farmers use fertilizers to compensate for losing nutrients, but can be a real problem on shallow soils. The loss of 20 cm of topsoil caused a massive decline in cassava yield, even with supplemental fertilizer. A different type of soil degradation called *desertification* occurs in many arid and semiarid regions. Some 50 years ago, observers noted gradual changes in the vegetation of the Sahel, a vegetation zone in West Africa, bordered to the north by the Sahara and to the south by dry savanna. The changes were described as the advance of the Sahara, because along its northern edge the Sahel was becoming like the Sahara, and along its southern edge the vegetation that characterized the Sahel was moving into the dry savanna.

The changes in vegetation were caused by a combination of climatic changes and poor land management resulting from population pressures. Overgrazing irreversibly changed the land by denuding it of protective cover. Overgrazing and the resulting desertification are not confined to the Sahel. Other semiarid regions such as Patagonia, Iran, Syria, India, and Kenya also show desertification, often because pastoralists grazed three to five times more grazing animals than the meager vegetation could support. When people irrigate land with water that contains dissolved salts (all groundwater and river water contains dissolved salts, but rainwater does not), there is always the danger that salinization, another form of land degradation, will occur.

Salinization, or the accumulation of salts in topsoil, happens in areas where mean annual potential evapo-transpiration (evaporation from soil plus transpiration by plants) exceeds precipitation by a factor of 1.3. The salts that accumulate can come from irrigation water, but salts can also rise to the surface in groundwater. In coastal areas, the intrusion of seawater below ground level can carry salts into topsoil. Salts accumulate because there is not enough rain to carry them out of topsoil into groundwater and/or rivers. If water is abundantly available, people can sometimes leach salts out of a salinized soil to be carried away into streams. However, this procedure may create problems downstream. Thus a greater abundance of saline soils occur in semiarid and arid regions than in tropical regions.

The salinity of a soil solution is expressed as molarity. A solution containing 58 g per liter of sodium chloride (sea salt) is said to be 1 molar (M) or 1,000 millimolar (mM). Saline soils are defined as soils that contain enough dissolved salts to osmotically stress plants and reduce growth. This is equivalent to approximately 40 mM NaCl (seawater contains approximately 500 mM NaCl or 29 g per liter). Some soils have reached salt concentrations twice that of seawater! Plants that are very sensitive to salinity may be affected by concentrations as low as 15 to 20 mM salts. The dissolved salts in saline soils are made up of many ions (just as in seawater); often Na^+ and Cl^- are the predominant ions but not always. Sulfates and carbonates may also significantly contribute to the saline solution.

Saline soils are increasing around the world, threatening the most productive, irrigated lands. Saline soils make up about 23% (340 million ha) of the world's cultivated land and are increasing by about 10 million ha per year. For example, two thirds of the saline soils are located in Asia, where two thirds of the world population depends on rice, a salt-sensitive crop produced primarily in low-lying areas. With global warming, the seas are expected to rise, which increases the threat of seawater intrusion into the low-elevation coastal areas.

Transpiration

Water is the medium in which soil minerals dissolve, and mineral nutrients must dissolve in soil water to enter the roots of plants. This important role of soil water, however, cannot explain the extremely high water requirements of many crops. The requirement that CO_2 needed for photosynthesis must be dissolved in water has resulted in a two-way gas exchange: The cost to the plant of allowing the CO_2 to enter the leaf is that water vapour escapes through the stomates. Water vapour in intercellular spaces is quickly replenished by evaporation of liquid water inside the leaf. However, liquid water within the leaf must be replenished. Therefore leaves obtain water from the conductive tissues, which in turn get their water from the soil via the roots.

Thus during the daytime, when the stomates are open, there is a continuous stream of water through the plant. This process is called *transpiration*. At night, when the stomates are closed, the transpiration stream stops, although the plant may continue to take up water from the soil until it has completely rehydrated. Transpiration dissipates heat in plants, similar to perspiration in humans. It also helps transport minerals and organic molecules from the root to the shoot. If the rate of water uptake from soil is inadequate to replenish the water lost from leaves by transpiration, leaves will lack water, and if the deficit is really severe the stomates will partially or fully close. Partial closure of stomates reduces further water loss, but it also reduces photosynthesis, because CO_2 movement into the leaf is reduced. On a hot summer day, plants may experience transient wilting because water loss from leaves is more rapid than water uptake into roots, even

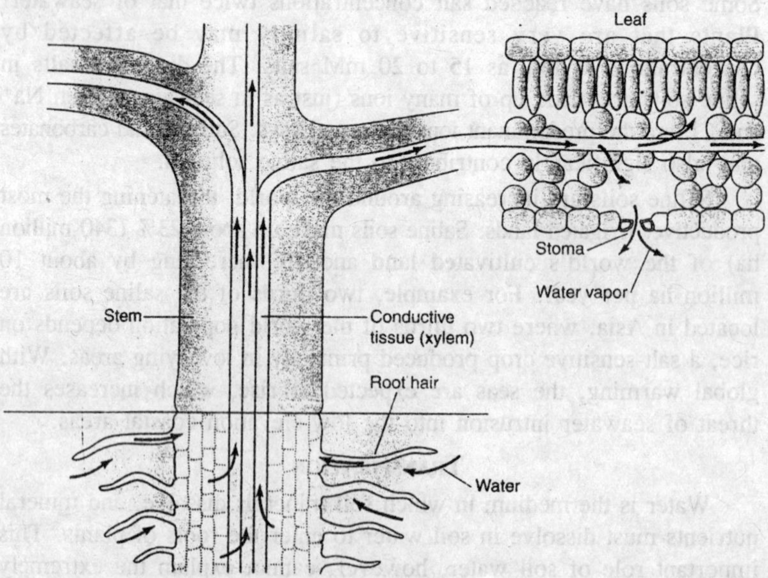

Fig. 3.4. The transpiration stream through a plant. Arrows indicate the direction of movement of water or water vapour.

though there is enough water in the soil. This transient wilting can severely depress photosynthesis in the middle of the day. Thus provision of adequate soil water is crucial for optimal plant productivity. Water is lost not only by transpiration but also by evaporation from the soil surface. The sun warms the soil and causes some of the water to escape as water vapour.

The term *evapotranspiration* describes the total amount of water lost from a plant-soil system. The total productivity of an ecosystem is very closely tied to the amount of water lost each year through evapotranspiration. Availability of water in soil determines plant growth more than any other factor. Thus it is not possible to "make the desert bloom" without irrigation. Although farmers can obtain small yields with drought-adapted plants, large yields always depend on supplying additional water. On a worldwide scale, water availability is a major limitation to crop production. Only a tiny fraction of the world's water is actually used by plants, but plants transpire daily an amount of water nearly equal to their total water content.

Water infiltrates the pores between soil particles and is held there with varying degrees of tenacity. Soil water can be measured directly by weight loss of the soil on drying, or by using special devices. The

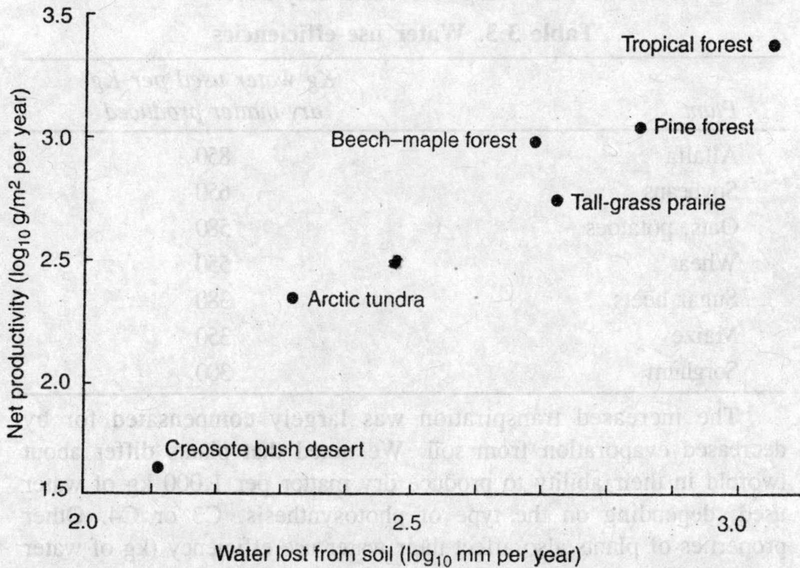

Fig. 3.5. The relationship between the water lost from the soil (by evaporation and transpiration) and plant productivity.

tenacity or soil tension with which soil particles hold water increases as soil water content decreases. Water tension in soil at any moment controls movement of soil water in soil and its use by plants. This water tension (a negative pressure) is expressed in units called megapascals or MPa. When tension is low—between -0.01 and -0.03 MPa (1 atmosphere is approximately equal to 0.1 MPa)—water moves to lower layers because of gravitational pull. Also, when soil water tension is -1.5 MPa or below the adhesive force is so strong that plant roots can hardly extract water from soil. At approximately this water tension, most plants permanently wilt and stop growing. Soil water between about -0.01 and -1.5 MPa is considered available to plants. The water requirements of any plant depend on its environment and nutrition.

Plants are grown in nutrient-rich soil are usually large and have extensive root systems. The total amount of water they lose by transpiration increases, but they use water more efficiently. Placing plants closer together increases total transpiration, because there are more plants, but decreases direct evaporation from soil, because sunlight does not hit soil directly. Where the number of maize plants per hectare was increased from 20,000 to 40,000, maize yield increased by 65%, but total water usage increased only 20%.

Table 3.3. Water use efficiencies

Plant	*Kg water used per Kg dry matter produced*
Alfalfa	850
Soybeans	650
Oats, potatoes	580
Wheat	550
Sugar beets	380
Maize	350
Sorghum	300

The increased transpiration was largely compensated for by decreased evaporation from soil. We noted that plants differ about twofold in their ability to produce dry matter per 1,000 kg of water used, depending on the type of photosynthesis, C3 or C4. Other properties of plants also affect their water use efficiency (kg of water used per kg of dry matter produced). On this list only maize and sorghum have C_4 photosynthesis. The others are all C_3 plants. Water use efficiencies are usually measured under ideal growth conditions, and do not predict how plants will perform with a normal (or abnormal) rainfall regime. Crop yield (usually seed production) is the bottom line, not dry matter production.

Mineral and Plants

The fact that plants need to take up minerals from the soil was shown for the first time in 1699 by the British botanist John Woodward. He grew small cuttings of mint in rainwater, river water, and water to which he had added some soil. When he weighed the plants some time later, he concluded that growth was related to the amount of dissolved substances in the water. But which chemical elements in the soil water are required for growth? Earth's crust consists of over 90 different chemical elements. An analysis of plant ash, the residue that remains after plants are burned, reveals that plants may take up as many as 50 or 60 different elements from the soil. Are all these elements essential, or does the plant take up whatever it happens to find in the soil? This question was first investigated in a systematic way in 1860 by the German plant physiologist Julius Sachs, who grew plants with their roots immersed in solutions of minerals (hydroponics). He found that many plants could grow satisfactorily in solutions containing only three mineral salts: calcium nitrate, potassium phosphate, and magnesium sulfate.

These salts provided the plants with six elements: calcium, potassium, magnesium, nitrogen, sulfur, and phosphorus—termed the *major nutrient elements* or *macronutrients*. If anyone of these elements was omitted from the culture solution, the plants did not grow well, so Sachs concluded that these six elements were essential for the plant. Later he also found iron to be essential for growth. These experiments showed that plants did not require any organic substances (such as vitamins), a question hotly debated at the time. Using hydroponics, it is relatively easy to demonstrate that these major nutrients are required for healthy plant growth.

Modern research suggests that the minerals used by Sachs were "contaminated" with small amounts of other minerals. Advances in chemistry made it possible to obtain much purer chemicals, and biologists later showed that plants also require trace amounts of seven other minerals. These seven, termed the *minor nutrient elements* or *micronutrients*, are boron, copper, chlorine, manganese, molybdenum, nickel, and zinc. Iron, which Sachs already knew to be essential, is also on the list of micronutrients. Together, these 14 nutrient elements are known as the *essential plant nutrients*. A mineral nutrient is said to be essential if without it the plant cannot complete its normal life cycle (flower and set seed). A few other elements are apparently needed by only some plants. Sodium is required by some plants that prefer salty environments for growth. Diatoms (a major component of oceanic phytoplankton) and some cereals, such as rice, need silicon. The microorganisms that live in the roots of leguminous plants and fix nitrogen need cobalt.

Plants take up from the soil the other 40 or 50 elements present in plant ash even though they may not require these chemicals for growth. This somewhat indiscriminate uptake of elements may benefit animals that eat the plants. For example, animals require sodium and iodine, two elements plants don't need. Plants may also accumulate elements that are toxic to animals. Certain plants in the genus *Astragalus*, called *locoweeds*, accumulate the element selenium. Selenium apparently does the plants no harm, but it kills the sheep, cattle, and horses that eat locoweeds.

Transport of Minerals

The 14 essential elements or nutrients just discussed must dissolve in soil water before plants can take them up. When minerals dissolve in water, they break up into smaller electrically charged particles called *ions*: positively charged *cations* and negatively charged *anions*. To understand the importance of this charge for the plant nutrients, consider again the surface properties of the soil particles, both mineral and organic, which determine not only the soil's ability to bind water but also its ability to bind many plant nutrients. Mineral and organic soil particles have an overall negative charge. Because opposite charges attract, cations bind electrostatically to these particles. As a result, most cations in a soil are more or less firmly bound to the particles, although some remain in the soil solution. The anions, in contrast, remain in the soil water. Because of their negative charge, they do not bind to soil particles.

As rainwater percolates through the soil, it carries anions and cations with it down to the groundwater and sometimes into streams and waterways. As noted earlier in this chapter, this process is called *leaching*. Because the anions are not bound to the soil particles, they are more easily lost by leaching than are the cations. However, phosphate is not leached, because it is rapidly tied up in the soil as insoluble precipitate. As soon as cations are removed from the soil solution, either by leaching or because they are taken up by plants,

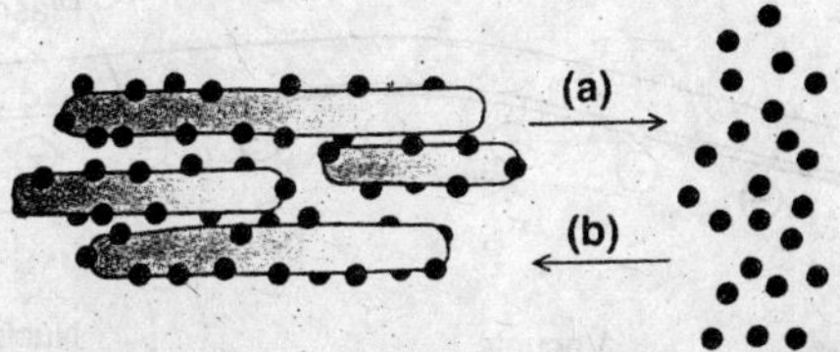

Fig. 3.6. Equilibrium between cations in the soil solution and those bound to the surface of clay particles.

they are replaced by others released from particle surfaces. Thus an equilibrium is kept between cations in soil water and cations bound to particles. Because the surface area available for binding nutrients depends on soil texture, a soil's texture greatly influences its fertility. The greater the surface area of the particles, as in clay soils, the more plant nutrients they can bind. As water percolates down through the soil, it carries some nutrients down with it, especially unbound anions such as nitrate.

In some agricultural areas, nitrate coming from fertilizers or produced by the decay of organic matter sometimes pervades groundwater at such high concentrations that the water is not suitable for use by babies. To take up nutrients from the soil solution, most plants must develop an extensive root system. The root system provides the plant with the surface area it needs to exploit a large volume of soil. In most plants, this surface area is increased even more either by root hairs—filamentous extensions of root epidermal cells—close to the tips of the roots, or by symbiotic fungi called *mycorrhizae*. Ions the root absorbs from soil come in contact with the root surface, either because they have moved through the soil with soil water (some nutrients move more readily than others), or because the root has grown into a previously unexploited area of soil.

If the soil is poor in mineral nutrients, the plant develops a much more extensive root system than if the soil is nutrient rich. However, this root system growth will be at the cost of shoot system, which usually produces the crop. Special proteins in the plasma membranes are needed for ion transport, forming either *ion channels* or *ion pumps*. The uptake of most ions into root cells requires energy because the concentration of the nutrient ions in the soil solution is much lower than within the cell. In terms of energy, this uphill process indirectly requires ATP. ATP maintains the plasma membrane in an "energized" state through the action of H^+-ATPases (proton or hydrogenion pumps). The pumping of H^+ through the plasma membrane to the outside of

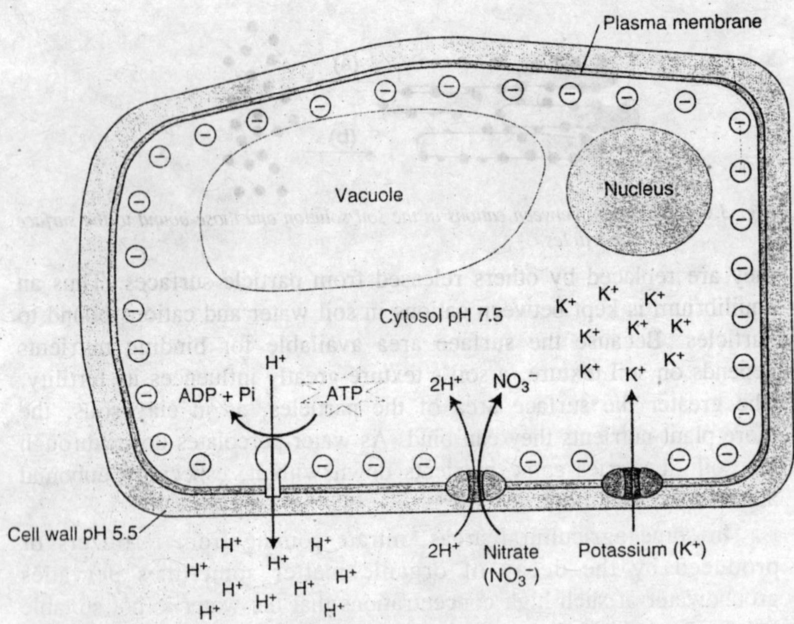

Fig. 3.7. Uptake of nutrients through the plasma membrane is powered indirectly by ATP.

the cell sets up a H^+ gradient: a greater concentration of H^+ and therefore greater acidity and lower pH on the outside than on the inside. This proton gradient powers the uptake of sugars, amino acids, and mineral nutrients such as nitrate through specific carrier proteins. The excess of negatively charged ions left in the cell creates a membrane potential that is a source of energy to move positively charged nutrient ions such as potassium to the inside. Once the mineral nutrients are inside the root cells, they pass from cell to cell until they reach the conductive xylem tissues in the center of the root. Once in the xylem, dissolved minerals can be carried upward with the transpiration stream. The various nutrients have specific roles in plants that are often similar to their roles in animals. For example, both plants and animals require nitrogen and sulfur to synthesize amino acids, and they need phosphorus to make phospholipids and nucleic acids.

Many minor nutrients, needed only in trace amounts, help enzymes carry out biochemical reactions. Some nutrients have unique roles in plants, such as calcium, which participates in maintaining cell wall integrity, and magnesium, which is part of the chlorophyll molecule. Nutrients, especially potassium, also play an important role in maintaining the turgidity of the plant's cells and consequently of the organs as a whole.

Potassium ions (and other solutes) become concentrated within the cell, and this drives water uptake into root cells. The influx of water causes cells to swell. Because cell walls resist stretching, pressure builds within each cell. Water pressure gives additional rigidity to cells and plant organs (analogous to air in a tire). When water loss by evaporation exceeds water uptake, pressure inside cells drops. As a result, plant organs (leaves, in many plants) do not retain their shape but collapse, and the plant wilts. Water pressure also drives cell enlargement in the plant's growing zones, thus playing a key role in plant growth.

Mineral Deficiencies

If the soil is deficient in even one plant nutrient, plant growth will be retarded and crop yield may be diminished. If the deficiency is severe, the plants will develop visible symptoms that are diagnostic of the lacking element. Treatment usually involves fertilizing the soil with the nutrient. Unless this is done at an early stage in plant growth, severe crop damage may occur. Furthermore, nutrient-deficient plants are more susceptible to pathogens. The appearance of symptoms such as stunted growth, leaf yellowing, "burned" leaf margins, or death of the terminal bud usually indicates a rather severe deficiency in one or more plant nutrients.

Plants do not develop these signs when the nutrients are only marginally deficient. However, marginal deficiencies may also reduce crop yield. This "hidden hunger" is difficult to diagnose. It can sometimes be uncovered by measuring amounts of plant nutrients in soil or in plant tissues; however, neither method is completely satisfactory. Although total amounts of particular nutrients present in soil can be measured, chemical tests cannot tell what proportion of the nutrients is available to the plants. Availability can only be measured by experiments with plants and measuring how crop yield responds to addition of nutrients (fertilizers) to a specific soil.

After yield responses are known, plants can also be analyzed to measure how much of the nutrients the plants took up from the soil. Harvesting crops removes large amounts of the major nutrients. These must be restored if soil fertility is to be maintained, and failure to restore them gradually reduces soil fertility and crop productivity. This is adequately demonstrated by the low yields obtained on experimental plots that have not been fertilized for many years. Plants may display nutrient deficiency symptoms even when a nutrient is present in the soil but unavailable to the plants, as when soil is too acid or too alkaline. There are many ways to increase or restore the fertility of the soil so that it will support crop production. A second widely practiced method is to *incorporate organic residues* and wastes into the soil.

Table 3.4 Nutrients contained in the total above-ground plant material in a hectare of corn yielding 10,000 kg of grain

Nutrient	*Kg/Ha*	*Nutrient*	*Kg/Ha*
Nitrogen	200	Iron	2.3
Phosphorus	42	Manganese	0.4
Potassium	205	Copper	0.1
Calcium	41	Zinc	0.42
Magnesium	48	Boron	0.19
Sulfur	24	Molybdenum	0.01
Chlorine	86.0		

A survey of agricultural practices around the world reveals that farmers incorporate all kinds of organic residues into the soil: manure, fish wastes, algae, human excrement, crop residues, sawdust, and composted kitchen scraps, among others. These organic fertilizers decompose in the soil, releasing their nutrients. Organic fertilizers provide a steady supply of nutrients but may not release enough nutrients during the period of rapid vegetative growth, when demand is greatest. Slower growth in the spring and early summer often means a reduced crop in the fall. A third method to restore the fertility of the soil is to add *inorganic fertilizers* (also called "chemical" or "synthetic" fertilizers). Chemical fertilizers release nutrients rapidly and can be made available when the plant needs them most.

A disadvantage is that some plant nutrients, especially nitrate, may leach out of the root zone before plants can use them. This problem is greatest if the fertilizers are applied just before or at planting time. The seedlings take several weeks to develop a root system large enough to take full advantage of the fertilizer. Minor nutrient elements are sometimes applied directly to the leaves as a spray. Weak solutions of iron, zinc, or copper are commonly used on crops such as pineapple, citrus fruit, or avocado. The three plant nutrients used most widely in inorganic fertilizers are nitrogen, phosphorus, and potassium. Adding fertilizer to the soil tends to

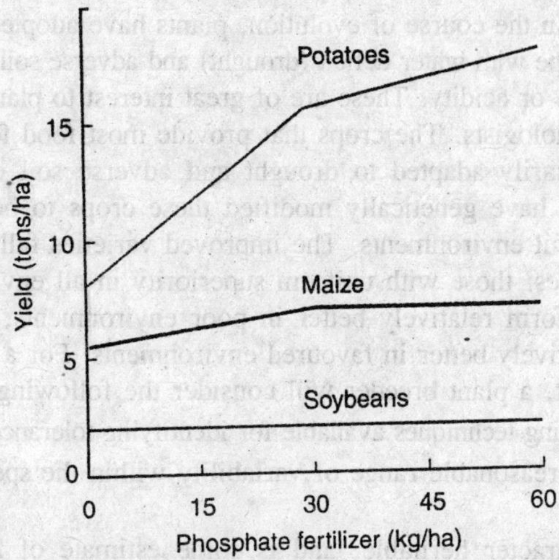

Fig. 3.8. Yield responses of three different crops to phosphorus fertilizers on a low-phosphorus silt loam soil.

stimulate overall plant growth and hence crop production. However, not all crop plants respond in the same way to adding a given amount of fertilizer to a certain soil.

Even within a given species there can be much variation from line to line, depending on genetic makeup. Integrated nutrient management (INM) is an approach to farming that uses both organic and inorganic plant nutrients for correcting nutrient imbalances in soil, to attain higher productivity, prevent soil degradation, and safeguard future food supplies. It relies on nutrient conservation and application, new technologies to increase nutrient availability to plants, and transfer of knowledge between farmers and researchers. Resource-poor farmers in the tropics can benefit from INM because it helps maintain or adjust soil fertility and plant nutrient supply to an optimum level for sustaining productivity. The appropriate combination of mineral fertilizers, organic manures, crop residues, compost, or N_2-fixing crops varies according to land use system and ecological, social, and economical conditions.

Adverse Soil Conditions

Of the total land in the world, only 11% of the land area is arable land and even on this land crops do not reach their genetic potential because of abiotic stresses. Climate (drought, cold, heat) and adverse soil conditions such as lack of nutrients, shallowness of the soil, and excess water limit crop productivity on the other 89% of the world's land. In the course of evolution, plants have adopted different strategies to cope with water deficit (drought) and adverse soil conditions such as salinity or acidity. These are of great interest to plant breeders and crop physiologists. The crops that provide most food for humans are not necessarily adapted to drought and adverse soil conditions. Plant breeders have genetically modified these crops to better adapt them to stressful environments. The improved varieties fall into three major categories: those with uniform superiority in all environments, those that perform relatively better in poor environments, and those that grow relatively better in favoured environments. For a successful breeding result, a plant breeder will consider the following:

- Are screening techniques available for identifying tolerance to stress?
- Is there a reasonable range of variability within the species to be bred?
- Is the character heritable, and is some estimate of heritability available?
- Are there any strong, undesirable genetic correlations with stress tolerance?
- Can an estimate be made of improvement to stress tolerance in the field?

With this information in hand a breeding program can be undertaken.

Drought

Drought is a complex syndrome involving three main factors—timing, intensity, and duration of water deficit—all of which vary widely in nature. The extreme year to-year and place-to-place variability in these three factors makes it difficult to define plant traits required for improved performance under all drought situations. Impact of drought on crop yield is a function of duration, crop growth stage, crop species or variety-within-species, and soil and management practices. A drought of about two weeks during flowering can result in complete loss of grain yields.

Droughts are an inevitable and recurring feature of world agriculture and, despite improved ability to predict their onset and modify their impact, drought remains the single most important factor affecting world food security. The problem of drought is most widespread in the developing world, where it has hastened the collapse of many fragile and unstable food production systems.

Water deficits cause morphological and physiological changes in plants. Morphological responses to water deficits consist of leaf shedding, leaf rolling, leaf angle changes, and an increase in the root-to-shoot ratio. Physiological responses caused by water deficits include changes in leaf wax thickness, transpiration, respiration, stomate behaviour, photosynthesis, translocation, mineral nutrition, and protein loss.

When the plant lacks adequate water, the stomates will close partially and eventually completely so that the plant does not dry out. This process is regulated by the hormone abscisic acid (ABA), which is transported by the sap from the roots to the leaves. The ABA triggers not only closure of the stomates, but a host of other processes that allow the leaves to survive this period of water deficit.

Plants have three ways of coping with drought: escape, avoidance, and tolerance. Drought escape allows a plant to grow and complete its life cycle before soil moisture becomes limiting. Drought avoidance is an alternate mechanism by which plants can maintain full hydration of their leaves and stems even under limited soil moisture conditions. They do so by decreasing water loss from the shoot or by more efficiently extracting moisture from the soil. The architecture of the plant is adapted to dry conditions, so the cells never experience water deficit. True drought tolerance, in contrast, functions at the tissue or cellular level. When the cells begin to dry out, they respond by synthesizing substances (sugars, proteins, and so on) that permit the cells to function normally even under water deficit.

Adverse Soil Conditions

Adverse soil conditions such as salinity, acidity, and nutrient deficiency and toxicity are problems for crop productivity throughout the world. In the past, farmers have amended such soils to meet plant needs. This approach requires farmers to purchase inputs: fertilizer, lime, irrigation water. These practices may no longer be economically feasible or environmentally sound. An alternative strategy is to select or improve plants for production on problem soils with fewer purchased inputs.

In many instances, adaptation to adverse soil conditions requires tolerance to excessive levels of mineral elements such as Al and Mn in acid soils, Mn and Fe in waterlogged soils, and sodium chloride in saline soils. Thus, tolerance to multiple stresses is often necessary for adaptation. In the case of rice, scientists from IRRI, the International Rice Research Institute in the Philippines have made impressive progress. In a large-scale screening and breeding program, they screened about 200,000 rice cultivars and breeding lines for tolerance to adverse soil conditions. They have developed adapted cultivars yielding up to 2,700 kg per hectare more than traditional cultivars in farmers' fields on unamended soils with adverse soil chemical conditions. In hybridization programs, plant breeders have successfully used as parents materials identified as tolerant. This has allowed marginal land to be brought into rice production without the need to take costly land reclamation measures.

Poor crop productivity and soil fertility in acid soils are mainly caused by a combination of Al and Mn toxicity and nutrient deficiencies (mainly P, Ca, and Mg). Among these problems, Al toxicity and P deficiency have been identified as the most important constraints to crop production. These two factors limit production of maize, sorghum, and rice in developing countries located in tropical areas of Africa, Asia, and Latin America. The most easily recognized symptom of Al toxicity is inhibited root growth. Because cell elongation is inhibited, root cells become shorter and wider. As a consequence, root elongation is impaired and the roots have a "stubby" appearance when grown in the presence of Al in solution; this has become a widely accepted measure of crop sensitivity to Al. There are two major mechanisms of plant resistance to toxic levels of Al in soil. First, the plant may exclude Al from being taken up by secreting organic acids that bind Al or by modifying the pH of the rhizosphere. Second, the plant may tolerate toxic concentrations of Al within the cell (for example, by binding Al internally after taking it up).

Salinity

Salinity can injure plants in three different ways: It can alter water relations, cause ion toxicity, and cause ion imbalance and nutrient deficiency within the plant. How each of these factors influences a plant depends on many environmental variables, such as humidity, temperature, soil structure and aeration, light intensity, composition and concentration of ion species in solution, and duration of exposure to salinity. Also, genetic and developmental variables influence the plant's sensitivity to salinity. Not surprisingly, therefore, plant responses to salinity can be very complex and there is a wide range of responses, which vary considerably with different genotypes.

Scientists who study responses and adaptations of plants to salinity want to understand the underlying physiological mechanisms and the genetic characteristics. Plants that have evolved in saline environments

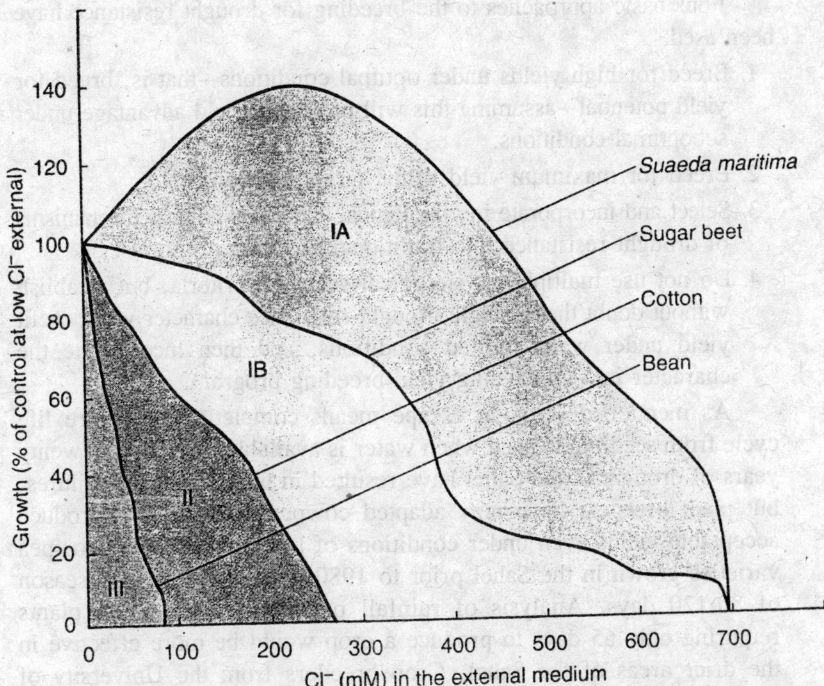

Fig. 3.9. The effect of salinity on the growth of different species of plants.

and tolerate salinity are called *halophytes*. Most plants, including virtually all crop plants, are *nonhalophytes*. Nonhalophytic plants are not adapted to saline environments, although some (such as sugar beets) may have halophytic ancestry. Based on their growth response, plants have been classified into three groups. Group I consists of halophytes, Group II represents a group of halophytes and nonhalophytes whose growth response to salinity overlaps, and Group III represents very sensitive nonhalophytes. The growth of most plants (Groups IB, II, and III), including many of the halophytes, is inhibited as the salinity levels in the soil solution increase. However, the growth of some halophytes (Group IA) can be stimulated by saline conditions.

Crop Resistance

Improving drought resistance of crops is an important aim of plant breeders. Three factors are involved in effective water use by plants: (1) capturing as much water as possible; (2) using the captured water as effectively as possible for growth; and (3) directing the photosynthate into harvestable crop (for example, grain).

Four basic approaches to the breeding for drought resistance have been used.

1. Breed for high yields under optimal conditions—that is, breed for yield potential—assuming this will provide a yield advantage under suboptimal conditions.
2. Breed for maximum yield in the target environment.
3. Select and incorporate morphological and physiological mechanisms of drought resistance into traditional breeding programs.
4. Do not use multiple physiological selection criteria, but establish without doubt that a single drought-resistance character will benefit yield under water-limited conditions, and then incorporate the character into an existing yield-breeding program.

As mentioned, drought escape means completing the entire life cycle from seedling to seed when water is available in the soil. Twenty years of drought in the Sahel have resulted in numerous crop failures, but plant breeders have now adapted cowpea varieties that produce acceptable yields even under conditions of low rainfall. Most cowpea varieties grown in the Sahel prior to 1980 required a growing season of 80-120 days. Analysis of rainfall patterns indicated that plants requiring only 65 days to produce a crop would be more effective in the drier areas of the Sahel. Crop breeders from the University of California crossed a variety that flowers early under the hot long-day conditions prevalent in the Imperial Valley of California with a day-neutral variety from Senegal, and selected from the progeny a variety that does well under conditions of low rainfall (short rainy season).

When rainfall was sufficient (452 mm), the new variety did not perform better than the parents or the local variety. But as the rainfall decreased and yields dropped dramatically, the new variety produced four times more bean than the local variety and 50% more than the parent from Senegal. At intermediate rainfall, the new variety outperformed both parents. It is important to note that when rainfall is low, crop production is also low, even for a drought-adapted variety.

Scientists from Pioneer Hi-Bred International Inc. Ltd. Of Iowa, in the United States, used 500 field sites to evaluate maize lines and tested whether plants selected to have improved yield under water-limited conditions also have improved yield under optimum conditions. Their experiments provide the strongest possible evidence that improvement under water-limited conditions need not sacrifice yield under favourable conditions. The number of replications, genotypes, and field sites was so large that this conclusion now seems beyond doubt. To develop drought-resistant maize, researchers at CIMMYT, the International Center for Maize and Wheat Improvement in Mexico, began selective breeding with the *Tuxpeno* variety that was already well adapted to tropical lowlands. They bred this drought tolerant *Tuxpeno* maize under drought conditions, selecting at flowering the plants with silks (female flowers) that appeared soon after the male flower emerged. When tassel and silk development most nearly coincided, grain production was highest.

The selected plants were then bred further over a decade for even greater resistance to drought. Scientists found that such plants allocate more carbohydrates, or energy, to the ear, which allows them to produce more grain with less water. This selective breeding led to the *Tuxpeno sequia* variety and others like it, which can be grown under a wide range of tropical conditions. The new breeds also show up to 10% increase in yield during non-drought seasons. In times of severe midseason droughts, these drought-resistant new maize varieties produce 2,800 kg per hectare, a 40% increase over regular maize yields under similar drought conditions.

Using molecular techniques, scientists have identified several classes of genes that confer resistance to water deficit. Some of these genes could be used to engineer plants for drought resistance and better crop yield under drought conditions. First they found enzymes that synthesize osmoprotectants, small molecules that accumulate in the cytoplasm of drought-stressed plants. Plants genetically engineered with the genes encoding these enzymes are more drought tolerant. Second, they found genes that encode transcription factors that regulate entire metabolic pathways leading to drought adaptation. By incorporating such genes into crops, scientists hope to ensure that plants respond rapidly and efficiently to any water deficit and continue all their developmental processes. Yet another approach may be to make stomates more responsive to ABA in plants so that they close more rapidly as soon as the plant senses water deficit in soil. Genes encoding proteins that function in the ABA signal transduction pathway may have this desired effect. Over expression of these proteins results in plants that remain green and active longer in a drying soil because they use water more slowly.

Conventional and GM breeding are complementary approaches and can be expected to enhance the drought resistance and yield of crops. People have entered a new era in which enhanced knowledge of both the physiology of yield accumulation and the physiological basis of genetic variation in drought resistance traits offer the potential for improving breeding efficiency for major food crops in different target environments. Using physiological knowledge and powerful tools of molecular genetic analysis requires a systems approach, with agronomists, physiologists, breeders, and biotechnologists working together with farmers to raise crop yields and farmer income in dry areas, particularly in developing countries.

Betterment of Nutrients

To achieve adequate levels of crop yields on various types of soils, nutrients must be applied. In addition, plants must take up these nutrients, which are present in very low concentrations in the soil solution, with high efficiency. Shortages and the high cost of fertilizer and other amendments, in association with environmental problems (soil and water quality), have intensified the need for more nutrient-efficient plants. Improved efficiency in recovering applied nutrients is becoming a prerequisite for lowering production costs, protecting the environment, and improving crop yield.

Nutrient efficiency is the ability of a crop system to convert nutrient inputs into desired outputs. Supply, availability, or amount of a mineral nutrient is the input, and plant growth, physiological activity, or yield is the output. Efficiency is the relationship of output to input. This can be expressed as a simple ratio, such as kg of yield per kg of fertilizer, or kg of dry matter per g of nutrient. In many soils, the efficiency of plants in recovering applied fertilizer is low.

There are two major approaches to improving nutrient efficiency of crops. First, developing appropriate nutrient management and supply strategies. Second, breeding for improved efficiency in acquiring and using nutrients. Nutrient uptake efficiency obviously depends on the characteristics of the root system, such as uptake rate of nutrients per unit root length, ability of a root system to quickly expand into a fertile zone of soil, ability of roots to modify the rhizosphere, and ability of the plant to grow symbiotically with mycorrhizal fungi. All these properties of the root system are determined by numerous genes. Certain properties of the shoot, such as ability to translocate nutrients from the roots or to redistribute nutrients from senescing leaves, are also important for efficient nutrient use.

Growing nutrient-efficient genotypes of crop plants on soils of low nutrient availability represents an environmentally friendly approach that would reduce land degradation by reducing use of machinery and by minimizing application of chemical fertilizers on agricultural land. The danger of exhausting soil nutrient resources ("land mining") would be negligible, at least for phosphate and micronutrients. The total supply of phosphate and micronutrients in soils is enough for hundreds of years of sustainable cropping by new, efficient genotypes that can gain access to nutrient pools generally considered plant unavailable.

Growing nutrient-efficient plants on soils low in plant-available nutrients represents a strategy of "tailoring the plant to fit the soil," in contrast to the traditional practice of "tailoring the soil to fit the plant." The approach of developing nutrient-efficient crops is very significant because out of all agricultural innovations, farmers most readily accept new cultivars. This is mainly because new cultivars can yield better, demand less fertilizer input, and need minimum or no changes in agricultural practices. However, relatively slow progress in defining the genetic, physiological, and biochemical basis of nutrient efficiency has hampered development of superior nutrient-efficient cultivars through conscious breeding efforts geared specifically toward that purpose.

Table 3.5. Genetic control of nutrient efficiency in crops

Trait	*Gene action*
Phosphorus efficiency	Multigenic (additive, dominance, epistasis, 3 chromosomes)
Phosphorus uptake	Multigenic
Potassium efficiency	Predominantly additive, but dominance and epistasis important. Single gene recessive
Iron efficiency	Single locus dominant, polygenic, additive, two complementary dominant genes, quantitative. Major gene governs uptake.
Boron efficiency	Single gene
Copper efficiency	Single gene

In alkaline soils, phosphate is much less available to plants because it becomes insoluble. Recently, scientists from Mexico expressed a bacterial gene that encodes the enzyme citric acid synthase in plants and found that the genetically engineered plants were better able to solubilize the insoluble phosphate present in alkaline soils. This work shows promise for enhancing the use of scarce nutrient resources by crops.

FURTHER IMPROVEMENT

Historically, farmers dealt with soil salinity by replacing salt-sensitive crops with more salt-tolerant ones; for example, barley replaced wheat thousands of years ago in Ethiopia. Thus people probably used crop substitution as a method of dealing with salinity long before they developed technologies to leach salts from soils and to avoid salinity problems using various management strategies. In all the saline growing areas of the world, farmers substitute salt-tolerant crops for sensitive crops. Crops such as sugar beet, barley, cotton, asparagus, sugar cane, and dates are very salt tolerant.

Improved salt tolerance in both sensitive and tolerant crops would allow more extensive use of brackish water supplies—an especially important consideration where water costs are high or water availability is low. Salt tolerance is a complex, quantitative, genetic trait controlled by many genes. Recently, scientists have identified a few genes that provide information useful in screening and selection programs for salt tolerance.

Scientists study the salt tolerance of plants with the goal of developing more salt-tolerant crops. Four major strategies have been proposed to develop salt-tolerant crops.

1. Gradually improve the salt tolerance of crops through conventional breeding and selection.
2. Introduce traits for salt tolerance from wild relatives into the crops by the process of back crossing.
3. Domesticate wild species that currently inhabit saline environments (halophytes) by breeding and selecting for improved agronomic characteristics.
4. Use molecular techniques to identify genes associated with salt tolerance, and enhance their expression in the crop species or transfer the genes from a noncrop species to a crop.

One example of breeding for salt tolerance is the development of salt tolerance in rice, a salt-sensitive species. Pokkali rice is grown in the state of Kerala in the southern tip of India. The region where this rice is grown has many inland waterways subject to tidal flow from the nearby sea. Because of seawater intrusion, the land is too salty to grow rice. However, just before the monsoon season, the farmers plant rice seeds that have been primed for fast germination. The seeds are planted into mounds, and when the monsoons come, rain washes some salt out of the soil, allowing germination. After a month, farmers transplant the seedlings throughout the field. Then the rice fields are flooded with freshwater and the rice grows as usual through the rest of the monsoon season. Most rice varieties would still not grow well in these salty soils even during the monsoon season, but over many generations local farmers have selected rice that will grow in these soils. After the monsoon season when farmers harvest the rice, they flood the rice stubble with salty river water and these conditions provide a very rich habitat for shrimp production. IRRI, the International Rice Research Institute in the Philippines, has used the salt-tolerant Pokkali rice extensively in its breeding program to develop salt tolerance in other, more desirable rice genotypes.

A promising approach attempted for tomato, barley, and wheat is to introduce traits of salt tolerance into these crops from their wild relatives through interspecific hybridization. This requires extensive backcrossing and may be difficult because a trait such as salt tolerance may be determined by several genes located on different chromosomes. Moreover, in the wild relatives these traits are favourable mainly for survival under natural conditions. For crop production in dry saline soils, high efficiency of water use (kg of dry matter produced per kg of water) is of key importance. Present wheat and barley cultivars have a higher water use efficiency and total biomass production on saline soils than most of their wild relatives that thrive in such environments. So far this approach has not yielded successful new crop strains.

Recent progress in genetic engineering has led to the development of salt tolerance in plants. Two notable examples are (1) increased expression of a gene that regulates stress responses in plants and (2) increased expression of a gene that encodes a sodium (Na^+) transporter in the tonoplast membrane of the vacuole.

On the molecular front, scientists are identifying the genes involved in sensing salt in the environment (signal transduction genes), transcription factor genes that turn on batteries of other genes that prepare cells to withstand a higher rate of salt influx, and genes that are part of the plant's adaptation to the presence of salt. An example of the latter category is the gene that encodes a vacuolar membrane sodium pump. Plants that can turn this gene on rapidly when the cells are exposed to salt, will be able to transport the salt from the cytoplasm and into the vacuole, thereby detoxifying the cytoplasm. Scientists in Canada expressed such a gene with a strong general promoter in tomato plants and found that the plants could be grown in 500 mM sodium chloride. The tomato plants apparently pump the leaf vacuoles full of sodium chloride and then allow those leaves to die, while at the same time producing new leaves. The plants produced an abundant crop of tomatoes in the greenhouse, but need to be field-tested.

Groups of Japanese and Belgian scientists independently discovered transcription factors that are specifically expressed when a plant is exposed to dehydration. The gene is turned on early in the drought response, and the protein activates expression of a large suite of genes involved in adaptation to stress. When the scientists engineered plants so that this gene was quickly expressed in response to dehydration, the genetically modified plants were more tolerant to water deficit in the laboratory. In addition, these plants were more tolerant to salinity and cold stress. All three stresses (drought, salinity, and cold) cause osmotic imbalances to which the plant must respond.

Tolerance to Acid Soils

Strategies for maintaining production on acid soils include applying lime to raise soil pH using lime from limestone or dolomite, and breeding for tolerance to acid soils. Liming increases nutrient uptake by plants and enhances organic matter decomposition, thus improving the conditions for plant growth. Concurrent application of lime and balanced nutrient replenishment is usually necessary to ensure continued long-term soil fertility. A problem associated with acid soils is the low availability of phosphate at low pH. However, the main problem on acid soils is aluminum toxicity, because at low pH the clay minerals are solubilized and aluminum oxide is released in the soil solution. Soluble aluminum inhibits root growth, causing the crops to be more easily water stressed, because they cannot mine a sufficient volume of soil for available water. Crops that are highly sensitive to toxic levels of Al in acid soils are maize, mung bean, cotton, and cocoa.

Table 3.6. Sensitivity of major crops to aluminum

Sensitivity to Aluminum	*Crops*
Low	Cassava, brachiaria, andropogon
Low to moderate	Millet, rice, cowpea, mucuna, rubber, oil palm
Moderate	Groundnut, centrosema, stylosanthes, kudzu
Moderate to high	Soybean, sorghum, bean, wheat, panicum
High	Maize, mung bean, cotton, crotalaria, cocoa

Breeding improved varieties for acid, infertile soils starts by identifying productive lines with general adaptation to biotic and climatic conditions of soils in the target environments. Scientists compare these varieties under conditions where soil fertility, but not pH, has been modified with modest quantities of plant nutrients. The immediate benefit for the farmer of cultivars adapted to acid soil is obvious and has been experimentally demonstrated for important crops such as maize, sorghum, and upland rice. For example, in South America the Colombian National Program released four Al-tolerant cultivars of sorghum that, in field tests, showed a two- to threefold yield advantage over other varieties. Also in Colombia, CIAT scientists successfully developed acid soil-adapted upland rice lines that were released to the farmers in the early 1990s.

These improved rice varieties permit commercial rice commercial production on the acid soil savannas of Colombia. Research has shown that certain plants that can tolerate high levels of soluble Al secrete organic acids into the rhizosphere. Researchers in Mexico transformed papaya plants with a bacterial gene that encodes the enzyme citric acid synthase and showed that the roots of these plants secreted citric acid and that the plants tolerated higher concentrations of soluble Al in their growth medium. Field evaluation of these plants is presently in progress.

4

Molecular Farming

'Molecular farming' is a term coined to describe the application of molecular biological techniques to the synthesis of commercial products in plants. The term can be applied to a broad spectrum of activities, from the enhanced production of products that are already extracted from plants through to the manufacture of compounds that are completely novel to plants. A wide range of products have already been identified as likely targets for molecular farming, these include a variety of carbohydrates, fats and proteins, as well as secondary products.

Many of the proteins being 'farmed' in plants are antibodies, vaccines or biopharmaceuticals aimed at improving human and animal health, and it is in the area of molecular farming that many of the most exciting and potentially beneficial developments in plant biotechnology are taking place.

Carbohydrates and Lipids

In this first section we will look at some examples of the production of novel or modified carbohydrates and oils (and their derivatives) in plants. Many of these examples have multiple applications (both as modified food- or feedstuffs and industrially) that will be highlighted here.

Carbohydrate Production

Plants produce a range of commercially valuable carbohydrates. The two most abundant carbohydrates are cellulose (used for fibres from cotton and flax, for paper-making from trees and for a range of industrial products such as paints and polymers) and starch (used for food, feed and industrial purposes). Some biotechnological effort is

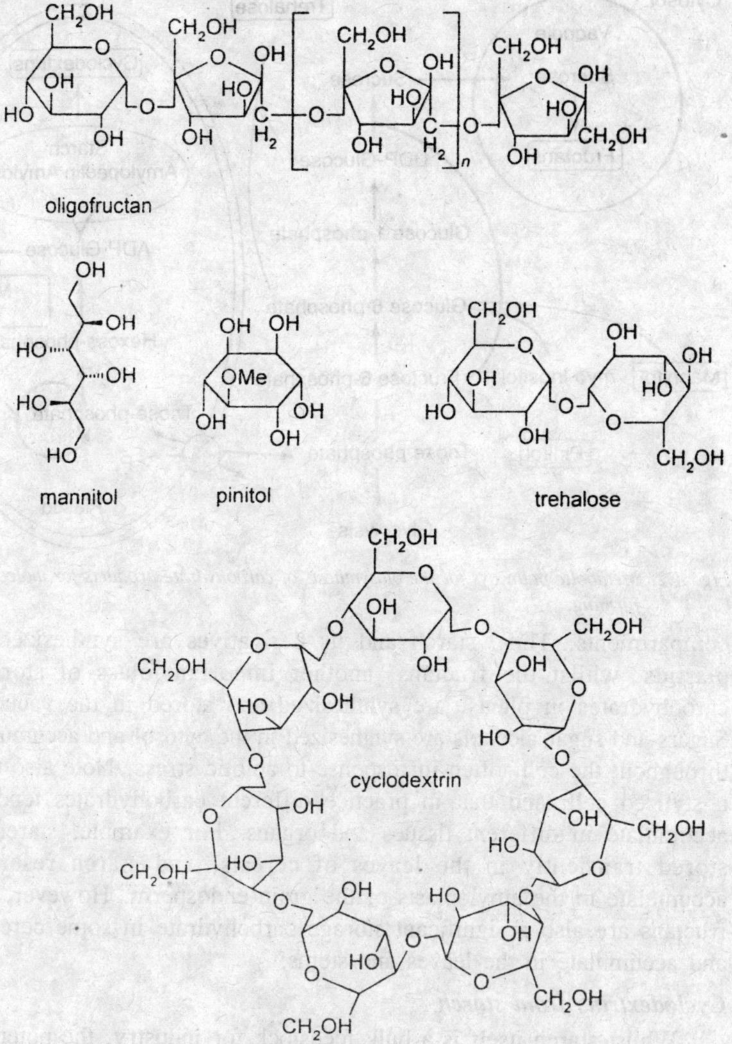

oligofructan

mannitol

pinitol

trehalose

cyclodextrin

Fig. 4.1. Target carbohydrate products for molecular farming.

directed towards improving the yield and quality of these bulk carbohydrates. However, there are a number of other carbohydrates that it could be attractive to produce in transgenic plants, including oligofructans, cyclodextrins and trehalose.

Fig. 4.2 indicates the biosynthetic pathways and cellular locations of these compounds in plants. One obvious point is that carbohydrates are synthesized and stored in a number of different cellular

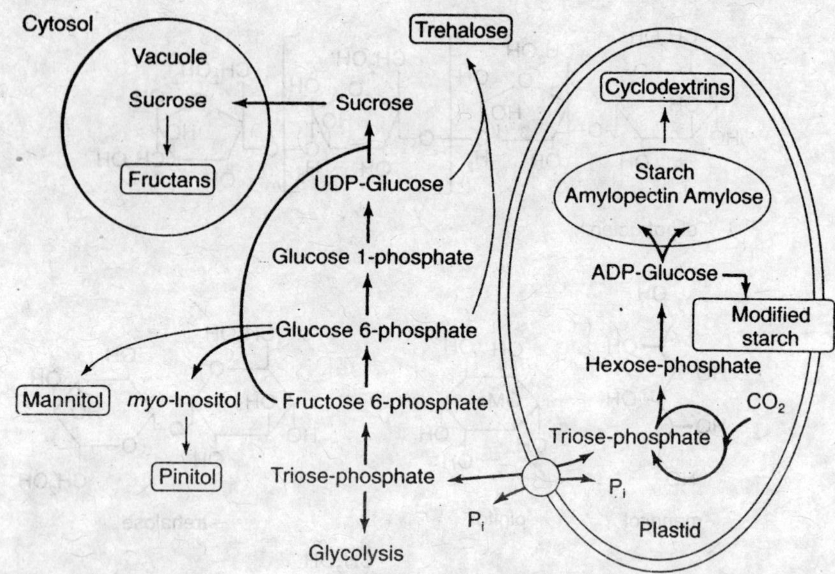

Fig. 4.2. Metabolic pathways for the biosynthesis of carbohydrate products for molecular farming.

compartments. Thus, starch and its derivatives are synthesized in plastids, whilst the fructans, another important class of storage carbohydrates in plants, are synthesized and stored in the vacuole. Sugars and sugar alcohols are synthesized in the cytosol and accumulate throughout the cell, often in response to abiotic stress. Note also that a stylized cell, and that in practice different carbohydrates tend to accumulate in different tissues and organs. For example, starch is stored transiently in the leaves of cereals, and starch reserves accumulate in the amyloplasts of the grain endosperm. However, the fructans are also a significant storage carbohydrate in some cereals, and accumulate in the leaves and stems.

Cyclodextrins from starch

Whilst starch itself is a bulk feedstock for industry, the potential exists for carrying out further biotransformations on the starch in the plant, rather than by chemical or fermentation processes after extraction. One type of high-value product that could be made from starch is the cyclodextrins. These compounds are typically 6-, 7- or 8-membered rings comprising glucopyranose subunits attached in $\alpha(1\rightarrow4)$ linkages. These compounds are normally produced by bacterial fermentation of maize starch, particularly for pharmaceutical applications.

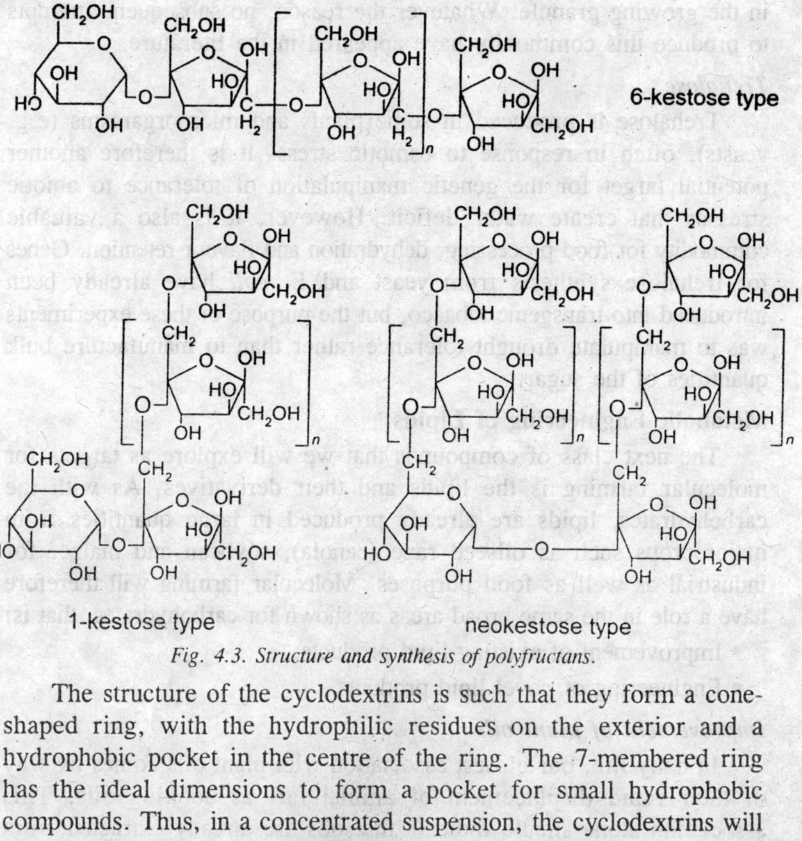

6-kestose type

1-kestose type

neokestose type

Fig. 4.3. Structure and synthesis of polyfructans.

The structure of the cyclodextrins is such that they form a cone-shaped ring, with the hydrophilic residues on the exterior and a hydrophobic pocket in the centre of the ring. The 7-membered ring has the ideal dimensions to form a pocket for small hydrophobic compounds. Thus, in a concentrated suspension, the cyclodextrins will effectively solubilize hydrophobic pharmaceuticals such as steroids. At lower concentrations, for example after injection into the bloodstream, the therapeutic agent is released. Having been identified as a suitable target for molecular farming, it is striking that only one report, in 1991, of an attempt to produce cyclodextrins has been published. A bacterial cyclodextrin glycosyl-transferase gene from *Klebsiella pneumoniae* was fused to a plastid-targeting sequence and placed under the control of the promoter from the patatin gene.

Patatin is a protein that accumulates in potato tubers, and the promoter directs high levels of expression in the tubers of transgenic potatoes. However, transformation of potatoes with this construct resulted in very little conversion (0.001-0.01 %) of starch to cyclodextrins. It was concluded that the insoluble starch granules may have been inaccessible to the bacterial enzyme, or that the enzyme became trapped in the growing granule. Whatever the reason, no subsequent attempts to produce this commodity have appeared in the literature.

Trehalose

Trehalose is produced in some plants and microorganisms (e.g. yeasts), often in response to osmotic stress. It is therefore another potential target for the genetic manipulation of tolerance to abiotic stresses that create water deficit. However, it is also a valuable commodity for food processing, dehydration and flavour retention. Genes for trehalose synthesis from yeast and *E. coli* have already been introduced into transgenic tobacco, but the purpose of these experiments was to manipulate drought tolerance rather than to manufacture bulk quantities of the sugar.

Metabolic Engineering of Lipids

The next class of compounds that we will explore as targets for molecular farming is the lipids and their derivatives. As with the carbohydrates, lipids are already produced in large quantities from major crops such as oilseed rape (canola), soybean and maize, for industrial as well as food purposes. Molecular farming will therefore have a role in the same broad areas as shown for carbohydrates, that is:

- Improvement of existing lipid products;
- Engineering of novel lipid products.

Improvement of plant oils

In daily life, our closest association with plant oils comes by way of their rapid displacement of animal fats as cooking oils. This association alone should indicate that oils are already extracted from a number of different types of crop. Oilseed rape or canola, and soybean are major sources of oil that are often unspecified, but sunflower oil, corn oil and olive oil are often identified because of the improved characteristics of these oils for particular purposes. Plant oils are also used in a wide range of processed foods, and for animal feeds. They also have an increasing number of applications in industry for the manufacture of soaps and detergents, lubricants and biofuel. However, the non-food applications currently comprise only about 10% of the total vegetable oils produced. One of the reasons for this relatively low level of industrial use is the complex and variable nature of plant oils, which generally makes them less suitable for oleochemical applications than cheaper alternatives from mineral oils.

A major long-term goal of this sector is therefore the improvement of plant oils for industrial applications, since they must eventually replace non-renewable, petroleum-based products. Storage oils that are used as carbon/energy reserves in plant seeds and some fruits (such as olives and avocados) are generally glycerol esters of fatty acids, called triacylglycerols (TAGs).

Table 4.1 The fatty acid composition of major oil crops

Fatty acid	*Soya oil*	*Palm oil*	*Rape-seed oil (LEAR)*	*Sunflower oil*	*Peanut oil*
16:0	11	42	4	5	10
18:0	3	5	1	1	3
18:1	22	41	60	15	50
18:2	55	10	20	79	30
18:3	8	0	9	0	0
20:1	0	0	2	0	3
Others	1	0	2	0	0

The differences between oils from different sources are largely due to the particular proportions of different fatty acids comprising TAGs. These differ in their length, and the number and positions of double bonds. Saturated fats contain no double bonds—the more double bonds there are, the greater the degree of unsaturation. The site of *de novo* fatty acid biosynthesis is exclusively in the stroma of the plastid, whereas most modification of the fatty acids-occurs in the cytoplasm and on the endoplasmic reticulum (ER). Assembly of TAGs takes place in the membrane of the ER.

The TAGs are stored in oil bodies, which are essentially oil droplets surrounded by a lipid monolayer formed from one half of the ER membrane lipid bilayer. The oil bodies of seeds (and other tissues that undergo extreme desiccation) also contain proteins, called 'oleosins', in the lipid monolayer. The first committed step in fatty acid biosynthesis in the plastid is the formation of malonyl-coenzyme A (CoA) from acetyl-CoA, in an ATP-dependent reaction catalyzed by acetyl-CoA carboxylase (ACCase). The CoA residue is then exchanged for the acyl carrier protein (ACP). Acetyl-CoA is then condensed to the malonyl-ACP, forming acetoacetate-ACP, and liberating CO_2, Three subsequent steps (reduction of the carbonyl group, removal of water to form a double bond and reduction of the double bond) produce an acyl-ACP, which is two carbons longer than the original.

The entire sequence of elongation reactions from the initial binding to ACP is catalyzed by a fatty acid synthase multienzyme complex. The elongated fatty acid chain is then transferred to another ACP

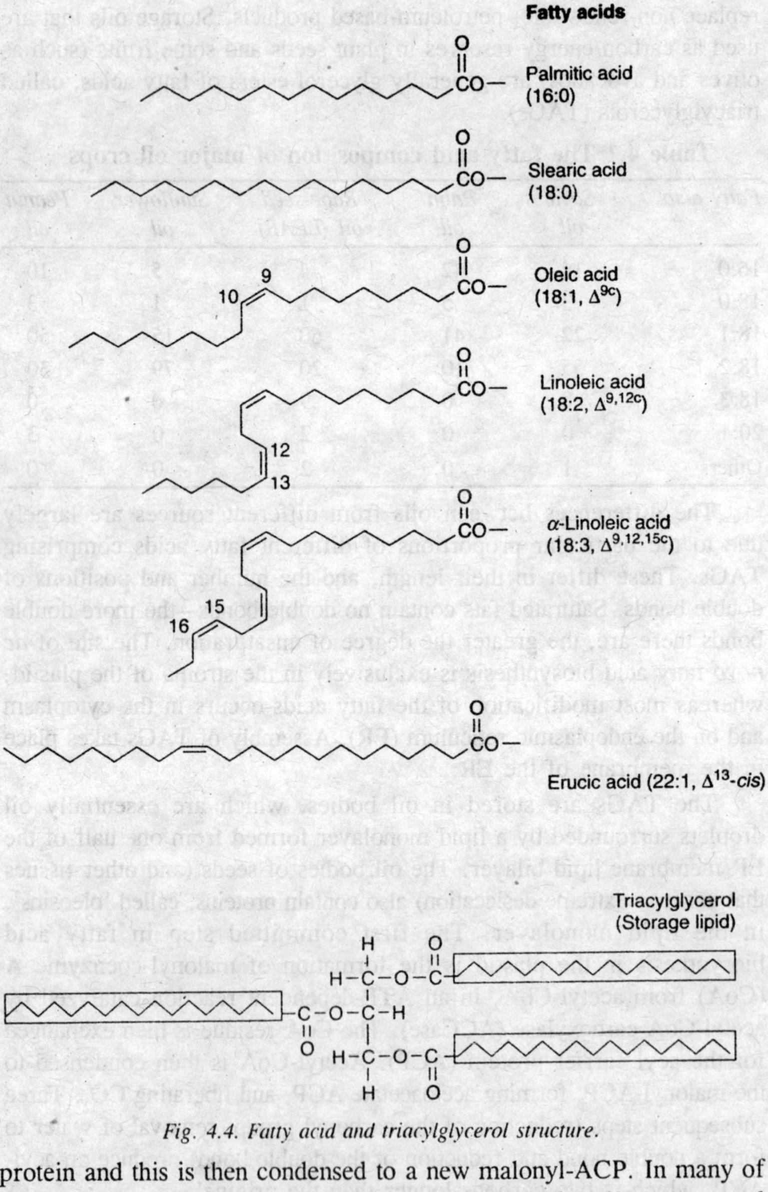

Fatty acids

Palmitic acid (16:0)

Slearic acid (18:0)

Oleic acid (18:1, Δ^{9c})

Linoleic acid (18:2, $\Delta^{9,12c}$)

α-Linoleic acid (18:3, $\Delta^{9,12,15c}$)

Erucic acid (22:1, Δ^{13}-*cis*)

Triacylglycerol (Storage lipid)

Fig. 4.4. Fatty acid and triacylglycerol structure.

protein and this is then condensed to a new malonyl-ACP. In many of the oil producing crops, this process stops at the 16-carbon stage, and the palmitoyl-ACP (16:0) is elongated to stearoyl-ACP (18:0) by a specific synthase. (The standard nomenclature of fatty acids indicates the number of carbon atoms and the number of double bonds.) At this stage, desaturation by a soluble Δ^9C-stearoyl desaturase in the plastid stroma converts most of the stearoyl-ACP to oleoyl-ACP (18:1Δ^9). (The Δ^9 indicates that the double bond starts at the 9th carbon counting from the carboxylic acid 'front end' of the molecule.) After termination of synthesis in the plastid, the fatty acids (mainly palmitic, stearic and oleic acids) are released from the ACP and exported to the cytoplasm, where they are converted to acyl-CoA esters. TAGs are formed by stepwise acylation of glycerol-3-phosphate in the ER membrane. Further modifications to the fatty acids normally occur after they have been attached to various glycerophosphatides. Additional double bonds may be added (typically to Δ^{12}, Δ^{15} and Δ^6 as well as Δ^9). Other modifications may include hydroxylation by the addition of water across a double bond.

Production of shorter chain fatty acids

The oils produced by the major oil crops of the world consist mainly of palmitic, stearic, oleic, linoleic and linolenic acids, which are all C16 or C18 in length. Coconut and palm kernel oils are largely C8-C14, and lauric acid (12:0), in particular, is an important raw material for the production of soaps, cosmetics and detergents (think of the widespread use of sodium lauryl sulphate, or SDS, in these products). The synthesis of fatty acids at a particular length is terminated by the hydrolysis of the acyl-ACP by a thioesterase. The California bay tree contains a very high proportion of lauric acid in its seeds, and an acyl-ACP thioesterase that specifically hydrolyses lauroyl-ACP has been cloned from this source. The introduction of this gene into oilseed rape causes fatty acid synthesis to terminate at the 12:0 stage, and a high proportion of lauric acid to accumulate in the seed oil. Most importantly, field tests show that these plants grow normally and produce normal yields.

Production of longer chain fatty acids

One of the important targets is to produce fatty acids longer than C18 for use as industrial oils. In oilseed rape and other *Brassica* species, there is a two-step elongation pathway from oleoyl-CoA (18:1Δ^9) to erucoyl-CoA (22:1Δ^{13}), such that erucic acid is one of the constituents of brassica oils. However, whilst erucic acid is valuable as an industrial oleochemical, it is nutritionally unsuitable for human consumption. Conventional breeding has led to the development of two distinct oilseed rape crops—high-erucic acid rape (HEAR) for industrial purposes, and low-erucic acid rape (LEAR) with virtually no erucic acid, for food products. (NB: the existence of two distinct crops that need to be kept apart means that systems are in place in the seed production, agriculture and processing industries to ensure minimal cross-contamination, a process called 'identity preservation'. Thus, requirements to keep GM and non-GM seeds and products separate do not create entirely new demands on this sector.) However, the highest erucic acid content of HEAR is about 50% of the total fatty acids, which makes the cost of separating out and disposing of the other fatty acids uncompetitive with mineral oil sources. Attempts are being made to overexpress the genes that encode the elongases, and also to transfer enzyme activities that preferentially incorporate erucic acid into TAGs.

Modification of the degree of saturation

As the erucic acid example shows, the ideal products for industrial purposes are as uniform as possible. This requires uniformity of de saturation as well as length. (The same does not necessarily apply to food constituents, where complex mixtures may produce the ideal product. In this case it is consistency, rather than uniformity, that is required.) One of the first attempts in this area was to increase the level of stearic acid (18:0) at the expense of oleic acid (18:1Δ^9) by inserting an antisense construct of the Δ^9 desaturase gene into oilseed rape. The antisense gene was under the control of the napin (a seed-specific protein of brassica) gene promoter. There was a marked decrease in the amount of the desaturase enzyme, resulting in a decreased formation of oleic acid, and a rise in the proportion of stearic acid from 1-2 % up to 40% of total fatty acids. This high stearic acid oil has potential as a cocoa butter substitute.

On the other hand, there is considerable potential in the production of a very high (>90%) oleic acid oil for food purposes and as a uniform oleochemical feed-stock. Conventional breeding has produced mutant oilseed rape lines with an oleic acid content close to 80%, but attempts to increase this level have produced plants with poor cold tolerance, presumably as a result of the lack of unsaturated fatty acids in the cellular membranes. Antisense repression and co-suppression of the Δ^{12} desaturase in the seeds of oilseed rape have, however, facilitated the raising of oleic acid levels to 87-88%, without affecting cold tolerance. Similar experiments in soybean increased the oleic acid level from 22% to 79%.

Production of rare fatty acids

There are 210 known types of fatty acids produced in plants, but most of these are not found in the major crop plants and would be difficult to produce commercially in their host plants. However, it is

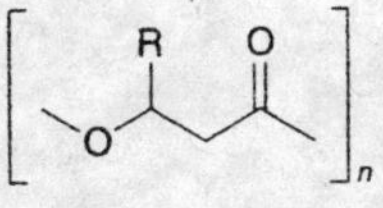

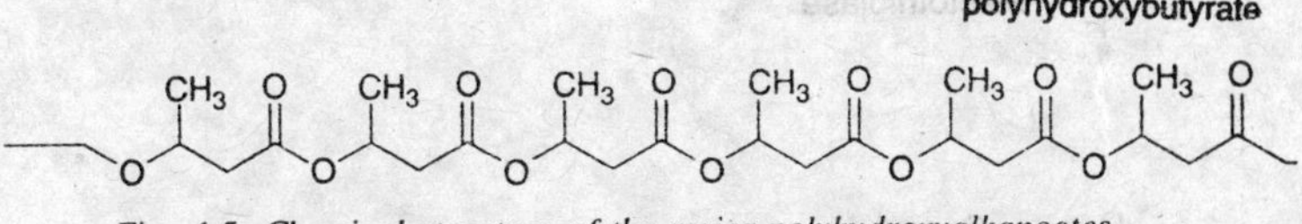

generic polyhydroxyalkanoate

polyhydroxybutyrate

Fig. 4.5. Chemical structure of the major polyhydroxyalkanoates.

possible to transfer genes from these plants in order to manipulate the profile of fatty acids produced in the major oil crop plants. One such fatty acid is petroselenic acid ($18:1\Delta^6$), which is found in certain Umbelliferae, such as coriander, and has potential as a raw material for industry. Oxidation of petroselenic acid by ozone produces lauric acid (12:0) (for soaps and detergents) and adipic acid (6:0), which can be used for nylon production.

Transformation of tobacco with a coriander acyl-ACP desaturase cDNA led to the production of petroselenic acid in calli to a level of 5% of total fatty acids. Certain polyunsaturated fatty acids have pharmaceutical or nutraceutical value. These include γ-linolenic acid ($18:3\Delta^{6,9,12}$) and arachidonic acid ($20:4\Delta^{5,8,11,14}$), which are essential fatty acids for humans and precursors of eicosanoids (including prostaglandins, leukotrienes and thromboxanes). Recently, it has become clear that the synthesis of these compounds involves a specific subclass of microsomal desaturases called 'front-end' desaturases. These enzymes insert additional double bonds between existing bonds and the carboxyl 'front-end' of the fatty acid, whereas most desaturases in plants add sequentially towards the methyl end.

The transformation of a front-end desaturase from borage (which is one of a few plant species to produce γ-linolenic acid) into tobacco resulted in the accumulation of high levels of Δ^6 unsaturated fatty acids. A Δ^5 desaturase has been cloned from the filamentous fungus *Mortierella alpina* (which accumulates arachidonic acid) and expression in oilseed rape produced significant levels of polyunsaturated fatty acids. Ricolenic acid (Δ^{12}-hydroxyoleic acid) is produced in castor beans to a level of 90% of the total fatty acids. However, the castor oil crop has a number of problems, including the presence of toxic compounds such as ricin in the residual meal.

The synthesis of ricolenic acid involves the direct hydroxylation of oleic acid bound to phosphatidylcholine on the ER membrane. The

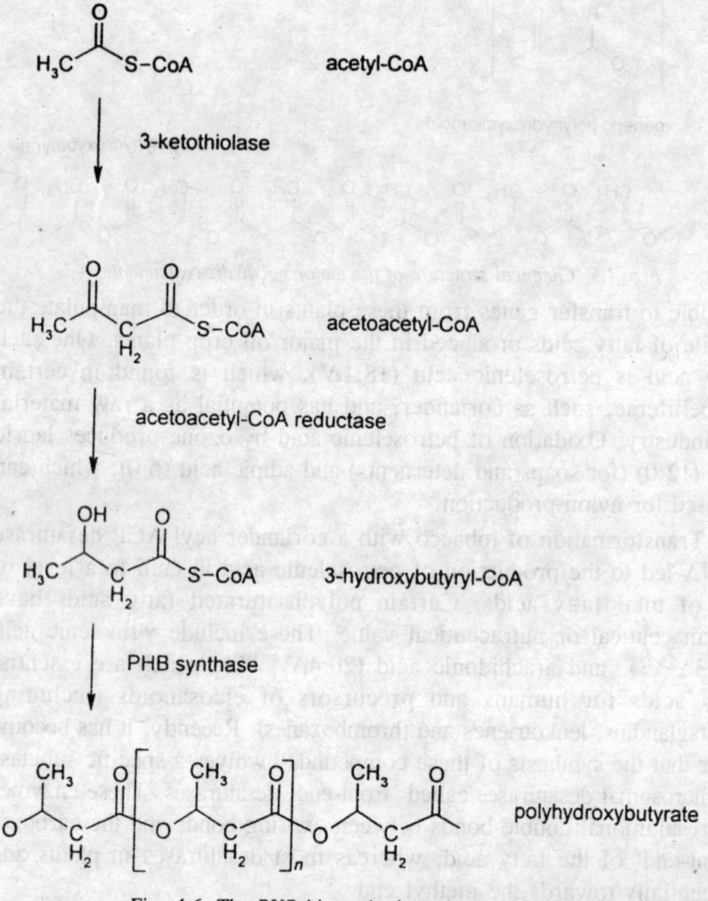

acetyl-CoA

3-ketothiolase

acetoacetyl-CoA

acetoacetyl-CoA reductase

3-hydroxybutyryl-CoA

PHB synthase

polyhydroxybutyrate

Fig. 4.6. The PHB biosynthesis pathway of bacteria.

cDNA for the 12-hydroxylase has been cloned from castor bean and could be used to produce 'castor oil' in major oil crops. Another fatty acid modifying enzyme has been cloned from *Crepis acetylenics*. This non-haem, di-iron protein catalyses triple bond and epoxy group formation in fatty acids, and would be a valuable way of inserting chemically reactive sites into oils.

Molecular Farming of Proteins

Throughout this book we have presented to you examples where transgene expression is used as a tool to, for example, modify a biosynthetic pathway, manipulate growth and development or alter some other property of the plant such as herbicide resistance. However, in some cases it is the transgenic protein itself that is important. It is already discussed that how attempts to improve the amino acid content of grain are being addressed by expression of proteins with a high content of lysine. In this section of the chapter, we will consider the production of bulk enzymes and potentially high-value proteins such as antibodies and vaccines.

The production of functional proteins in plants on an industrial scale does not, at first sight, suggest too many difficulties. However, the production of a complex protein that is folded, processed, easily purified and can be shown to be safe for pharmaceutical use, all at low cost, is a real challenge. The economics of protein production in plants is complicated. The actual cost will depend on many factors, amongst them being the cost of growing the plant, transport costs and processing and protein purification costs.

The costs of proteins produced in plants may undercut the costs of producing proteins by existing methods, but if too much (in economic terms) is produced, then the market price will fall. This has already proved to be the case for some vaccines produced by existing methods. In this remaining section, the protein products that will be considered fall into two main types. First, enzymes for industrial and agricultural uses will be looked at. Second, we will consider medically related proteins such as:

1. Antibodies-full immunoglobulins and engineered types such as scFvs (single chain antibodies);
2. Subunit vaccines-vaccines based upon short peptide sequences that act as antigens rather than whole organism-based vaccines; and
3. Protein antibiotics.

Production Systems

The interest in producing such proteins in plants comes in part from the problems associated with existing fermentation or bioreactor systems. Mammalian cell systems are expensive and cannot easily be scaled up (input costs can approach $US1 million per kg of crude product). Bacterial systems can be scaled up, but often the recombinant proteins are not properly processed (they are not properly folded and disulphide bridges are not formed), so intracellular precipitation of non-functional proteins can occur. Plant systems can be scaled up so that large amounts of material can be harvested and processed, allowing industrial-scale amounts of protein to be purified. In some cases it may be possible to omit purification, as is the case with vaccines from plant sources, and where plant material containing recombinant enzymes can be added directly to animal feed or industrial processes. There are some differences between plant and animal post-translational modification systems, and the difference in glycosylation patterns between plants and animals has caused some concerns, particularly when expressing antibodies in plants. However, plants are able to fold, cross-link, and post-translationally modify (by, for example, glycosylation) non-plant proteins sufficiently well to ensure that, in most cases, functional proteins are obtained.

The production of recombinant proteins in plants also benefits from the ability to target the proteins to cellular compartments where processing occurs, or where the proteins may be more stable. Another advantage of using plant systems is that the potential for contamination of the protein products with toxins or pathogens of animals and humans is very much reduced. The plants initially chosen for expression studies related to the ability of scientists to transform the plant material. A lot of work was first done with tobacco, but this was not a suitable plant for feeding to animals. For work in which edible products are being developed, plants such as potatoes, tomato, maize and lettuce have been used. For bulk production, the leaves of plants such as tobacco and alfalfa remain a favourite, because they can be harvested several times a year. However, plants such as rice, wheat, maize and soybean have also been used, although their biomass yields are lower.

Table 4.2 Potential biomass yields from various plants considered as potential vehicles for protein production

Crop	*Potential annual biomass yield (tonne per hectare)*
Tobacco	>100
Alfalfa	25
Maize	12
Rice	6
Wheat	~3

Strategies for protein production

Two major strategies have been developed for the production of proteins in plants: the stable integration approach, which has been used for many systems; and the use of plant viruses as transient vectors.

Stable expression

The stable transgene expression approach, in which the transgene is regulated by a strong, constitutive promoter (such as the 35S promoter), is perhaps the most suitable for the bulk production of soluble proteins in leaves, although yields can be low using this approach. A more sophisticated approach has been to target gene expression and protein production to specific tissues. Restricting gene expression to particular organs or tissues often leads to higher yields of recombinant proteins.

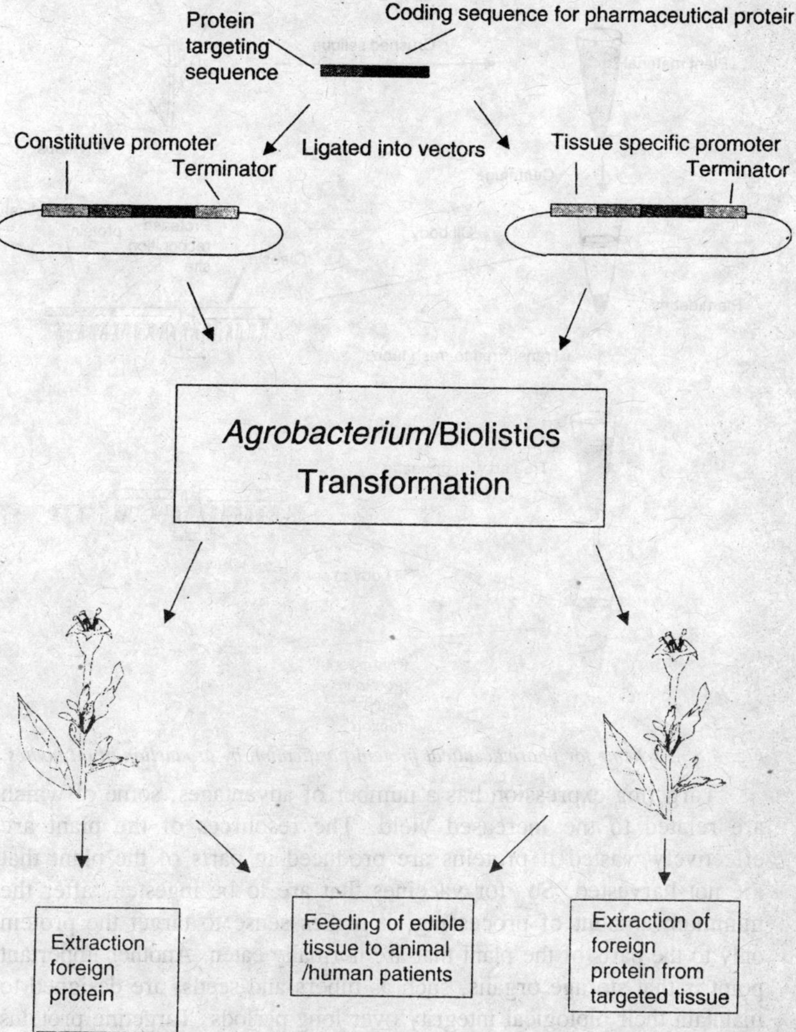

Fig. 4.7. Different routes of protein manufacture in stably transformed plants.

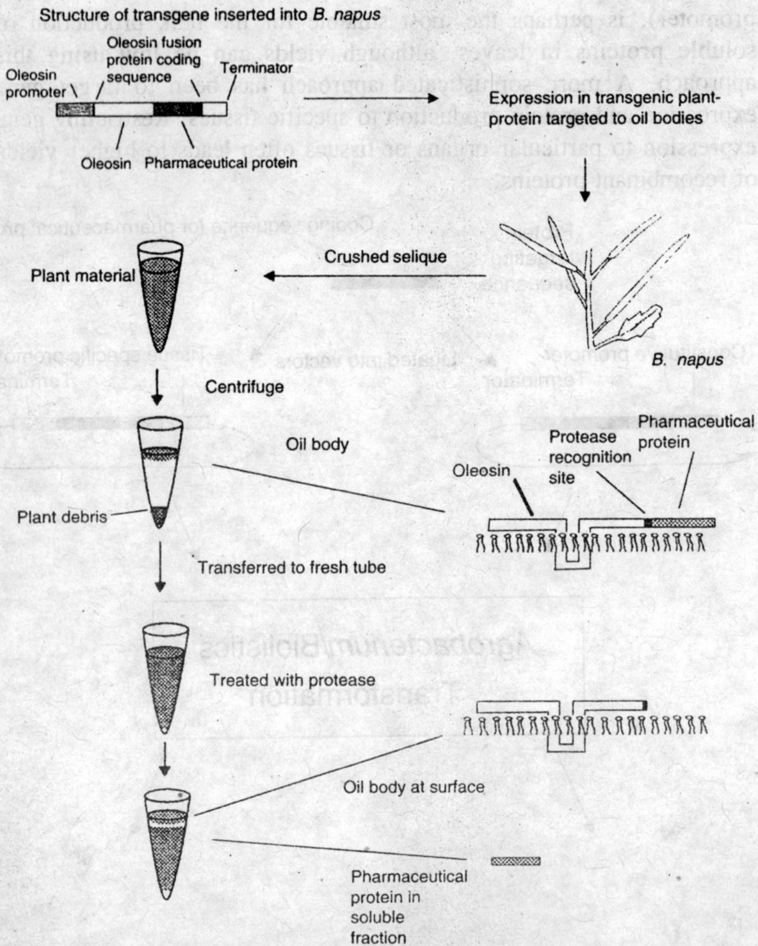

Fig. 4.8. A scheme for pharmaceutical protein purification by production in oil bodies.

Targeting expression has a number of advantages, some of which are related to the increased yield. The resources of the plant are effectively wasted if proteins are produced in parts of the plant that are not harvested. So, for vaccines that are to be ingested, after the minimum amount of processing, it makes sense to target the protein only to the parts of the plant that are normally eaten. Another important point is that storage organs, such as tubers and seeds, are designed to maintain their biological integrity over long periods. Targeting proteins to these structures has been shown to increase the stability of recombinant proteins from weeks to years.

Seeds are also of great interest because they can offer simple strategies for protein purification. In tobacco, seeds contain a simpler mixture of proteins and lipids, with fewer phenolic compounds, than do green leaves. As well as targeting expression to particular organs or tissues, intracellular targeting has been used to optimize the production of active proteins. Post-translational modification has been) shown to be associated with the ER and/or apoplastic export, and the inclusion of transit peptides has been shown to improve both yield and activity. Tissue and intracellular targeting are brought together in the use of oleos ins as vehicles for protein purification.

Transient expression

Plant virus capsids have also been used as carriers of recombinant proteins; particularly vaccines. In one approach, coding sequences for epitopes or proteins have been introduced into the coat protein gene of the virus genome so that it is 'presented' externally when virus particles are made. Another approach that has been used is to construct viral vectors to produce recombinant proteins that are targeted to the ER for processing. The virus can be replicated in the host plant, and through serial passage enough protein can be generated for clinical use.

Production of enzymes for industrial uses

The global industrial enzyme market has been estimated to be worth something approaching $US2 billion per annum. Proof of concept incursions by plant biotechnology into this market were on a small scale. The first commercialized 'industrial proteins' produced from transgenic plants were two proteins familiar to the molecular biologist—avidin (from chicken) and β-glucuronidase (from *E. coli*)—both of which were produced in maize. Following on from these successes, ProdiGene Inc., the company responsible for their development, is leading the race to large-scale production of plant-derived transgenic enzymes for commercial use. They announced in Spring 2002 that they would be going into large-scale production of trypsin, a proteolytic enzyme, which by its very nature is difficult to produce in conventional recombinant systems. This protein is currently harvested from bovine and porcine pancreases, and is used for a number of applications, including the production of pharmaceuticals such as insulin, vaccine production and wound care. The company has predicted that the worldwide demand for the enzyme will increase fivefold in the next 5 years.

Existing sources may be unable to supply this increase, and there is the added worry of the spread of pathogens (like BSE) with products derived from animals. ProdiGene Ine. aims to meet this demand by producing kilogram quantities of the protein from plant sources. Some of the other important enzymes that may be produced in plants are, interestingly, normally targeted against components of the plant cell. Cellulases and xylanases, which are normally produced by the microorganisms associated with digestion in ruminants, have been produced in a number of different plants.

Table 4.3 Examples of enzymes produced in plants

Enzyme	*Use*
Avidin	Diagnostic kits
β-Glucuronidase	Diagnostic kits
Trypsin	Pharmaceuticals, wound care
Cellulase	Ethanol production from cellulose waste
Thermostable xylanase	Biomass processing
Phytase	Phytate breakdown, improved phosphate utilization
α-Amylase	Food processing
(1-3) (1-4) β-Glucanase	Brewing
Lignin peroxidase	Paper manufacture

There is a large market for these enzymes in the bioethanol, textile, pulp and paper industries and for the production of animal feed. It may, at first sight, seem a strange concept to express such enzymes in plants, due to the risk of 'autodigestion', but engineered, thermostable, forms of the enzymes, with high temperature optima were used. Therefore, no deleterious effects were detected at the temperatures at which a plant normally grows, but high levels of enzyme activity were detected when plant extracts were heated to the temperature optima of the enzyme. These were only small-scale initial trials, but the process could easily be scaled up. An interesting point to note is that these enzymes would require only the minimum of purification (an important consideration when determining applicability and economic viability). This is also the case for another important hydrolytic enzyme, phytase. This enzyme catalyses the hydrolysis of inositol hexaphosphate (phytate) to its constituents, the sugar-alcohol inositol and inorganic phosphate.

Phytase is a very useful enzyme as it releases phosphate from a substrate that is normally indigestible to monogastric animals. As phytate is found in significant quantities in seeds used to formulate

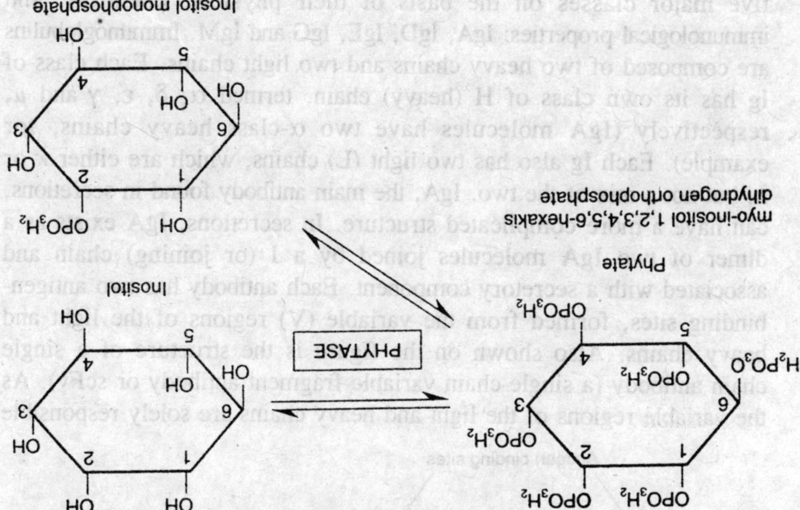

Fig. 4.9. The action of phytase.

feed meal for pigs and poultry, these animals do not benefit from this source of phosphate, which is required for proper skeletal growth. Problems with phytate are exacerbated by the fact that monogastric animals excrete the indigestible phytate in their faeces, which contributes to an excessive phosphate build-up in ground water, leading to eutrophication. Transgenic plants expressing phytase in, for example, seeds have been produced. When these transgenic seeds are added to the diet of pigs and poultry, phosphate utilization is increased (negating the need for supplements) and phosphate excretion is decreased. Thus, the addition of the enzyme in a simple plant-derived formulation solved both a nutritional and an environmental problem.

Medically Related Proteins

This is an important, expanding, area for plant biotechnology, in which there is great hope that recombinant proteins made *in planta* will be accepted and widely utilized. This section of the chapter is divided into three sections:

1. Antibodies;
2. Vaccines; and
3. Other medically related proteins.

Antibodies or 'plantibodies'

In mammals there are a number of different types of antibodies, or immunoglobulins (Ig). The immunoglobulins have been divided into five major classes on the basis of their physical, chemical, and immunological properties: IgA, IgD, IgE, IgG and IgM. Immunoglobulins are composed of two heavy chains and two light chains. Each class of Ig has its own class of H (heavy) chain, termed α, δ, ε, γ and μ, respectively (IgA molecules have two α-class heavy chains, for example). Each Ig also has two light (L) chains, which are either κ or λ, but not a mix of the two. IgA, the main antibody found in secretions, can have a more complicated structure. In secretions, IgA exists as a dimer of two IgA molecules joined by a J (or joining) chain and associated with a secretory component. Each antibody has two antigen-binding sites, formed from the variable (V) regions of the light and heavy chains. Also shown on the figure is the structure of a single chain antibody (a single-chain variable fragment antibody or scFv). As the variable regions of the light and heavy chains are solely responsible

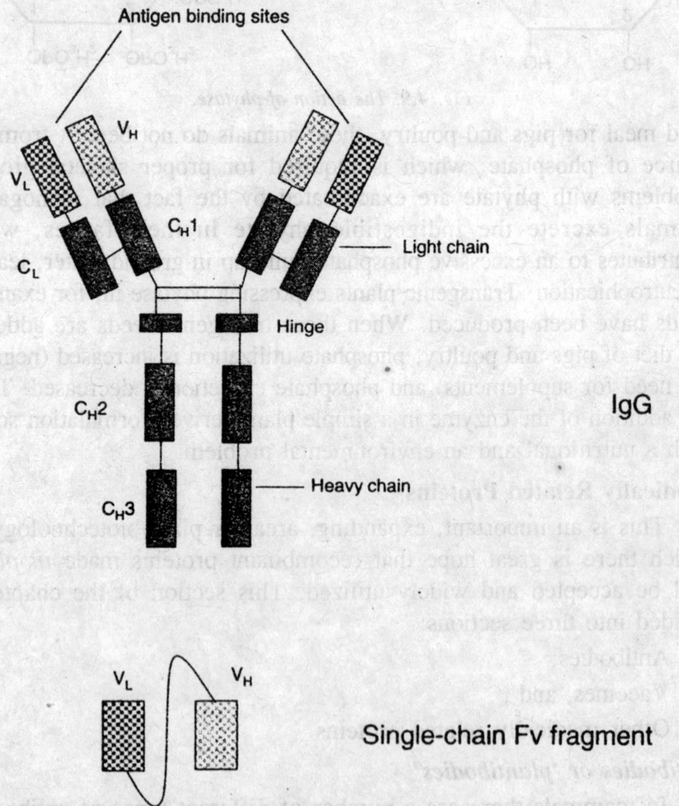

Fig. 4.10. Structures of IgG and an engineered single-chain variable fragment antibody.

for antigen binding, simpler, smaller molecules that still bind antigens can be produced.

Single-chain variable fragment antibodies are produced from synthetic genes made by fusing the sequences for light- and heavy-chain variable regions. Although not the only example of engineered antibodies, scFv antibodies are the most commonly used and most successful. The first example of a functional antibody (a mouse immunoglobulin IgG1) being produced in plants was reported in 1989. The construction of this antibody was a two-step process. In the first stage, two separate transgenic tobacco lines were produced, one contained the gene for the γ-heavy chain the other contained the gene for the k-light chain. The second stage of the process was to cross these plants to generate progeny that expressed both genes.

It was found that in the F1 progeny expressing both chains, significant amounts (up to 1.3% of total leaf protein) of assembled antibody were produced. These transgenes contained native immunoglobulin signal sequences, which targeted the nascent light and heavy chains to the ER and were found to be necessary for efficient assembly of the antibody in plants. Several important observations, which underpinned much of the future work in this area, were made in this original research.

Targeting of the antibody chains to the lumen of the ER was found to be necessary for efficient chain assembly and stability. Conditions in the lumen of the ER favour the correct processing of the antibody chains, especially disulphide bridge formation, and molecular chaperones also help to direct antibody chain folding and assembly. Another important observation was that the formation of the complete antibody in the F1 plants increased the protein yield over that of either of the parents that contained single chains, reflecting the stable nature of the complete antibody in the plant system. Although a pioneering study, demonstrating that functional antibodies could be assembled in plants, the antibody in question had no practical applications.

Antibodies produced in plants are thought to be particularly suitable for topical immunotherapy. In an extension of the strategy of producing the various antibody subunits in different plant lines that are subsequently crossed to produce a functional antibody, a secretory IgA (sIgA) that protects against dental caries produced by *Streptococcus mutans* has been expressed in plants. The antibody does this by recognizing the native streptococcal antigen (SA) I/II cell-surface adhesion molecule, thereby preventing colonization. This multiple-component sIgA antibody required four parental transgenic tobacco lines to be produced, in which cDNAs for either the H, L or J chains (the H and L chains were derived from a murine antibody termed 'Guy's 13' that recognizes the SA I/II cell-surface adhesion molecule), or the secretory component, were expressed under control of the 35S promoter, and then crossed. The success of the approach has been quite an impressive technical achievement.

In mammalian systems, both plasma cells and epithelial cells are involved in the synthesis and assembly of secretory antibodies, yet

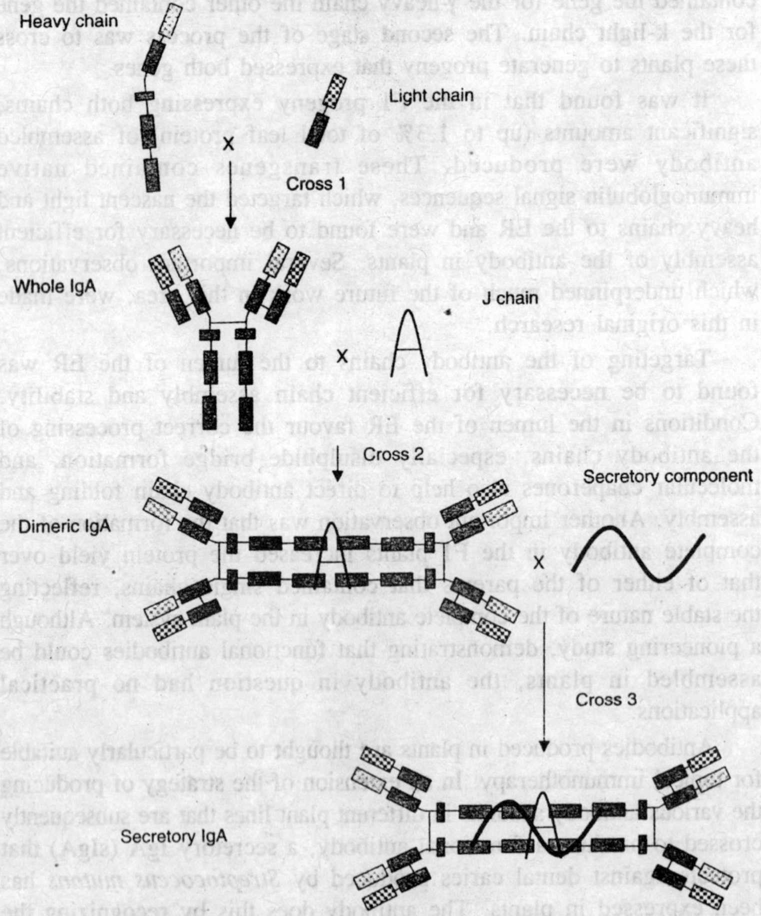

Fig. 4.11. Synthesis of a secretory IgA (sIgA) molecule in transgenic plants.

single plant cells containing all four transgenes are able to produce them efficiently. Secretory antibodies have many theoretical advantages in situations like the one described here. Secretory IgAs are the predominant antibody type that protects against microbial infections at mucosal sites. They are also more resistant to proteolysis (which contributes to high yield), and they bind antigens with a higher efficiency than do the original antibodies produced in plants. These secretory antibodies have now been tested on humans. They have been topically applied to teeth and found to be effective at preventing recolonization by *S. mutans* for up to 4 months, which is as effective as an IgG produced in a murine hybridoma.

Despite the differences in structure, there appear to be no differences in binding properties between the proteins. Plants have been used to produce a variety of antibodies, including whole antibodies, antigen-binding fragments and single-chain variable fragment antibodies. Despite the now well-characterized differences in the glycosylation pattern seen between antibodies produced in plant and mammalian expression systems, antibodies produced in plants generally seem to exhibit similar properties to antibodies produced in other, mammalian, systems.

Plant-derived vaccines

Both stable and transient expression systems have been used to produce vaccines in plants. One of the first successful attempts in this area used a transient expression system. Recombinant cowpea mosaic *comovirus* (CPMV) was used as a vector for a linear epitope (antigenic determinant) from the VP2 capsid protein of mink enteritis virus (MEV). The recombinant CPMV was replicated in black-eyed beans (*Vigna unguiculata*), from which chimeric virus particles (CPMV-VP2), displaying the VP2 epitope, were isolated and used for the subcutaneous injection of mink. The vaccine was able to induce resistance to clinical infection by MEV. The main aim of many of these studies is the generation of edible vaccines. The driving force behind the use of edible vaccines has been the need for cheap, effective treatments for enteric disease in the developing world.

Plant-based systems have the potential to provide low-cost, easily administered, locally produced edible vaccines that could provide mucosal immunity against the infectious agents that are responsible for the deaths of millions of people, particularly children, annually. To bring this about it is necessary to present, in a practical, edible form, a quantity of vaccine that will stimulate the GALT (gut-

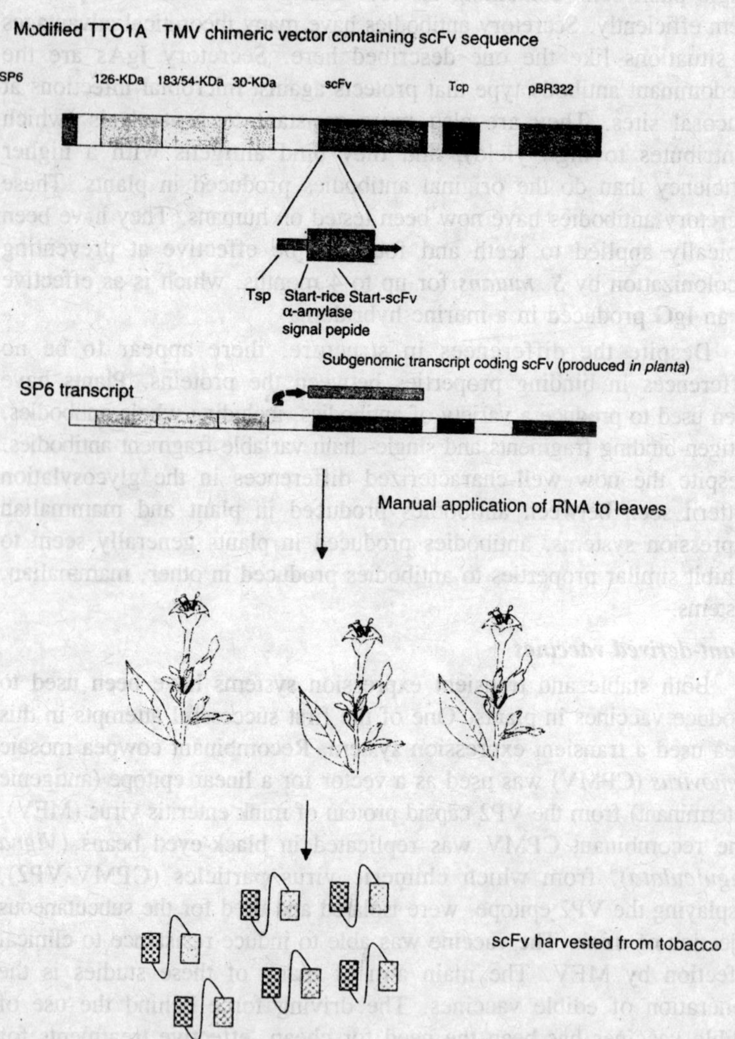

Fig. 4.12. The use of engineered tobacco mosaic virus (TMV) vectors for the expression of pharmaceutical proteins.

associated lymphoid tissue) system in the gut to generate secretory antibodies. The questions that have had to be considered relate to delivery of the vaccine to the gut through the stomach. As proteins, these vaccines are likely to be degraded in the stomach. The experiments carried out so far indicate that orally administered plant material expressing a vaccine can induce an immune response, with particulate material the most effective. Interestingly, when virus coat proteins are expressed in plant leaves they can form virus particles that may also improve the stability and effectiveness of the vaccine by presenting the protein in its native form. Much of the initial work on producing vaccines in plants was carried out in inedible plants, such as tobacco. There has been a move by some scientists towards producing vaccines in plants that could be eaten raw, such as tomato and banana.

Bananas are considered to be the ideal production system by some scientists as they are grown widely in the countries of the developing world and can be eaten raw. For use in animals, it is possible to consider the use of fodder crops or other food crops. These so-called 'edible vaccines' do work in the laboratory, but would they really work as originally envisaged? The major flaw in the process is that one needs to control very carefully the dose being given to patients if effective and safe protection against disease is to be ensured.

One of the drawbacks of edible vaccines is that it may prove very difficult to control the dose administered if consumption as part of a foodstuff is the means of delivery. Charles Arntzen, one of the main lights in this field, has recently withdrawn from the concept of edible vaccines. He believes they should be considered vaccines that come from a plant source. His group produces vaccines in tomato, but they are partially processed so that food-based tablets containing a known dose of vaccine can be produced. Not only is this better science but it will be easier to license the use of these vaccines from plant sources. Bananas may, of course, still prove to be a useful system that allows locally based production facilities to be established in developing countries.

Other proteins-biopharmaceuticals

A wide range of proteins of pharmaceutical interest has been expressed in transgenic plants in the hope of producing an economically viable system for large-scale production. Although active recombinant proteins have been produced, one problem associated with production in plant systems is that these often give a relatively low product yield and recovery. Various strategies to overcome this problem are being developed, amongst them the use of novel purification systems (see the section on oleos ins for an example) and chloroplast transformation.

Trichosanthin

Trichosanthin (TCS) is a component of the tuber of the plant *Trichosanthes kirilowii*, which is used in Chinese medicine. A ribosome-inactivating protein, TCS inhibits tumour growth and the immune response, and may be useful as a treatment for HIV/AIDS. Relatively high levels of TCS accumulation (2% of total soluble protein 2 weeks after inoculation) have been achieved in *Nicotiana benthamiana* using a viral RNA-based transfection system. This system offers the possibility of rapid, large-scale production of TCS. The rapid production also offers the opportunity to rapidly produce mutated forms of TCS for screening in trials aimed at identifying forms with increased efficacy or reduced side-effects.

Glucocerebrosidase

Gaucher's disease is an inherited disorder in which glucocerebroside (a component of red blood cells that is normally broken down into glucose and ceramide as old and damaged red blood cells are removed by the body) accumulates in lysosomes, due to a deficiency in the enzyme glucocerebrosidase. Gaucher's disease, which results in swelling of the spleen and liver and severe bone damage, can be extremely debilitating and painful. Currently, treatment is based on managing the symptoms or by treatment with a drug developed from glucocerebrosidase purified from human placentas. However, a huge amount required (10-12 tons of placentas a year for one patient), making this one of the world's most expensive drugs. Recombinant approaches to production in mammalian cell cultures have reduced the cost to some extent, although it remains an extremely expensive drug. A process to produce glucocerebrosidase in tobacco has recently been patented, which will hopefully result in cheaper glucocerebrosidase being available for patients with Gaucher's disease.

Human serum albumin

Human serum albumin (HSA) has many potential medical applications (the treatment of burns and in liver cirrhosis, for example). HSA has been expressed in tobacco and potato under the control of the 35S promoter. Two forms of the HSA protein were expressed, with different signal sequences to ensure secretion of the HSA. One form had the human prepro-sequence, whilst the other had the signal sequence from the tobacco extracellular PR-S protein. Although both forms of the HSA were secreted, analysis showed that the form with the human prepro-sequence was not properly processed, leading to proHSA being secreted. However, the form with the PR-S signal sequence was correctly processed, and mature HAS, that was indistinguishable from the human protein, was produced.

Arginine vasopressin

The examples we have looked at demonstrate that both plant and non-plant pharmaceutical proteins, which are of high value both clinically and economically, can be successfully produced in plants. However, in what is a typical situation in developing technologies, not all attempts at protein production in plants have been successful. Attempts to express arginine vasopressin (AVP) in tobacco cultures, for example, have proved unsuccessful due to protein instability and retarded cell growth.

ECONOMIC CONSIDERATIONS FOR MOLECULAR FARMING

Whilst there has been much talk in this chapter of suitable targets for molecular farming, few of the examples described have yet been produced commercially. This may result from problems with the technology itself, or it could reflect the long time between initial laboratory-based demonstration and subsequent optimization and regulatory approvals for field release and approval for food or drug use. However, it may be the case that levels of production that represent major scientific breakthroughs are none the less insufficient to be economically viable. This may seem surprising, because it is tempting to make the naive assumption that the production of compounds in a field, with free energy from the sun, must be cheaper than production in a chemical factory or industrial fermenter. This is often not the case, because farming is not a cost-free activity, the extraction of the product from the plants may be a difficult and expensive process and the price of the product must be significantly higher than that of the crops it has replaced. This may mean that the current levels of synthesis in transgenic plants coupled to the costs of production make a number of the examples described above uneconomic when compared to current production methods.

On the other hand, continuing improvements in plant biotechnology, coupled to the inevitable rise in costs of non-renewable resources, such as petrochemicals, make it likely that many of these products will eventually be made in plants. The economics of producing medically related proteins in plants has been much discussed, particularly because of the high unit price of many of these proteins and the expansion of the market for medically related proteins. It is in this area that many companies and people have invested money, in the hope of achieving economically viable molecular farming in plants. Fortunately for these companies and investors, the figures seem to suggest that there is a future for molecular farming. It has been estimated that the cost of producing immunoglobulins in alfalfa, in a 250-m^2 greenhouse, are \$US500-600g^{-1} compared with the figure of \$US5000g^{-1} for a hybridoma-produced antibody. The company Planet Biotechnology has compared the costs of producing an IgA antibody in four different systems.

The comparison was between production in a cell culture system, transgenic goats, grain (assuming a yield of 7.5 tonnes ha^{-1}) and green biomass, such as leaves (assuming a yield of 120 tonnes ha^{-1}). The cost of an IgA produced in leaves was estimated to be below \$US50g^{-1}. This compared very favourably with the costs for production in a cell culture system (\$US1000g^{-1}) and in a transgenic animal (\$US100 g^{-1}). It must be remembered that these figures are fairly crude, being based on projected yields, etc. However, they do illustrate that costs of protein production in plants are at least comparable with other systems of production and further developments are now coming on stream. As we have seen in chapter 3 transformation systems are now available for the introduction of foreign genes into chloroplast, giving an estimated 10,000 copies per cell.

Using such systems it is claimed that recombinant protein content has reached as much as 47% of the total soluble protein. Somatotrophin (growth hormone), produced in chloroplasts, reached about 7% of the TSP. This compares favourable with that of a nuclear transgene (0.01% TSP). Using Chloroplast based plant systems have other potential advantages over existing systems. Despite the very high number of gene copies there have been no reports of gene silencing that could effect expression levels. Also, protein processing is one of the most expensive parts of producing proteins in existing systems such as *E. coli*. It seems that chloroplasts can assemble and fold complex foreign proteins, but whether they can cope with the other post translational modification problems remain to be seen.

Another advantage of chloroplast is that they are maternally inherited. This means that they will not be spread via pollen to non-transgenic systems. This type of clean gene technology is most important for getting approval for growth in non-contained facilities. Other Health related proteins produced in chloroplasts include: human proinsulin for the treatment of autoimmune disease, Interferon IFNa 2b, human serum albumen, synthetic human haemoglobin genes, Guys 13 Monoclonal antibody, and a *Bacillus anthracis* protective antigen.

5

Crop Quality

The variety of ways in which the productivity of crop plants can be improved by enhancing their ability to resist or tolerate biotic and abiotic stresses is already discussed. These strategies can all contribute to an improvement in crop yield by allowing the plants to better withstand external factors that reduce the amount and quality of harvestable plant material. However, the performance of crop plants is also determined by endogenous factors that affect the yield and quality of the harvestable material produced. The yield of a crop is ultimately determined by the amount of solar radiation intercepted by the crop canopy, the photosynthetic efficiency (i.e. the efficiency of conversion of radiant to chemical potential energy) and the harvest index (the fraction of dry matter allocated to the harvested part of the crop). Thus, manipulation of crop yield requires some understanding of the fundamental processes that determine photosynthetic efficiency and dry-matter partitioning. On the other hand, the quality of a crop is determined by a wide range of desirable characteristics such as nutritional quality, flavour, processing quality and shelf-life.

The nature of yield and quality are therefore much more varied and wide-ranging than the other traits described to this point, and the structure of this chapter is therefore somewhat different to reflect this. Rather than try to cover the entire range of targets for different crops, a small number of yield and quality traits have been selected to exemplify the types of approaches that have been adopted in this area. The chapter therefore starts with an extended discussion of one system—tomato ripening—because it is possible to draw several general lessons from the techniques and concepts that were deployed in this project.

Genetic Manipulation of Fruit Ripening

The predominance of herbicide and pest resistance amongst the commercially developed GM traits has been discussed previously. However, amongst the first GM products to reach the market, around 1994, were Calgene's FlavrSavr fresh tomatoes, and processed tomato products containing delayed-softening tomato fruit developed in the UK by Zeneca and the University of Nottingham. This may seem rather puzzling, because the tomato crop is relatively small compared to the major GM crops of Northern America (maize, soybean, oilseed rape and cotton). Furthermore, it is not immediately clear why a large biotechnology company like Zeneca would be involved, when they had no direct interest in tomato seeds or processed products when the project was started (when the company was still ICI). Part of the answer lies in the early recognition by ICI and Calgene of the value of using tomato ripening as a model system on which to develop expertise and test the potential of plant biotechnology to modify crop quality.

In order to understand why tomato ripening was seen as such a good model system, it is necessary to look at some of the basic biology

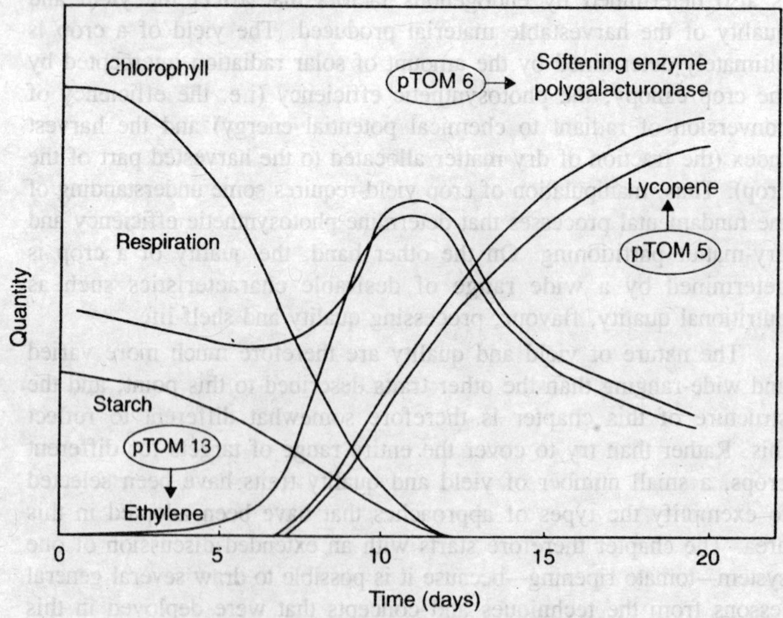

Fig. 5.1. Biochemical changes during tomato ripening.

of fruit ripening. Fruit ripening is an active process that, in climacteric fruit such as tomatoes, is characterized by a burst of respiration (the respiratory climacteric), ethylene production, softening and changes to colour and flavour. The transient peaks in respiration and ethylene production at the start of tomato ripening, which are accompanied by softening, a change in colour from green to red and enhanced flavour.

The peak of ethylene production is significant, because ethylene is known to be the phytohormone that triggers ripening in climacteric fruit. The colour change results from the degradation of chlorophyll and the production of the red pigment, lycopene. Flavour changes occur as starch is broken down and sugars accumulate. A large number of secondary products that improve the smell and taste of the fruit are also produced. The softening of the fruit is largely the result of the cell wall degrading activity of the enzymes polygalacturonase (PG)

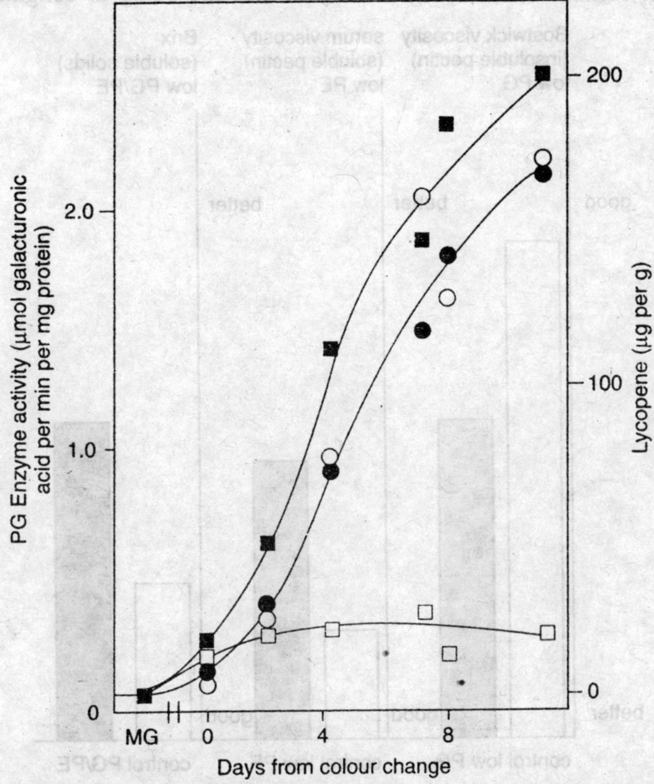

Fig. 5.2. Polygalacturonase (PG) activity and lycopene content during the ripening of antisense PG fruit.

and pectin methylesterase (PME). The PG enzyme is synthesized *de novo* during ripening and acts to break down the polygalacturonic acid chains that form the pectin 'glue' of the middle lamella, which 'sticks' neighbouring cells together. The realization that the ripening process involved the activation of specific ripening-related genes, such as the one encoding polygalacturonase, enabled these genes to be cloned.

The genetic modification of fruit softening demonstrated the general validity of the concept that a specific trait could be manipulated by downregulating the expression of an endogenous gene using antisense and co-suppression techniques. The specific example of reducing cell wall degradation by targeting endogenous PG and PME has been applied to a number of other fruit, such as mango, peach and pear. Furthermore, it is now apparent that pectin depolymerization is not the only determinant of texture, and the roles of other enzymes such as pectate lyases, cellulases and xyloglucan hydrolases are being investigated.

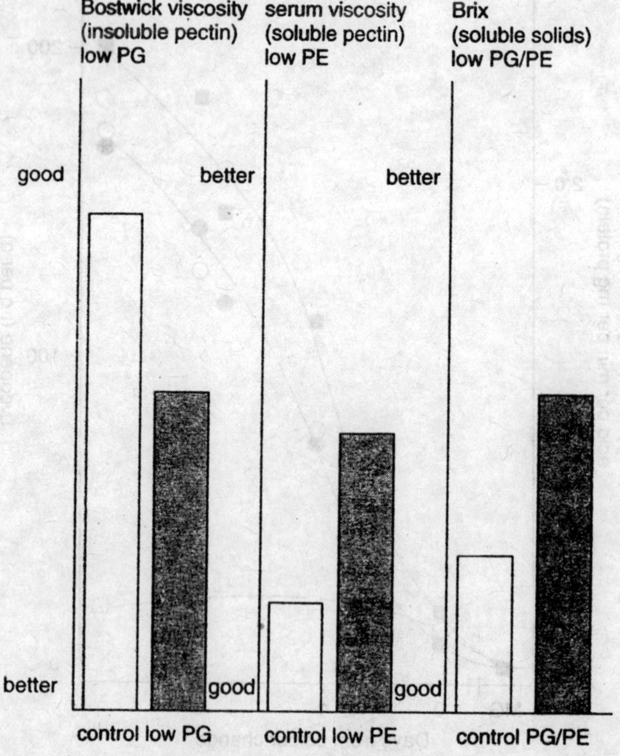

Fig. 5.3. Relative processing performance of low-PG and low-PE fruit.

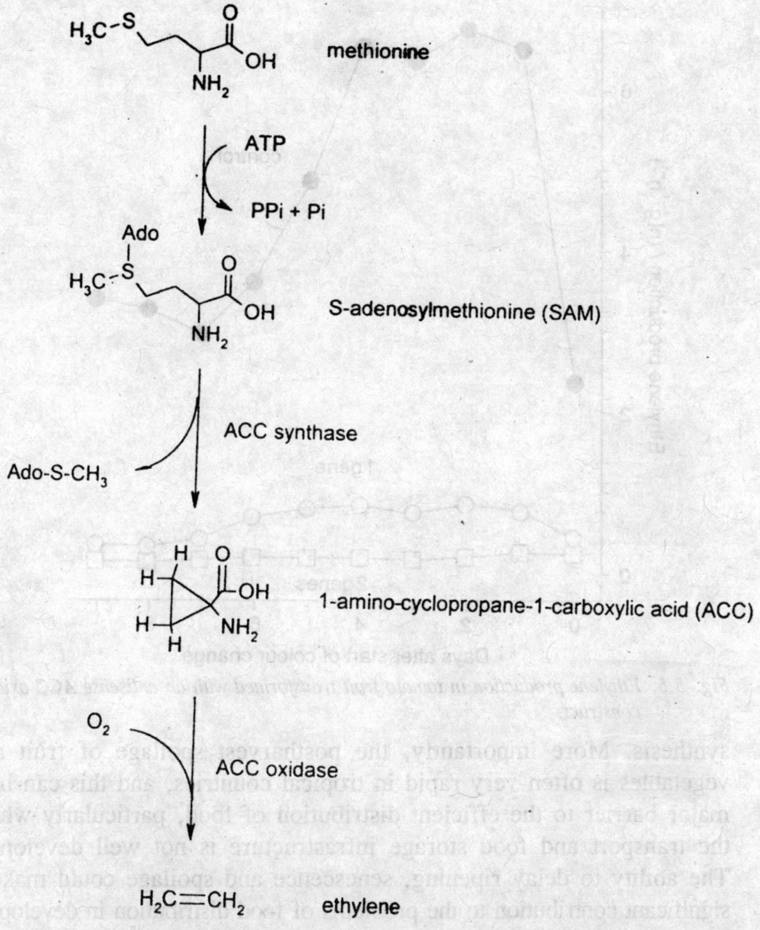

methionine

ATP

PPi + Pi

S-adenosylmethionine (SAM)

ACC synthase

Ado-S-CH_3

1-amino-cyclopropane-1-carboxylic acid (ACC)

O_2

ACC oxidase

$H_2C{=}CH_2$ ethylene

Fig. 5.4. Ethylene biosynthesis and its regulation.

Plant cell wall structure and metabolism are still poorly understood, but research in these areas holds out the prospect of being able to manipulate many aspects of plant growth and development. "Antisense ethylene" technology is applicable to all climacteric fruits, and to other systems triggered by ethylene, such as the spoilage of vegetables and the senescence of picked flowers. For example, it has been known for some time that the vase-life of cut flowers can be extended by adding silver salts to the water, which blocks the response to ethylene.

The introduction of antisense ACC synthase or ACC oxidase would have a similar effect of extending vase-life by suppressing ethylene

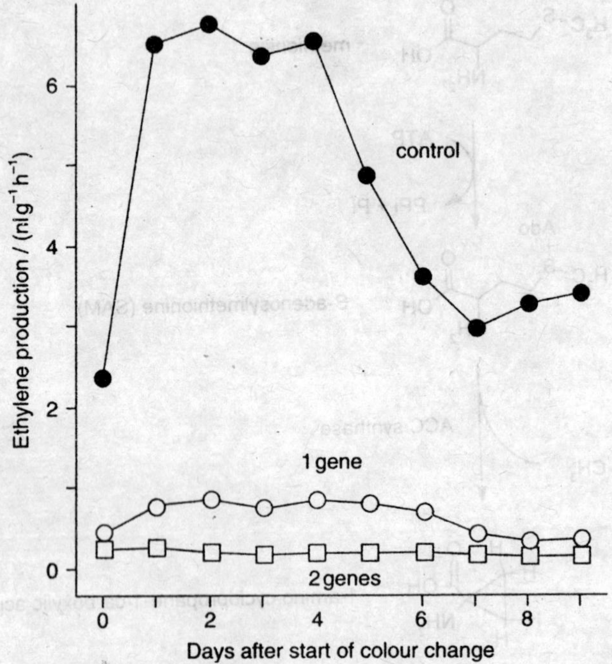

Fig. 5.5. Ethylene production in tomato fruit transformed with an antisense ACC oxidase construct.

synthesis. More importantly, the postharvest spoilage of fruit and vegetables is often very rapid in tropical countries, and this can be a major barrier to the efficient distribution of food, particularly where the transport and food storage infrastructure is not well developed. The ability to delay ripening, senescence and spoilage could make a significant contribution to the problems of food distribution in developing countries. In addition, the successful manipulation of one plant growth regulator indicates the potential for engineering plant development by manipulating other plant growth regulators. The early availability of auxin and cytokinin biosynthetic genes from the *Agrobacterium tumefaciens* Ti plasmid ha meant that both phytohormones have been synthesized in a range of plants under the control of various promoters.

The more recent cloning of genes involved in gibberellin and abscisic acid synthesis/response has also permitted the genetic manipulation of important development processes. The manipulation of lycopene synthesis, and its effect on gibberellin synthesis, leads on to three different examples of the manipulation of yield and quality. The

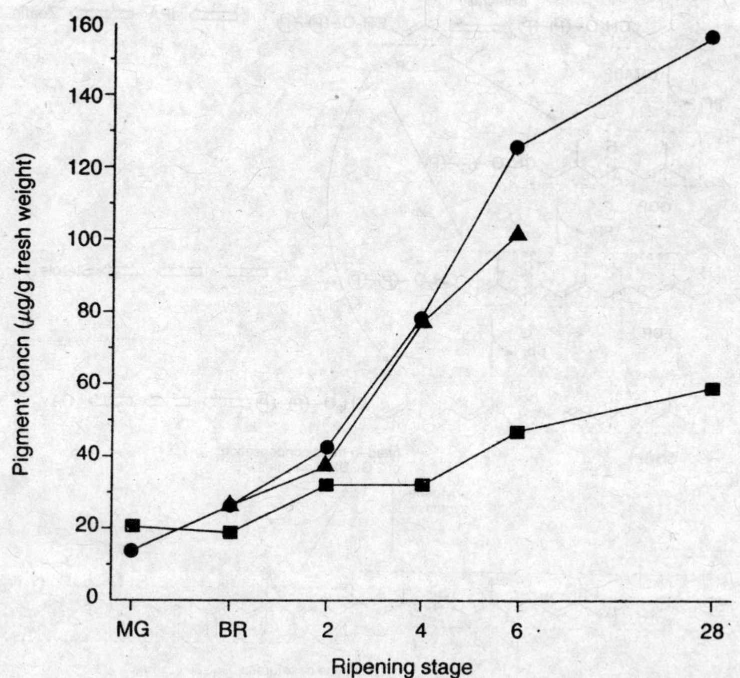

Fig. 5.6. Pigments in antisense ACC oxidase fruit during ripening.

first general area to consider is the modification of colour. There are many examples of this, particularly amongst the ornamental flowers.

Another area that relates to the genetic manipulation of tomato colour is the effect that overexpression of phytoene synthase had on the gibberellin pathway, and the resultant dwarfing. This is of interest because dwarfing is an important agronomic trait. Dwarf plants require less of the plant's resources to be committed to the growth of the stem, allowing more dry matter to be partitioned to the grain (the main commercial product) rather than the straw. This increase in harvest index directly improves yield, and, in addition, the dwarf crops are less prone to damage by wind and rain. The dwarf character was a major feature of the high-yielding varieties of wheat and maize that gave rise to the 'Green Revolution'.

The 'Green Revolution' strains of wheat were shown to be short because they respond abnormally to gibberellin. This trait is conferred by mutant dwarfing alleles at two reduced height-1 loci (*Rht-B1* and *Rht-D1*). Recent work has shown that the *Rbt-B1/Rht-D1* and the maize dwarf-8 genes are orthologues of the *Arabidopsis* gibberellin insensitive

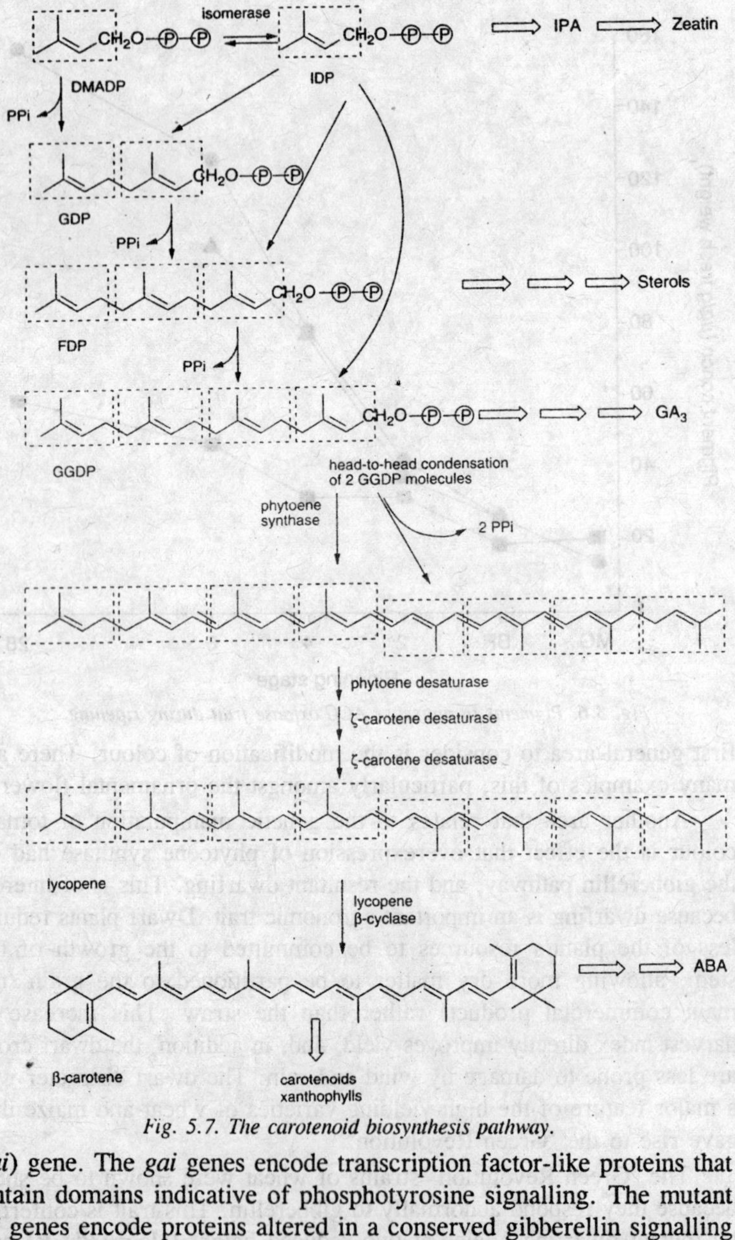

Fig. 5.7. The carotenoid biosynthesis pathway.

(*gai*) gene. The *gai* genes encode transcription factor-like proteins that contain domains indicative of phosphotyrosine signalling. The mutant *gai* genes encode proteins altered in a conserved gibberellin signalling domain. Transgenic rice and wheat plants into which a mutant *gai* gene was introduced were shown to have a reduced response to

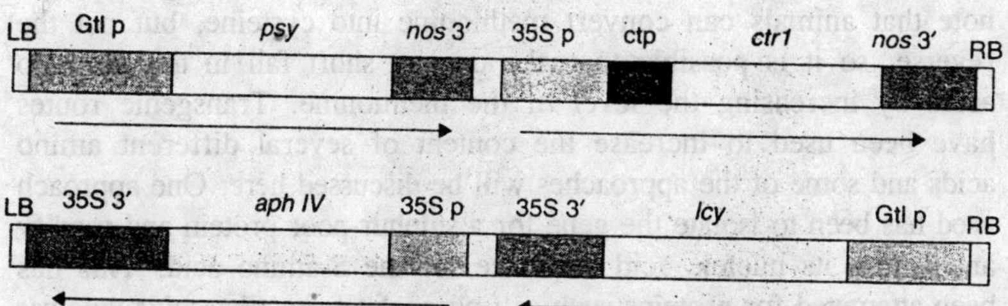

Fig. 5.8. Constructs for the production of Golden Rice.

gibberellin, and to have a dwarf phenotype. This result is particularly significant because, whilst the dwarf character has been found in rice, attempts to breed a dwarf strain of basmati rice have failed to date because the resulting short plants have lost their characteristic flavour.

The success of the *gai* gene experiments suggests that the dwarf character could be introduced into a wide range of crop species to improve crop yield, without having to shuffle all the other genes that contribute to the desirable characteristics of the current elite lines. An important lesson here is that transgenic approaches to crop improvement, particularly those that use well-characterized genes, can be much more precise and predictable than conventional breeding. However, as we look at effects of modifying phytochromes in a later section, it will become clear that this is not always the case: The final example of crop quality enhancement that flows from the study of carotenoid biosynthesis is that of provitamin A production in rice grains—so called 'Golden Rice'.

Engineering Plant Protein Composition for Improved Nutrition

The example of Golden Rice has dealt with one aspect of nutritional quality. Another that has become a target for genetic engineering is the amino acid content of plant foods. Humans are only capable of synthesizing 10 of the 20 naturally occurring amino acids. The other, 'essential', amino acids are obtained from the diet. Obtaining balanced quantities of the amino acids can be problematic if certain foods predominate within a diet. This is the case with cereal grains, which are commonly used as the principal energy source, and with legume seeds, which are important sources of proteins in the diet of humans and livestock.

Cereal grains are often limiting for lysine, while the legume seeds have an adequate level of lysine but are limiting for the sulphur-containing amino acids, methionine and cysteine. It is important to note that animals can convert methionine into cysteine, but not the reverse, so it is possible to make up any short fall in the S-amino acids by increasing the level of the methionine. Transgenic routes have been used to increase the content of several different amino acids and some of the approaches will be discussed here. One approach used has been to isolate the gene for a sulphur-poor protein and modify and enrich its nucleic acid sequence for the S-amino acid. This has been attempted for proteins such as (-phaseolin from *Phaseolus vulgaris* and vicilin from *Vicia faba*, but the modified proteins were either unstable or contained too little methionine to make them useful.

A more successful approach has been to construct totally artificial genes that code for proteins containing a high S-amino acid content. One such totally synthetic protein, containing 13% methionine residues, has been successfully expressed in sweet potato (*Ipomoea batatas*). Several methionine-rich proteins have been identified in maize (21-kDa zein—28% methionine, 10-kDa zein—23% methionine); rice (10-kDa prolamin—20% methionine); sunflower (2S sunflower seed albumin—16% methionine, 8% cysteine); Brazil nut (Brazil nut albumin—18% methionine, 8% cysteine). The genes for these proteins have been introduced into a number of crops (maize, soybean, lupin, canola) to increase the level of S-amino acids in blended stock feeds containing cereal and legume grains.

One major problem associated with moving proteins between species has become apparent with the Brazil nut albumin (BNA), i.e. the protein responsible for the potent allergenicity of Brazil nuts. This property is maintained in the seeds of BNA-containing transgenic plants, which makes them unacceptable for human consumption and has been a timely warning that safety issues should be paramount when dealing with food crops. Some of the problems associated with these studies relate to the stability of the proteins and their localization within the plant. This point can be illustrated by considering the expression of high-lysine proteins in the seed endosperm of cereals such as rice. These studies raise important issues about the types of protein that can be used and how they are localized in the grain, issues that should be considered when ectopically expressing proteins for whatever reason.

In cereal endosperm, storage proteins accumulate primarily in protein storage vacuoles (PSVs) of terminally differentiated cells and as protein bodies (PBs) assembled directly within the endoplasmic reticulum (ER). Specific storage proteins are stored in these structures. In rice for example, there are two main types of storage protein: prolamins, which form accretions within the lumen of the ER, and glutelins (related to 11S globulins). Glutelins comprising up to 80% of the total seed protein are synthesized on rough ER and then transported to PSVs. It has recently been found that mRNAs for both types of protein are found in rough ER polysomes, but that the prolamin transcripts are preferentially located on membranes surrounding prolamin-containing protein bodies; while glutelin mRNAs are predominantly associated with polysomes on the cisternal ER. One interesting study has been carried out with rice, where a lysine-rich protein from soybean has been expressed in the endosperm.

To optimize expression and localization, the soybean protein chosen was a glycinin, which is a member of the 11S globulin family. The important feature of the glycinin is that it is processed and stored in soybean by a process very similar to that found for the rice glutelins. When the transgenic rice plants were analyzed it was found that both globulins (glutelin and glycinin) were expressed and formed complexes of discrete sizes that were targeted correctly. If this had not been the case then it would not be possible to obtain optimum levels of proteins. The other interesting point of glycinin proteins in soybean is that they are associated with a hyporcholesterolaemic effect (reduction in cholesterol) if the dietary intake of these proteins is in .excess of 6 g/ day. Therefore, expressing the soybean gene in rice potentially has two beneficial effects: it increases the amount of lysine in the endosperm protein; and it would contribute to the dietary levels of soybean globulins that lead to a reduction in the level of cholesterol.

Genetic Manipulation of Crop Yield by Enhancement of Photosynthesis

Crop yield is totally dependent upon light and its conversion (light harvesting and electron transport) into the usable energy (ATP, NADP) that drives the dark reactions of photosynthesis (carbon dioxide conversion into carbohydrates). These reactions usually take place in chloroplasts within the leaves, the source, but the efficiency of the process is also dependent upon the capacity of sink tissues and organs to assimilate this fixed carbon. These are complex processes and outside the scope of this book, but some studies have been carried out to investigate the potential for enhancing photosynthesis by biotechnological means. Although these have been very preliminary in nature, several examples will be briefly discussed to describe the potential for manipulating complex physiological interactions. Direct manipulation of light harvesting, electron transfer or the processes of photoprotection and photoacclimation will not be considered here.

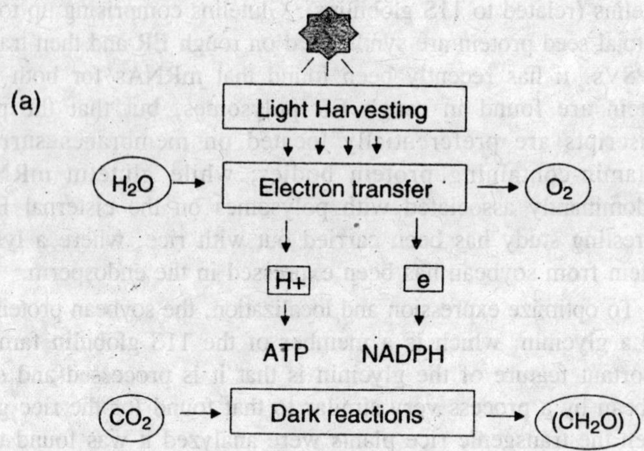

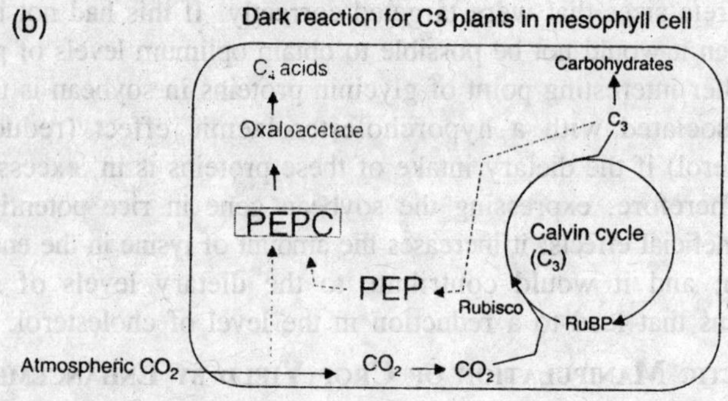

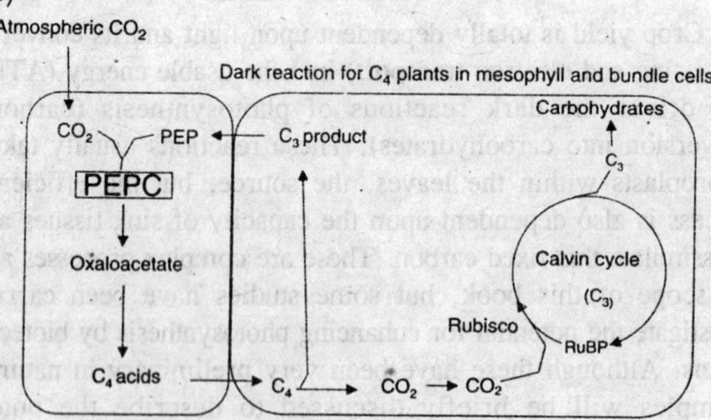

Fig. 5.9. Photosynthesis.

Manipulation of Light Harvesting and the Assimilate Distribution —phytochromes

When grown in the field or in their natural environments, plants are not always under optimum light conditions, even during daylight hours. They are faced with changes in the amount, direction, duration and quality of incident light radiation. Plants have evolved mechanisms that optimize the acquisition of available light energy for photosynthesis. In dense populations or in shady conditions, plants display the shade response. The plants use available resources to increase stem and petiole elongation to outgrow any shading plant; there is also a reduction in chlorophyll synthesis, leaf thickening and an increase in apical dominance. From an agricultural point of view, this shade response has important implications for yield, as assimilates are used to re-establish optimal light conditions for photosynthesis rather than being stored. The shade response is regulated through a series of photo receptors.

The phytochrome family of proteins are the best characterized of these proteins (there are at least five Phy proteins in dicots). They are able to detect the level and quality of the incident light; and then control growth and development via a series of signal transduction pathways that regulate gene expression. Phy proteins are photochromic proteins, they have a protein moiety connected to a tetrapyrrole chromophore, and they can exist in two forms: physiologically inactive (P_R, red-light absorbing) and active (P_{FR}, far-red light absorbing). These forms are interconvertible by red or far-red light, respectively. The photosynthetic pigments in leaves absorb most visible radiation (400-700 nm) but reflect light beyond 700 nm (far red, 700-800 nm). This leads to a high proportion of FR radiation being found in dense plant stands. Plants are able to perceive this radiation and the proximity of other plants because the equilibrium of the phytochromes is shifted towards the inactive form.

Two of the major phytochromes involved in this response are PhyA and PhyB. PhyA accumulates in the dark and is rapidly degraded upon conversion, by the absorption of red light, into the labile PFR form. Despite this, it is responsible for the detection of continuous FR light and dampens the shade avoidance response. PhyB is responsible for the detection of red light, so under high light conditions (high relative proportions of red light) it is converted to the active form, in which form it suppresses the shade-avoidance response. The manipulation of phytochromes is a general approach that can have pleiotropic effects on photosynthesis and assimilate distribution. Analysis of these systems suggests that phytochrome manipulation is likely to have important consequences on photosynthesis (the phytochromes coordinately regulate many of the genes involved in photosynthesis) and the sink-source relationship. So, a number of different transgenic plant types, ectopically expressing either PhyA or PhyB, have been constructed and some interesting effects have been observed.

Plants expressing the phytochromes at very high levels were seriously affected in their growth, but it was possible to identify plants in which the phytochromes improved the harvest index under high densities in the field. In one set of experiments, tobacco plants, engineered to overexpress an oat *PHYA* cDNA, were grown at five different planting densities. At the higher planting density the wild-type tobacco demonstrated the shade response, in that the plants were normally taller than those planted at low densities. With the transgenic lines, they elongated at lower rates—so, at the high planting densities, relative to the lower planting densities, the result was dwarfed plants.

The ectopically expressed PhyA was not only damping the shade response in response to far-red light, but was actually reducing stem elongation to below the 'normal levels' found in plants grown in optimal light conditions. Overall yield was not affected, but assimilates showed an enhanced allocation to leaves with increases in the harvest index approaching 20%. So the beneficial outcome of this approach was that at high planting densities assimilates were not being utili sed in stem growth but were being utilized in leaf production or were being stored. Similar dwarfing effects were found in an experiment with transgenic potato plants containing an ectopically expressed *Arabidopsis PHYB* sequence. However, overexpression of the phytochrome also led to an increase in chloroplast number and an overall increase in photosynthesis per leaf area. Photosynthesis was also less sensitive to photo inactivation under prolonged light stress. Although leaf senescence was not delayed, chlorophyll degradation was, leading to a prolonged period of photosynthetic activity.

The increase in the rate of photosynthesis and the prolonged period of photosynthetic activity resulted in an increase in biomass. This was demonstrated by extended underground organs (roots/shoots) and with a large number of small tubers. In fact, tuber yields were increased by as much as 50% in weight. The pleiotropic nature of this approach has some overlap with strategies to prolong the photosynthetic period of plants by delaying leaf senescence. One such strategy has involved the production of transgenic plants containing the senescence-related *SAG12* promoter (*SAG,* senescence activated gene) driving expression of the *Agrobacterium* cytokinin biosynthetic gene *ipt*. As the leaves begin to senesce, the expression of the *ipt* gene is switched on, thereby increasing the cytokinin level in the leaf. This delays senescence and maintains chlorophyll content. The prolonged period of photosynthetic activity is expected to increase assimilate production. However, it is not clear if this strategy will be successful as there may be problems with nitrogen metabolism. As the nitrogen released from senescing leaves is required for other developmental processes, such as grain filling in cereals,, its maintenance in the leaf is unlikely to be beneficial.

Direct Manipulation of Photosynthesis—Enhancement of Dark Reactions

The sheer complexity of photosynthesis and assimilate storage suggests the task of increasing yield will require whole suites of genes—nuclear and chloroplast-to be enhanced. The phytochrome and the *ipt* approaches allow for this, in that they have pleiotropic effects. It is likely that different crops will require different strategies to be adopted. This is the case for C_3 and C_4 plants (this 'refers to whether the plant uses a 3-carbon or a 4-carbon route in the fixation of CO_2), which will be discussed in the next section. However, identifying limiting steps in pathways can allow increases in yield to be obtained. This has recently been found with tobacco where two enzymes in the Calvin cycle (C_3 cycle) have been shown to be limiting. Most of the enzymes in the Calvin cycle are present at levels in excess of those required to sustain a continued rate of CO_2 fixation. However, levels of fructose-1,6-bisphosphatase (FBPase) and sedoheptulose-1,7-bisphosphatase (SBPase) are extremely low compared with those of other enzymes in the Calvin cycle.

Transgenic tobacco plants expressing a single gene for a chloroplast-targeted, dual-function cyanobacterial fructose-1,6/sedoheptulose-1, 7-bisphosphatase show enhanced photosynthetic efficiency and growth characteristics. Dry matter and photosynthetic CO_2 fixation were 1.5- and 1.24-fold higher in the transgenic plants compared with wild-type tobacco. It was also found that ribulose-1,5-bisphosphate carboxylase/oxygenase (Rubisco) activity was increased, and that various Calvin cycle intermediates and the accumulation of carbohydrates were also higher. These results seem to relate to an increased metabolic flux, and support the notion that the enzymes were limiting the photosynthetic pathway in tobacco. Perhaps the most challenging approach attempted has been to introduce various enzymes from the C_4 photosynthesis pathway into C_3 crop species.

The C_3 plants—which include wheat, oats and soybean—have chloroplast-containing mesophyll cells in the leaves. CO_2 is assimilated in these chloroplasts, via Rubisco, as part of the Calvin cycle, and the first product from this is the C_3 molecule 3-phosphoglyceric acid. This is not a very efficient system as the CO_2 concentration in C_3 plants can be a rate-limiting factor. Also, Rubisco can use O_2 as an alternative substrate for reactions with ribulose-1,5-bisphosphate, which ultimately leads to photorespiration (metabolism of C_3 molecules leading to the release CO_2 in mitochondria). In essence, O_2 is a competitive inhibitor with respect to CO_2 fixation by Rubisco. In addition, high temperature decreases the availability of CO_2 to Rubisco, and water stress (by closing stomata) increases the resistance to CO_2 diffusion into the leaf.

In C_4 plants, such as maize and sugar cane, two different types of chloroplast-containing cells are involved in CO_2 assimilation. CO_2 is first fixed in mesophyll cells by phosphoenolpyruvate carboxylase (PEPC). C_4 acids then migrate into associated bundle-sheath cells where the C_4 pathway generates a high concentration of available CO_2 which is then used by Rubisco. This system is very efficient under all the conditions where the C_3 system fails. Recently, a complete maize PEPC gene (maize is a C_4 plant) has been introduced into rice (a C_3 plant) with surprising results. Many of the plants obtained showed high levels of expression, two or three times that found in maize, and the enzyme accounted for up to 12 % of the total leaf protein.

The most important observation was that the O_2 sensitivity of photosynthesis in rice was decreased in the transformed plants. Why this should be the case was not clear, but it suggested that carbon fixation by the maize enzyme increased the supply of CO_2 to Rubisco. This experiment raises the possibility of manipulating other stages of the photosynthetic pathways. However, it is also likely to be a very difficult process, even the expression of the *PEPC* gene had drawbacks. Expression of PEPC in the absence of the other enzymes of the C_4 pathway led to a drain of carbon into C_4 acids rather than the carbohydrates, so reducing plant yield.

6

Crop Production

People are very much a part of Earth's ecosystems, and every ecosystem has both producers and consumers. The producers are the plants: They convert solar energy into chemical energy stored in food that is available to the consumers. The consumers are the animals—from the smallest to the largest—and the microorganisms that break down organic matter and recycle nutrients. Humans are therefore consumers who directly or indirectly depend on plants for their food. Before humans first practiced agriculture, ancient hunter-gatherers had evolved a complex relationship with their environment. They had an intimate knowledge of the plants and animals in their surroundings and used a wide variety of plant and animal foods.

Aboriginal peoples relied largely on plants, but most had diets that included animal products. With the development of agriculture some 10,000 years ago, people narrowed their food selections. Instead of the many plant species once gathered in the wild, 24 cultivated plants now account for much of the food people eat. Three of them—the cereals: wheat, rice, and maize—make up about two thirds of the human diet. Various cultures rely on different plants as their main food crop, or staple. The major food plants evolved at the same time as the societies that use them. Thus the Japanese have many different soybean-based foods and sauces, and Westerners eat wheat under many guises. Many Latin-Americans and Africans eat cassava, a tuber crop that is virtually unknown in North America and Europe.

Plants in Human Food

A look at food sources in various regions of the world reveals another interesting feature: Some societies eat almost entirely plants, whereas others use a substantial amount of animal-derived foods. For example, in India plants supply 80% of dietary protein (cereals and legumes), but in the United States plants provide only 25% of dietary protein. The rest comes from animals and animal products.

Table 6.1. Humanity's most important food crops

Category	*Crop Name*	*Amount produced (millions of metric tons)*
Cereals	Maize	596
	Rice	593
	Wheat	582
	Barley	136
	Sorghum	60
	Millet	27
Rootcrop	Potatoes	302
	Cassava	170
	Sweet potatoes	138
Legumes	Beans, dry	20
	All others combined	40
Vegetables	Tomatoes	100
	Cabbage	50
	Onions	50
Oil crops	Soybeans	162
	Oil palm fruit	98
	Coconuts	48
	Rapeseed (canola)	40
Starchy fruits	Bananas	58
	Plantains	30
Fruits	Oranges	66
	Apples	60
	Grapes	60

There are also differences among regions, social classes, and people of various religions. In developing countries, city dwellers typically eat more meat—historically regarded as a sign of affluence—than do farmers. For political reasons, governments often encourage meat consumption by providing agricultural price supports for feed grains, tax incentives for feedlot operators, guaranteed minimum prices, or government storage of surpluses. Researchers estimate that 85% of the calories and 80% of the protein in the human diet now come directly from plants. However, this situation is changing as people become more affluent. Rising affluence therefore puts additional pressures on the food system. We have already noted that urbanization and economic development, and the rise in income and expectations associated with them, make increasing demands on the food production system. People want to eat a more varied diet, and they want and can afford to eat more animal products.

Table 6.2. Comparison of diets in India and United States

	Source of Calories		*Source of Protein*	
Food	*India*	*United States*	*India*	*United State*
Cereals, starchy foods	61%	23%	60%	21%
Sugars	6	12	—	—
Beans, lentils	11	4	18	4
Fruits, vegetables	2	6	1	4
Fats, oils	4	19	—	—
Milk, milk products	7	14	12	24
Meat, poultry, eggs, fish	9	22	9	47

Global meat production has increased dramatically in the last 50 years, quadrupling since 1950. Until about 1950 in the industrialized countries, and even today in many parts of the developing world, farmers who practiced a sustainable mode of farming integrated livestock rearing

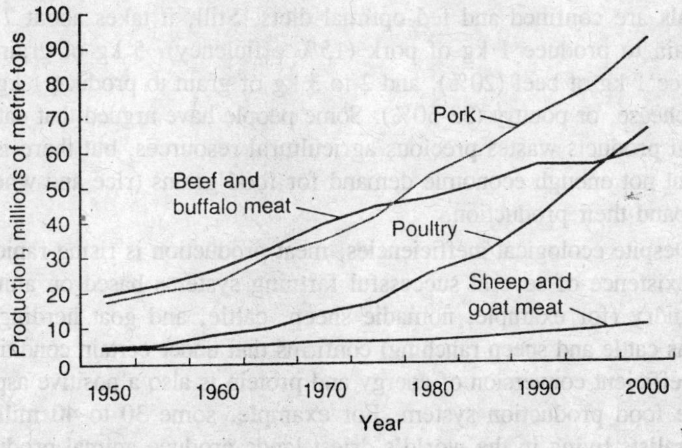

Fig. 6.1. Increase in world meat production, 1950-2000.

with food crop production. They rotated food crops (wheat, potatoes and sugar beets) with feed crops (hay, clover, and alfalfa). People ate the former; farm animals ate the latter. Farmers spread animal waste (manure) over the soil as fertilizer, and leguminous feed crops (clover and alfalfa) added nitrogen to the soil. Unfortunately, this integrated system is hard to sustain when demand for meat is high. Specialized animal production facilities, often located close to the population centers, feed animals soybeans and grains such as maize and sorghum. This system requires large areas of land just to raise food for animals.

To understand why eating animal products puts additional demands on the food production system, also consider how matter and energy transfer from one organism to another in the ecosystem. In the food chain plants are called *primary producers*. When an animal eats the plants, only about 10% of the plant matter and energy ends up as part of the growing animal's body. The other 90% is "lost": The animal uses a lot as energy to fuel its digestion and to refashion plant material into animal material. So if humans eat plants directly, the transfer efficiency is about 10%. Now consider what happens if an extra step occurs between plant and human: Let's say a cow eats grain, and then a human eats the cow.

The efficiency from plant to human via cow is 10% ÷ 10%, or 1%. So people must grow 10 times the amount of grain to supply a meat-eating human with the same energy and matter as would be needed if the person ate the grain directly. Actually, in some cases modern methods have improved the efficiency of transfer from grain to animal. Efficiency is higher in feedlots and chicken farms, where animals are confined and fed optimal diets. Still, it takes about 7 kg of grain to produce 1 kg of pork (15% efficiency), 5 kg of grain to produce 1 kg of beef (20%), and 2 to 3 kg of grain to produce 1 kg of egg, cheese, or poultry (33-50%). Some people have argued that eating animal products wastes precious agricultural resources, but there is at present not enough economic demand for food grains (rice and wheat) to expand their production.

Despite ecological inefficiencies, meat production is rising rapidly. The existence of several successful farming systems based on animal husbandry (for example, nomadic sheep, cattle, and goat herding as well as cattle and sheep ranching) confirms that under certain conditions the inefficient conversion of energy and protein is also a positive aspect of the food production system. For example, some 30 to 40 million pastoralists living in the world's driest lands produce animal products (milk, meat, and wool) in areas that are essentially unsuitable for crop agriculture. Nevertheless, meat consumption can sometimes have unintended negative consequences as well.

Use on Arable Land

With suitable temperature and rainfall, people can successfully use land to grow crops and produce foods. But where is the arable land (cropland) located, and how much exists? Certainly not all of Earth's land is used for crops. Of the total land area of Earth, people use about 11 % for crops (arable land) and use another 24% for pastures; 31 % is forested. The arable land is heavily concentrated in Canada, the United States, Europe (including Russia and the Ukraine), India, China, and Southeast Asia. The rest of the land surface—about one third—is too cold, too dry, or too steep for plant growth.

The unequal distribution of agricultural resources over the globe is shown clearly if we express the amount of cropland per inhabitant and the available freshwater per inhabitant for each continent. Asia, which already has the most people, has the least agricultural land and freshwater per person. The location of arable land is itself a function of climate, soil type, and type of vegetation that grew in the area before people cleared the land. Climate, soil type, and vegetation all interacted during the soil formation process and the most productive soils formed in areas of permanent grassland or hardwood forests (most of Europe and India). Grasslands produce rich soils with abundant organic matter that can remain productive for many decades after farmers convert them to cropland. These soils are quite suitable for annual row crops (maize, wheat, beans, and so on). The process of converting natural ecosystems into pastures or cropland is now being repeated most dramatically in the tropical rainforests of South America. However, converting a tropical rainforest into an agricultural field can be very difficult.

Most rainforests grow on relatively infertile and highly weathered soil. When people clear such land and burn the vegetation, they expose bare earth to the sun, soil temperatures rise, and the soil organic matter quickly decays, because the vegetation does not replenish it. Because of the high rainfall, the soil erodes easily. Asia, Latin America, and Africa contain many examples of massive soil degradation caused by clearing away tropical forests and imposing agricultural practices on soils that are totally unsuitable for row cropping. People need new approaches for these soils. Plowing more land may not achieve a greater total of cropland than the 1.5 billion hectares presently used. The

FAO estimates that 5 to 7 million hectares are being lost each year to land degradation. One problem is that the areas with the greatest population pressures, such as the east coast of China and most of India, have the least land reserves that could be cleared for crops. Other areas—in the former Soviet Union, for example—have good arable land but not enough rainfall. Improved agricultural technologies could greatly expand food production in these regions. The development of agriculture in different regions of the world has intensified land use. Each stage of intensity has its own characteristics and productivity:

Forest Fallow

Cutting and burning parts of a forest release nutrients contained in the plants; this fertilizes the land, allowing it to support crops. But this fertility lasts only until crops deplete the soil of its nutrients, and the farmers then move on to another part of the forest, repeating the cycle. In the meantime, they leave the original land fallow for up to 20 years, to restore its fertility.

Short Fallow

As human population density grows, grassland replaces the forest and farmers shorten the fallow period to one or two years. This is not long enough to fully restore the soil, to let it support crops, so added fertilizer—often animal manure—is required. Farmers often plant special legumes to accelerate restoration of soil fertility. In addition, they achieve higher yields by manual pest control (for example, weeding).

Annual Cultivation

With a still-increasing population, now coupled with market demand, farmers drop the fallow period and grow a crop every year. Now they need constant fertilization and pest control to keep up the yield. Crop rotations increase the sustainability of the system.

Multiple Cropping

This is the most intensive use of the land. Not only do farmers grow several crops on the same land sequentially in one year, but also they grow two or more crops at the same time. Farmers often follow this latter pattern, called *inter cropping*, for the same reason as they rotate crops: One species removes one type of nutrients, and the intercrop removes other nutrients.

Modern Agriculture

Historically, the transition from stages 1 to 3 in the preceding schema has been gradual, as populations grew slowly and the necessary technologies evolved over time. Farmers now purchase the following inputs:

Farm Machinery

The process of mechanization started in the mid-19th century. New farming techniques depended on scientific advances and the industrial production of inputs. Manual labour and animal power were replaced by steam power and later by the internal combustion engine. Farmers used tractors to pull plows or the reaper-binder, which they later replaced with the combine harvester. Milking machines, cotton gins, cotton pickers, sugar beet and potato harvesters, and tomato-harvesting machines all have greatly reduced the need for manual labour. In developing countries small tractors have revolutionized agricultural production and greatly reduced the input of labour.

New and Improved Varieties

Initially, farmers selected suitable varieties of crops for planting the next year. These selected strains were the mainstay of agriculture until the science of genetics emerged in the 20th century. Then it became possible to deliberately interbreed different strains of a crop to consolidate desirable characteristics in a single strain. The first step is always to evaluate the field performance of known varieties. Scientists have systematically applied plant-breeding principles to improving rice and wheat, and have widely introduced new varieties in Latin America and Asia, displacing local varieties. This process, often called the Green Revolution, has significantly raised crop yields in an era when the amount of land cultivated has remained more or less constant. The introduction of hybrid rice in China and Southeast Asia is raising rice yields even further. About roughly 40% of all increases in crop productivity during the past 50 years stemmed from breeding new varieties. The other 60% improvement has come from managing the crop environment by inputs such as energy, fertilizer, and pesticides.

Inorganic Fertilizers

Originally European farmers relied on manure and crop rotation to maintain fertility. Later they found that ground bones and rock phosphate enhanced crop production, as well as nitrates (long imported from Chile, in the form of fossil guano). The invention of a chemical process that combines nitrogen gas with hydrogen gas to form ammonia, allowed the widespread production of nitrogen fertilizers. This led to a tremendous increase in nitrogen fertilizer use worldwide. Fertilizer applications have been slowly decreasing in developed countries, but in developing countries they are still rising.

Herbicides and Pesticides

Herbicides replaced manual labour for weeding and farmers used insecticides and fungicides to minimize crop losses. These changes in food production had a great impact on commercial grain farming (for example, in the United States, Canada, and Australia), on mixed crop and livestock farming (in Europe), and on intensive wetland rice farming in Asia and Africa. More complex approaches to weed control and pest control are now replacing heavy use of chemicals in agriculture.

Irrigation Technology

Irrigation has become a very important agricultural input in the past 40 years and accounts for much success in raising food production. Indeed, although only 18% of the world's arable land is irrigated, this land produces 40% of our food. Irrigated land is highly productive. However, this trend cannot be maintained: There simply is not enough water in many areas. Experts now see that water is the resource that will be most limiting for food production in the 21st century.

Information Technology

New innovations in agriculture that use information technology are often called *precision agriculture*. To obtain maximal yields, farmers of very large farms use remote sensing of their land and their crops by satellite or airplanes to adjust irrigation water, fertilizer application, and genetic varieties. Large tractors equipped with computers can now control row spacing and crop planting densities. Inputs do not exist in isolation, but interact. Scientists often breed improved varieties to fit other production-enhancing inputs. For example, after scientists introduced hybrid maize in the United States in the 1930s and nitrogen fertilizers became widely available, agronomists bred maize to respond to nitrogen fertilizers.

After the invention of the automatic tomato harvester, tomatoes were bred to withstand the rougher handling to which such machines subject these fruit. These technological advances have led to a style of agriculture that depends on purchased inputs. A survey of U.S. production methods for maize from 1910 to the present clearly shows this dependence. The technological advances resulted from government-sponsored and industrial research, the main thrust of which has always been to develop technological inputs that let farmers minimize per unit production costs. Many of these advances diminished work

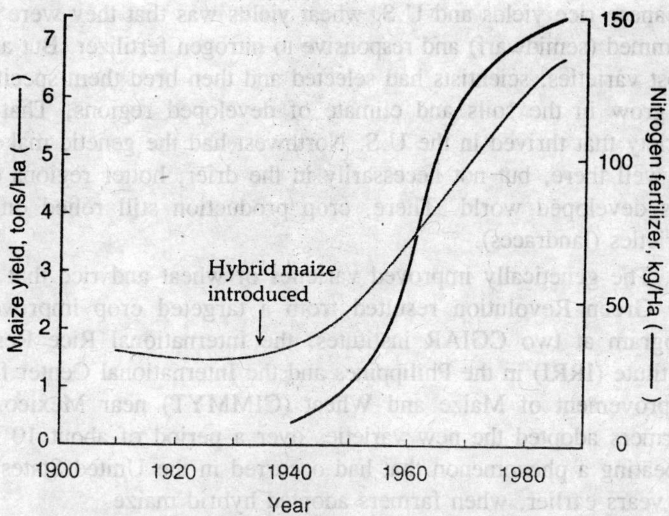

Fig. 6.2. Interaction of inputs in crop production: Maize yields and the use of nitrogen fertilizers.

opportunities on the farm in favour of jobs in towns and cities where the inputs are manufactured. The benefits of the lower production costs (cheaper food) accrued to the consumers and to agribusiness, whereas society as a whole had to bear the penalties (pollution and land degradation).

Green Revolution

Worldwide food production has been rising by 2.3% annually as a result of the development of high-input agriculture, which in turn results from applying agricultural research and technology. This use of scientific knowledge started about 150 years ago in the developed countries and has resulted in yields taking off spectacularly, as can be charted quite accurately for individual crops. Japan is a classic case of a country turning to high-yield agriculture. By 1900, this island nation was growing crops, mostly rice, on all its arable lands. But the population was still growing.

The government was reluctant to become dependent on foreign food imports lest a hostile nation cut off food from Japan during an international crisis. Short of lowering the people's nutritional standard, the only thing to do was to increase the yield of the rice crop. The government mobilized the country's political, social, and scientific resources, and as a result the yield per hectare tripled between 1900 and 1965. One characteristic of the new crop varieties that raised

Japanese rice yields and U.S. wheat yields was that they were short-stemmed (semidwarf) and responsive to nitrogen fertilizer. But as with most varieties, scientists had selected and then bred them specifically to grow in the soils and climate of developed regions, That is, a variety that thrived in the U.S. Northwest had the genetic makeup to do well there, but not necessarily in the drier, hotter regions of the less developed world. There, crop production still relied on local varieties (landraces).

The genetically improved varieties of wheat and rice that drove the Green Revolution resulted from a targeted crop improvement program at two CGIAR institutes: the International Rice Research Institute (IRRI) in the Philippines and the International Center for the Improvement of Maize and Wheat (CIMMYT) near Mexico City. Farmers adopted the new varieties over a period of about 10 years, repeating a phenomenon that had occurred in the United States some 30 years earlier, when farmers adopted hybrid maize.

In the case of high-yielding varieties of wheat and rice, the farmers had to adopt not only the new seeds, but an entire technology package, including fertilizers, insecticides, weed killers, equipment for irrigation, and tractors to till the land. Indeed, the new varieties let farmers grow two or even three crops per year, potentially increasing the demand for labour. Thus labour-saving technologies (tractors and herbicides) had to accompany the new strains. Furthermore, farmers needed nitrogen fertilizer so the new varieties would yield up to their potential (without fertilizer, they produced the same yields or less, than the landraces they replaced). Adopting these new varieties of wheat doubled and tripled production in Mexico, India, and other countries in Asia. Rice production increased similarly.

But the yield increases on farms were never as great as those at research stations. Scientists at IRRI showed that the new varieties could outproduce the old ones by a factor of 3 or 4, yet most farmers in Asia realized increases of only 1.5. This discrepancy in production is not surprising and also occurs in developed countries, where farms seldom match yields obtained in experiment stations. Surveys of farmers who were using the new technology showed that the constraints on higher productivity were poor water control, inadequate nitrogen fertilizers, and diseases. Put another way, the major constraint was inadequate capital to purchase the input package needed to obtain the high productivity of which the seeds are capable. The fundamental fact of the Green Revolution is that the farmers must manipulate the environment to get the most out of the plants, and too often they have not been able to afford it. Nevertheless, the Green Revolution allowed many developing countries to raise food production and even become food exporters.

International agricultural experts saw breeding the high-yielding varieties and applying the industrial inputs as the quickest way to increase production. As a result, agriculture in developing countries came to resemble agriculture in developed countries: more heavily dependent on purchased outside inputs, more capital intensive, and less dependent on labour. Earlier, we noted that families needed the money generated from labour to buy food. Merely raising production is not enough to eliminate hunger. (But raising production is intended to create profits, not to eliminate hunger, as noted earlier.) Because the need for farm labourers decreases as agriculture becomes more technological, the displaced labourers must find other employment. When farmers first apply a new technology—for example, fertilizer or pest control—yield immediately increases as the amount of input rises. A little bit is good, and more seems better! During this phase, the money spent on the technology results in a substantial return. Once the plant's inherent capacity to respond to the input is reached, however, the yield increase slows drastically.

Adding fertilizer and other inputs follows the law of diminishing returns: less output for a given amount of input. In the United States, farmers now apply 90-100 kg of fertilizer per ha of cropland, and it is not easy to raise the yield further by adding more. This figure is 65 kg/Ha for India, so there is still hope for improvement, but fertilizer use is only 4 kg/Ha in Ethiopia and 12 kg/Ha for Costa Rica. Gains in crop productivity should be possible in these countries if researchers can find the proper cropping systems and crop varieties that can take advantage of additional fertilizer. When all technologies are maximal, agriculture may reach a yield plateau. Some crop physiologists argue that people will always be able to find new crop varieties to escape the yield plateau and take advantage of new and existing technologies. Biotechnology has the potential to allow continual crop production increases. For example, scientists at IRRI and elsewhere are now producing hybrid rice varieties to raise rice yields even further.

Catalyzing Agricultural Research

The success of the Green Revolution, and the perceived need for more research on tropical and subtropical food crops, led to the establishment of other research institutes in different countries of the developing World. These institutes have a triple mandate: (1) to improve the crops assigned to them, (2) to study farming systems for these crops, and (3) to create gene banks in which the landraces of the crops under study can be preserved (in the form of seeds stored under conditions that ensure their longevity). Initially the institutes focused on the approach that was successful with high yielding wheat and rice varieties in the 1960s and 1970s. That is, they concentrated on monoculture and on improving crop production by a package of improved varieties and associated technologies to modify the environment. More recently, this focus has begun to change, showing a greater concern for integrating agriculture more with the natural environment.

For example, scientists at ICRISAT in India have identified strains of groundnut (a drought-resistant legume) that produce more than 1,000 kg grain per hectare, even when stressed by drought, compared to the 500-800 kg per hectare the normal varieties produce. Similarly, they have found strains of sorghum that produce two to three times more grain than the present commercial types. One of their goals is to select drought-resistant crop plants to use not only in Asia, but also in the Sahel region of Africa.

Scientists at CIAT in Colombia have identified genes in wild varieties of beans that let them resist attack by bruchid beetles. These beetles do enormous damage to the seeds after harvest because they multiply in the dry, stored beans. By crossing a wild bean variety with a cultivated variety, CIAT scientists were able to introduce the bruchid-resistance genes into the cultivated variety.

When the CGIAR system of research institutes was first established, people saw biological and technological research problems as separable from political and social issues. The research institutes generated the necessary innovations and offered these as packages to various national and regional extension services or development authorities. Fine-tuning these packages to local conditions, and solving social and political problems that resulted from implementing new technologies, were the responsibilities of the implementing agencies or the governments of the countries in question. In this system, people saw agricultural development as an activity that came from the top (the CGIAR institutes) and trickled down to the bottom (the farmers).

This approach has worked reasonably well in some countries or areas, and poorly in others. Agricultural development has not often been a high priority for many governments of developing nations, except for developing cash crops that earn foreign exchange. In many African countries, national agricultural research efforts have been weak, in part because governments mistakenly assumed the CGIAR institutes would solve all their agricultural problems, and also because these high-profile institutes attracted all the international funds and the best scientists.

The Green Revolution has done more than raise overall food production. It has also caused the rapid disappearance of the landraces and of indigenous farming systems that may have had much to offer to agricultural researchers. The top-down approach of development implies that agricultural researchers can learn little of value from subsistence farmers. For example, although the institutes have often stressed monoculture, many farmers have been growing more than one crop on the land (either by crop rotation or by intercropping) for centuries. These practices led to a more sustainable agricultural system than did monoculture.

Scientists at the CGIAR institutes know they must do more than create new strains to hand over to extension agents. They need to become involved in understanding the tropical agroecosystems at the farm level and to evaluate with the farmers the local varieties (genotypes) and breed varieties for different environments. For example, IRRI is located at a place where rainfall is reliable and adequate—in the Philippines. But much rice in Asia is grown on land where rainfall is intermittent and unreliable. So how can a technology developed at IRRI be useful to a farmer who has very different needs? This type of problem led to the formulation of a second mandate for these institutes: to study local farming and make sure that proposed development uses the strengths of those systems. Biotechnology is an important tool, but only one of many, to raise crop productivity in the less developed countries.

Crop Yield Biotechnology

Between 1960 and 1985 the yield of the major cereals (wheat and rice) rose dramatically in many developing countries, resulting in a 25% rise in the per capita food availability. However, in more recent years (1985-2000) this trend has leveled off. As noted in Chapter 1, the rate of population growth is slowing, so it is hard to predict whether world crop production is on track to meet the needs of the future human population. But two relatively new factors have arisen to challenge crop producers. First, increased demand for feed grains has followed increased demand for meat. Second, people realize that high-input agriculture can degrade the environment, so that plants must be even better adapted to the natural system. Although people still need ever better genetically improved varieties in combination with new technologies, they also need to more thoroughly understand the sustainability of crop production. Researchers need to devise different strategies for the highly productive regions that are already intensely cropped and where yields have risen dramatically and for the more marginal soils where, as in sub-Saharan Africa, productivity has stagnated and per capita food availability has declined.

Classical plant breeding has been a very important factor in the past successes. A new set of tools has become available in the past 20 years, that—combined with plant breeding—will allow people to produce the genetically improved varieties of the future. This set of tools, which comes under the general title of *biotechnology*, encompasses a variety of laboratory methods that include

1. Cell, tissue, and embryo culture
2. Clonal propagation of disease-free plants
3. Identification of chromosome regions (quantitative trait loci) that carry important multigenic traits
4. Gene identification and isolation
5. Genetic engineering for agronomic traits such as pest and disease resistance or better adaptation to environmental stresses (such as salinity, or water deficit)
6. Genetic engineering for greater nutritive value (such as vitamin A, minerals, and better amino acid balance)
7. Genetic engineering to reduce postharvest losses
8. Genetically engineered male sterility to facilitate hybrid seed production

Embryo culture is used to rescue the embryos produced when plants of different species are crossed. For example, a cross between the wild rice *Oryza nivara* and domesticated rice *Oryza sativa* is not normally viable. Researchers at IRRI used embryo rescue to introduce a trait for resistance to the grassy stunt virus present in this wild rice into domesticated rice. The horticulture industry routinely uses meristem culture and plant regeneration from small parts of the meristem to propagate virus-free planting materials.

The discoveries that the genetic material (DNA) of all organisms basically has the same structure and that genes from one organism can function normally after transfer to another organism have opened up the field of genetic engineering. In plants, gene transfer is made possible by a natural process discovered in a soil-dwelling plant pathogen that transfers a few of its genes to plant cells. Molecular plant biologists have manipulated this process so that the pathogen will transfer one or more genes of the scientist's choosing. This technique opens up unlimited possibilities for modifying crop traits. Plant breeders were previously limited to using the genes of the crop's closest relatives, but with genetic engineering they can now use any gene from any organism. It is important to emphasize repeatedly that the genetic engineering and biotechnology can make important contributions but are not the silver bullet that will solve all food production problems. It is but one of the many technologies that people need.

Genetically engineered plants that resist certain insect pests or can tolerate herbicides so that farmers can destroy weeds more efficiently, are being grown in more than half a dozen countries already (Argentina, Australia, Canada, China, India, Mexico, and the United States). The rapid adoption of these new varieties has pleased the biotechnology companies that produce them and the farmers who adopt them because it lowers their production costs, but has aroused some anxiety in the general public. Some countries in western Europe and a few localities in the United States have banned genetically engineered (or manipulated, GM) crops. The four major GM crops being grown are rapeseed (canola), cotton, maize, and soybean. Molecular techniques will also greatly facilitate chromosome mapping of important agronomic traits. This will make it easier to follow these traits in the progeny of crosses when the traits have no easily discernible phenotype (characteristic) in the field. It will also greatly speed up classical breeding, which is a rather slow and cumbersome process. Gene replacement (gene therapy) is not yet a reality, but is on the drawing board. It eventually will be possible to take a small chromosomal region that encompasses a whole set of genes, and simply replace it with another region from a plant that is more distantly related.

Effects of Agriculture on Ecosystem

Growing populations and increasing affluence demand ever-increased output from agricultural systems. This is true whether we are talking about the high-input agriculture of the United States and Western Europe, about nomadic herds in Africa, or about intensive rice cultivation in Asia. Concerns are now being raised about the sustainability of this ever intensifying productivity.

The World Commission on Environment and Development defined sustainable development as "development which meets the needs of the present without compromising the ability of future generations to meet their own needs." To understand sustainability, we must first look at the rate at which people are losing productive soils as a result of present agricultural practices. The FAG of the United Nations has estimated that salinization, soil erosion, and desertification have degraded a quarter of the world's arable land. Second, there may not be enough energy resources to maintain high-input agriculture. Third, the increasing use of pesticides is arousing concern that people are polluting ecosystems with synthetic chemicals. Fourth, the trend toward genetically uniform crops increases the potential for serious disasters by eliminating the many different strains of a given crop that farmers previously used. Fifth, government policies perpetuate conventional agriculture and discourage farming practices that could make agriculture more sustainable.

Land Degradation

The term *land degradation* denotes the physical and chemical changes that reduce long-term soil productivity. Such changes are sometimes very obvious but are often difficult to measure because they occur over long periods of time. Other improvements, such as more fertilizer or new genetic strains, often compensated for them, so the effects of land degradation do not always show up as decreased yield—at least, not in the short run. Physical degradation takes the form of erosion, the carrying away of soil particles by wind and water. In an annual cropping system, the soil is alternately covered with plants, and then almost totally bare.

When the soil is bare, it is exposed to higher wind velocities and to the force of raindrops, which destroy the structure of the uppermost layer of topsoil. The net result is muddy runoff and/or dust storms. Chemical degradation can take several forms, including (1) acidification from acid rain, (2) alkalinization and salinization (build up of salts) as a result of irrigation or the intrusion of saline groundwater into topsoil, (3) exhaustion of mineral nutrients when farmers don't use enough fertilizer to replace minerals removed with the crop, and (4) leaching of excess mineral fertilizers into streams, lakes, and groundwater.

Pesticides and Herbicides

Agriculture uses pesticides and herbicides extensively in the developed countries, and their use in developing countries has also increased rapidly. Many problems accompany their use, including the emergence of pesticide-resistant pests, adverse health effects on farm workers, pesticide residues on crops, and pollution of lakes, streams, and groundwater with pesticides and herbicides. Furthermore, the realization that pesticides have not diminished the proportion of the crop lost to pests is prompting many people to try alternate pest control methods. There is renewed emphasis on pest-resistant varieties and biological pest control.

Loss of Crop Biodiversity Because of Genetically Uniform Crops

A main feature of the high-input system of agriculture is genetically uniform crops. These reliably produce high yields if used with appropriate inputs. But their use has several drawbacks. First, when farmers start planting seeds furnished by seed companies or state agencies, they stop using and usually discard their own varieties, which were well adapted to the microenvironment on their farm. Local varieties contain valuable genes that are thus lost to plant breeders. Second, genetic uniformity means crops are more susceptible to sudden outbreaks of disease. When a pathogen mutates and overcomes the plant's resistance, the entire crop over a wide geographic area can be ravaged. An excellent example is the potato blight that ravaged the Irish potato crop in the 1840s.

Government Policies

Sometimes government policies actively promote high-input agriculture and discourage sustainable practices and technologies. Such policies are difficult to change because large and powerful lobbies, including farm organizations and industry advocates, have vested interests in such policies.

To protect their resource base—the soil—and to decrease costs, farmers have begun to adopt alternative practices. Taken together, these practices constitute sustainable agriculture, which differs from conventional agriculture not so much by the practices it *rejects* (for example, heavy use of inorganic fertilizers), but by the practices it *incorporates* into the farming system. The more widely used term *organic farming* (or *biological farming*, in Europe) describes two major aspects of alternative agriculture: (1) substituting manures and other organic matter for inorganic fertilizers and (2) using biological pest control instead of chemical pest control. The objective of sustainable agriculture is to sustain and enhance, rather than to reduce and simplify the biological interactions on which production agriculture depends. Alternative agriculture is not a single system of farm practices, but encompasses many farming systems variously called *organic*, *biological*, *low input*, *regenerative*, or *sustainable* systems. Such systems emphasize management practices as well as biological relationships between organisms; in addition, they take advantage of naturally occurring processes such as nitrogen fixation. Alternative or sustainable agriculture is any food or fiber production system that systematically pursues the following goals:

1. More thorough use of natural processes such as nutrient cycles, nitrogen fixation, and pest-predator relationships in agricultural production
2. Reduced use of off-farm inputs with the greatest potential to harm the environment or the health of farmers and consumers
3. Greater productive use of the biological and genetic potentials of plant and animal species
4. Improved match between cropping patterns and the productive potential and physical limitations of agricultural lands, to ensure long-term sustainability of current production levels
5. Profitable and efficient production, with emphasis on improving farm management and conserving soil, water, energy, and biochemical resources

Studies of alternative agriculture show that the farms derive sustained economic benefits and are not necessarily at a disadvantage. Although yields are often lower, costs are also considerably lower.

Environment Affecting Crop Production

The weather is the one factor with which people are familiar that profoundly affects the growth of plants. Everyone knows that to grow plants need sunlight and rain and that both are quite variable around the world. Weather—defined as short-term (less than a few weeks) variations in the atmosphere—affects the year-to-year yield of crops. The longer-term variations that comprise climate determine the geographic conditions where crops can grow at all. Short-term variations caused by the weather that affect crop yields occur even in the modern era of technology-based agriculture:

1. *Moisture stress* is caused by insufficient, too much, or ill-timed rainfall. Plants often have precise water requirements, in terms of both amount and time. If these requirements are not met, yields and/or quality suffer.
2. *Temperature stress* occurs commonly when the temperature is either too low or too high for optimal growth. A short period of stress can severely depress growth later. Sudden cold snaps injure plants because they do not have enough time to become acclimatized to

the cold weather. Moisture stress often accompanies high-temperature stress.

3. *Natural disasters* occur, such as cyclones, hurricanes, and hailstorms. Although these are more common in such Asian countries as Sri Lanka and Bangladesh, they can also cause severe damage in developed countries by physically harming crops and the soil that supports them.

Yields of many crops vary considerably year to year in many locations. The probability of a staple crop yielding the same year after year can be quite low. Over much of North Africa and the Middle East, for example, yields fall significantly below expectations in half of the crop years. A similar situation exists in the countries of the former Soviet Union. Complex statistical models, proposed to explain crop yield variability in other situations, show that in many cases climate is the major factor. When things go well (the yield trend is a smooth upward or constant curve), it is because the weather is unusually benign. Different regions of the world are affected differently by climatic variation. Some of these effects are as follows:

Temperate Regions

The north and south temperate regions produce 75% of the world's wheat and maize. The most prominent climatic factors affecting yield are moisture stress and temperature extremes. Wheat and soybeans are more drought resistant than maize.

Humid Tropics

Ranging from fertile valleys to jungles to floodplains, these regions are heterogeneous and show great variability of year-to-year yields. The major determining factor here is rainfall. There are pronounced wet and dry seasons, with the wet season often being in the form of a monsoon, with very heavy, sustained rainfall. But the unpredictability of the intensity of the monsoon, combined with poor water-holding capacity of soils and high rate of evaporation, lead to unstable water supplies for agriculture. In addition, torrential rains on lowland areas, such as river deltas, inundate the crops.

Semiarid Tropics

These regions (for example, the Sahel south of the Sahara, in Africa, and much of India) have a long history of periodic crop failures and resulting famines. The major problem here is rainfall or, more precisely, the lack of it. Most of the rain falls in a two- to five-month period when temperatures are at their yearly peak. This leads to extensive evaporation and less water availability. The sequence of events in growing a crop here is exquisitely sensitive to the annual rainfall cycle. Farmers time the growth of the crop to coincide with the maximum available water. If this period is very short, so is the growing season. In these areas farmers often plant two crops: one that will yield some food if the rains fail to come, and one that will yield abundantly if they do come. This relationship of agricultural practices to climate is an ancient one. Every culture has its deities and stories relating the cycle of the seasons to food production.

Effect of Climate

Human activities can substantially alter the global environment and thereby negatively impact our ability to grow crops. The public is now keenly aware of the effects of pollution on society, and generally considers industry to be the main source of pollutants. However, in the past 30 years agriculture has also become an important environmental polluter. The book *Silent Spring*, written by biologist and science writer Rachel Carson in 1962, first awakened public opinion to unintended effects that pesticides can have on the environment. Although organochlorine insecticides then used have since been phased out, they did considerable damage both in developed and developing countries. These pesticides prevented crop damage by insects, but in Asian rice paddies pesticides kill fish, shrimp, and crabs, important sources of protein for poor people. (However, it is good to remember that these pesticides, used for mosquito control, also saved millions of people from contracting malaria.)

Burning vegetation to clear forestland or to encourage grass growth in savannas is an important source of global air pollution. When vegetation burns, some nutrients (potassium, phosphate, and calcium) return to the soil, but nitrogen, sulfur, and carbon disappear into the atmosphere as gases. In addition, burning releases particulates into the air. The sulfur dioxide and nitrogen oxides, in combination with those same two gases released from the burning of fossil fuels, help acidify soil via acid rain. In Europe and North America acid rain is killing some forests, and in China it has shrunk rice and wheat harvests, with most of the damage occurring at the beginning of the rainy season.

Burning biomass (vegetation) annually releases about 1 to 2 billion tons of carbon into the atmosphere as carbon dioxide. This amount adds to the 2 billion tons of carbon released by burning fossil fuels. Carbon dioxide is a greenhouse gas, which means it contributes to warming of Earth. Anaerobic decay of organic matter in marshes produces methane, another greenhouse gas. Rice paddies are an important source of methane worldwide, and increased rice cultivation in Southeast Asia has much increased methane production. These gases are responsible for the greenhouse effect, which warms the surface of Earth. If greenhouse gases were completely absent from the atmosphere, Earth's surface would be a chilly -18°C. Their presence maintains Earth at a higher temperature, making human life possible.

The temperature of the planet has been rising slowly—about 0.3 to 0.6°C during the past 100 years, and is expected to rise another 0.4°C by 2020. A 2°-C rise by the end of the 21st century is possible. Although this slow global warming could be a natural phenomenon, most experts believe it is the direct consequence of the rise in emissions of greenhouse gases. A likely result of global warming is greater variability in the weather and more extreme weather conditions (droughts, floods, typhoons, frosts, heat waves, and so on). Rainfall may be more variable, making crop production even more variable than it is now. Other gases, especially nitrous oxides and the now banned chlorofluorocarbons formerly used as refrigerator coolant, destroy the protective ozone layer in the upper atmosphere. Ozone shields Earth from destructive ultraviolet solar radiation. Closer to Earth, nitrous oxides emitted from cars or released by burning vegetation help create photochemical smog, of which ozone is a significant component. Even at quite low concentrations, ozone inhibits plant growth and recent experiments suggest that the ozone levels around cities and in some rural areas are now high enough to impair crop yields. Scientists may need to take this into account when breeding new crop varieties.

Converting Solar Energy into Crop Production

The energy contained in the food that people eat is ultimately derived from sunlight through photosynthesis. Agriculture and crop production are basically all about capturing solar energy and converting it into food and fiber with the highest possible efficiency in a sustainable manner. The amount of solar energy received at the site where a crop is grown sets the upper limit on potential photosynthetic production. The solar energy at a given location on Earth on a clear day may be predicted for any minute of the day and time of the year from Sun-Earth geometry. At 10°N latitude of the equator, as with all sites within the tropics (0-23.5°) there is only modest annual variation in irradiance (sunlight) contrasting with 70°N, the northern limit of any crop production, where there is strong seasonal variation.

At the polar latitudes (66.5-90°), there is no sunlight for a period of days in midwinter, but in midsummer total solar radiation can exceed that of the tropics because of very long summer days. When the sky is clear, solar energy over the course of a single day shows symmetrical variation around midday, but during intermittent cloud cover there are frequent variations. Both clouds and atmospheric pollution can significantly lower the actual amount of solar radiation reaching Earth's surface. Whereas solar energy input sets the upper limit on the energy that can be transformed into crop yield, the actual yield of energy in food *(Y)* depends on the product of solar input (*S*) and the efficiency with which the solar energy is transformed into the harvested product. This efficiency is expressed as a number between 0 and 1 that shows what proportion of the solar energy reaching the field is transformed into food. This overall efficiency is the product of the efficiencies of the interception of sunlight energy by the leaves (ε_i), the conversion of the intercepted energy into plant matter (ε_c), and the partitioning of this plant matter into the harvested food product (ε_p):

$$Y = S \cdot \varepsilon_i \cdot \varepsilon_c \cdot \varepsilon_p$$

The last 50 years have seen large increases in crop yields worldwide as a result of the joint efforts of plant breeders and agronomists. Using wheat and rice—the world's two most important food sources—as examples, we can see that two factors have contributed to these yield increases. First, increased use of fertilizers and pesticides has resulted in larger and longer-lived leaves, improving ε_i. Second, plant breeders have selected genotypes that invest more of their total production into grain and less into other plant parts, raising ε_p. For the major food crops, these two efficiencies now seem close to the maximums achievable. In contrast, there has been very little change in ε_c the efficiency of conversion of solar energy intercepted by leaves into crop biomass through photosynthesis. Improving ε_c must be a central focus in meeting the challenge of doubling global food production that will be needed to feed the human population by the middle of this century. The following sections explain photosynthesis and explore the prospects for improving its efficiency in crops.

Photosynthetic Energy Conversion

Photosynthesis is a two-stage process that occurs in chloroplasts. In the first stage, the plant converts light energy into chemical energy in the form of two high energy compounds called ATP and NADPH. These reactions are intimately associated with the photosynthetic

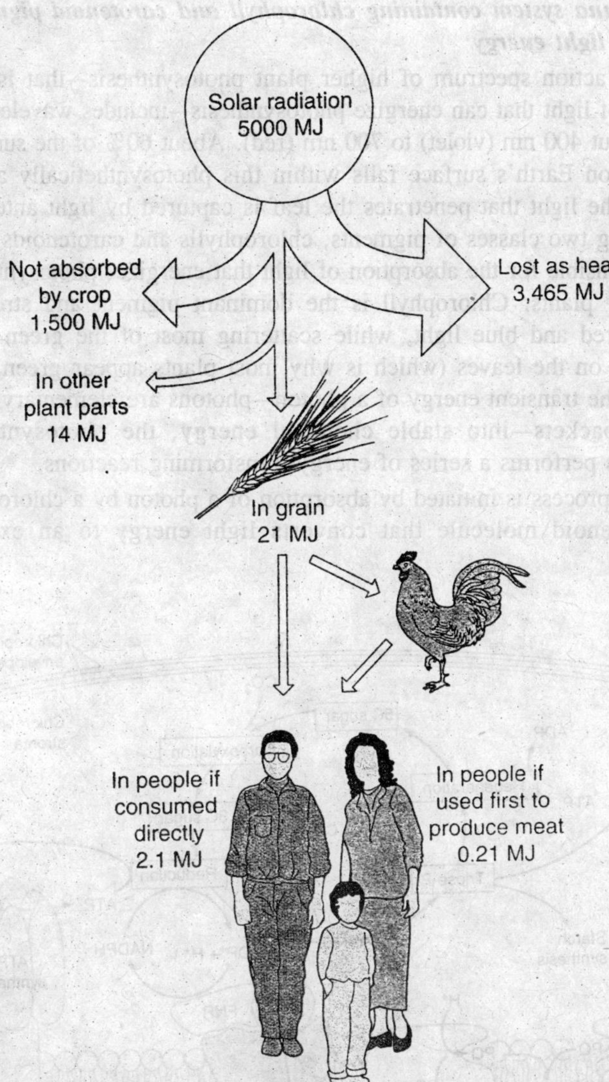

Fig. 6.3. Efficiency of food production from solar energy.

membranes. In the second stage, the plant uses these energy-rich biochemicals to convert carbon dioxide (CO_2) and water into simple sugars. These two processes occur simultaneously in the chloroplast stroma, and simple sugars are exported from the chloroplast into the cytoplasm, where they are used to make sucrose.

An antenna system containing chlorophyll and carotenoid pigments captures light energy

The action spectrum of higher plant photosynthesis—that is, the colours of light that can energize photosynthesis—includes wavelengths from about 400 nm (violet) to 700 nm (red). About 60% of the sunlight incident on Earth's surface falls within this photosynthetically active range. The light that penetrates the leaf is captured by light antennae containing two classes of pigments, chlorophylls and carotenoids, that are responsible for the absorption of light that energizes photosynthesis in higher plants. Chlorophyll is the dominant pigment and strongly absorbs red and blue light, while scattering most of the green light that falls on the leaves (which is why most plants appear green). To convert the transient energy of a photon—photons are elementary light energy packets—into stable chemical energy, the photosynthetic apparatus performs a series of energy-transforming reactions.

The process is initiated by absorption of a photon by a chlorophyll or carotenoid molecule that converts light energy to an excited

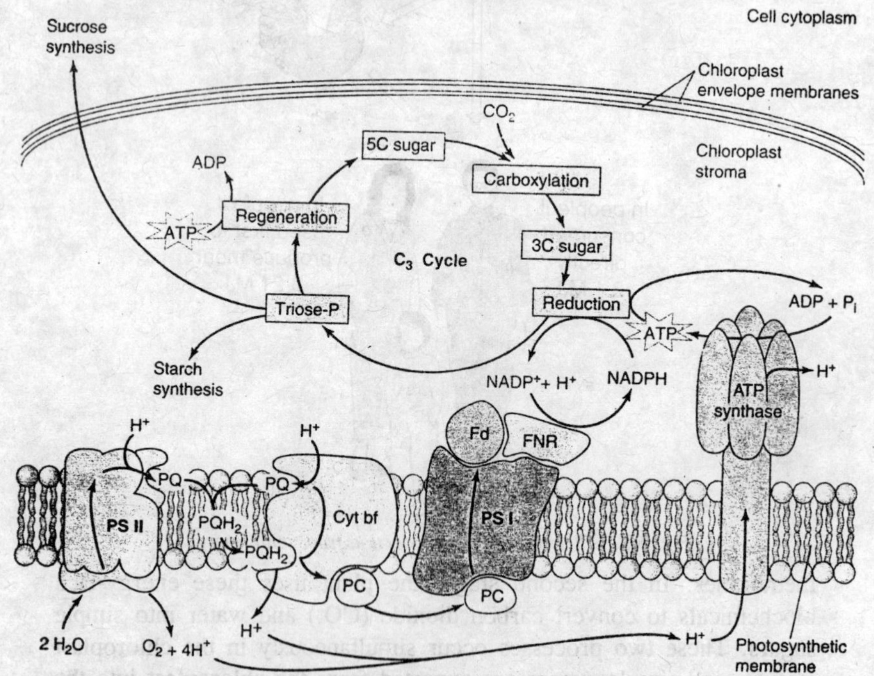

Fig. 6.4. Schematic drawing of a chloroplast.

electronic state of the pigment molecule. These pigment molecules are arranged in the photosynthetic membrane of chloroplasts in groups of 250-300 anchored there by specialized proteins, which provide a scaffolding for the precise arrangement of each molecule within the antenna. Because of the proximity of other unexcited pigment molecules with the same or similar energy states, the excited state is rapidly transferred over the antenna system. Under optimum conditions, over 90% of the photons captured by this antenna are successfully used to drive the next step in photosynthesis: primary charge separation.

The energy from absorbed light is used to separate positive and negative charge

Photosystem II (PSII) and photosystem I (PSI) of plants are the sites of the primary photochemical reactions of photosynthesis, which involves separation of a positively charged molecule from a negatively charged one. Each photosynthetic membrane in a chloroplast contains thousands of these two photosystems, which are themselves complexes of multiple proteins embedded in the membranes. In PSII the excited state energy is transferred to a specialized group of six chlorophyll molecules associated with proteins at the core of PSII. In the primary photochemical reaction, a negatively charged electron is transferred from an excited chlorophyll molecule to an electron-accepting pigment, creating positively and negatively charged molecules that are adjacent to each other.

The negatively charged electron cannot flow back to the positively charged molecule, and this separation of charge "captures" the energy and drives the subsequent electron transfer reactions in the photosynthetic membrane. In PSII the energy is used to remove electrons from water and to add electrons, as well as protons, to plastoquinone (PQ). In PSI, the energy captured in the primary charge separation drives the oxidation of plastocyanin, a soluble copper-containing protein located in the lumen, and the reduction of ferredoxin, a soluble iron-containing protein located in the stroma. As with PSII, the reactions of PSI produce an electric potential across the photosynthetic membrane and generate a strong reductant. PSI differs from PSII in that it does not deposit protons into the lumen but uses its energy to reduce NADP- producing NADPH.

Electron transport between PSI and PSII creates a pH difference plus an electric potential across the photosynthetic membrane

The light-driven electron and proton transfer reactions of the two photosystems are interconnected through the activity of another membrane protein complex (cytochrome bf complex, abbreviated cyt bf), which catalyzes the energetically downhill reaction of reducing plastocyanin. Plastoquinone, a special membrane lipid, serves as a mobile hydrogen carrier, transporting hydrogen (that is, a proton plus its electron) from PSII to cyt bf, and plastocyanin serves as a mobile electron carrier (that is, does not carry a proton), transporting electrons from cyt bf to PSI. In addition to linking the activity of PSII and PSI, the cyt bf complex plays a central role in energy transformation and storage by converting energy available in reduced plastoquinone (PQH_2) into a transmembrane pH difference as well as an electric potential difference. This feat is accomplished as cyt bf oxidizes PQH_2 at a site near the inside (lumenal side) of the membrane vesicle, resulting in the release of protons into the lumen. Because plastoquinone is reduced near the outside (stromal side) of the membrane, the protons for its reduction are taken up from the stroma.

In addition to oxidizing plastoquinone, cyt bf also reduces plastoquinone at a second site that is near the stromal side of the membrane. The net result of these reactions is the transfer of protons from the stroma to the lumen, creating a transmembrane pH difference and an electric potential difference. The energy stored in the pH difference and electrical potential is used for the energy-requiring reaction of converting ADP to ATP by the addition of a phosphate group. This conversion is done by the enzyme ATP-synthase located in the photosynthetic membrane.

Carbon Metabolism in Photosynthetic

Although the energy stored in ATP and NADPH is chemically stable, plants do not accumulate high levels of these chemicals. The ATP and NADPH are rapidly used in the biosynthesis of carbohydrates from CO_2 and water. This intricate biosynthetic pathway, known as the C_3 *photosynthetic carbon reduction cycle* (C_3 *cycle*), takes place in the stroma of chloroplasts, and involves more than a dozen different enzymes.

The C_3 photosynthetic carbon reduction cycle fixes CO_2 and produces sugar

The C_3 cycle begins with a carboxylation reaction in which CO_2 is attached to the five-carbon acceptor molecule ribulose bisphosphate, abbreviated as RuBP. This reaction is catalyzed by Rubisco (ribulose bisphosphate carboxylase/oxygenase), an exceptionally abundant enzyme in leaves, often accounting for 40% or more of the soluble leaf protein. The resulting 6-carbon compound is not stable and immediately splits into two 3-carbon molecules, which accounts for the C_3 name of the cycle. The next stage of the C_3 cycle requires energy in the form of both ATP and NADPH to form triose phosphate (glyceraldehyde-3-phosphate). Some of the triose phosphate is transported out of the chloroplast and used in synthesizing sucrose. Triose phosphate is used within the chloroplast for starch synthesis and is reinvested into the C_3 cycle to regenerate the five-carbon CO_2 acceptor RuBp' thereby completing the photosynthetic carbon reduction cycle.

After rubisco fixes O_2 rather than CO_2, the C_2 photorespiration cycle retrieves some of the lost carbon, but at a cost

The fossil record indicates that photosynthetic organisms have been producing oxygen for over 3 billion years, although stable molecular oxygen did not appear in Earth's atmosphere until about 2 billion years ago. This means that photosynthesis evolved in an oxygen-free or oxygen-poor atmosphere. This fact probably explains a curious feature about Rubisco that is the origin of a major inefficiency in photosynthesis. In addition to the carboxylation of RuBP by CO_2, Rubisco also catalyzes its oxidation by atmospheric O_2 (oxygenation) to yield one molecule of a three-carbon compound (PGA) and a molecule of a two-carbon compound. This oxygenation reaction creates a significant inefficiency in the photosynthetic process, because the two-carbon compound cannot enter the C_3 cycle.

In a typical C_3 crop (such as wheat), the rate of Rubisco-catalyzed oxygenation is about 20% of the rate of Rubisco-catalyzed CO_2 fixation. The inability of Rubisco to distinguish between molecular O_2 and CO_2 appears to be an unavoidable consequence of having evolved in an oxygen-free atmosphere. To compensate for the oxygenation of RuBP by Rubisco, a metabolic pathway evolved in plants that recovers the two-carbon skeletons that are diverted from the C_3 cycle by the oxygenation reaction. However, this scavenging operation, known as *photorespiration*, is energetically expensive. In the biochemical reactions of photorespiration, two of the two-carbon molecules are converted into a single three-carbon molecule with the release of one CO_2 and at the expense of ATP and reducing power. The cell thus succeeds in returning 75% of photorespiratory carbon to the C_3 cycle, with the remainder released as CO_2.

C_4 photosynthetic metabolism defeats photorespiration but at a considerable energy cost

The photorespiration reactions just discussed make the best of a wasteful situation caused by Rubisco's oxygenase activity. An alternative "strategy" taken by plants in the course of evolution is to prevent or greatly reduce Rubisco's oxygenase activity by exploiting the competition between CO_2 and O_2 as substrates of Rubisco. Plants such as maize, sorghum, and sugar cane greatly suppress or completely inhibit the oxygenation reaction of Rubisco by concentrating CO_2 in specialized leaf cells that contain Rubisco. These species are known as *C_4 plants* because the initial carboxylation reaction produces a four-carbon acid. They have a unique leaf anatomy with two distinct photosynthetic cell types in which chloroplast-containing mesophyll cells surround chloroplast-containing bundle sheath cells, which in turn encircle the

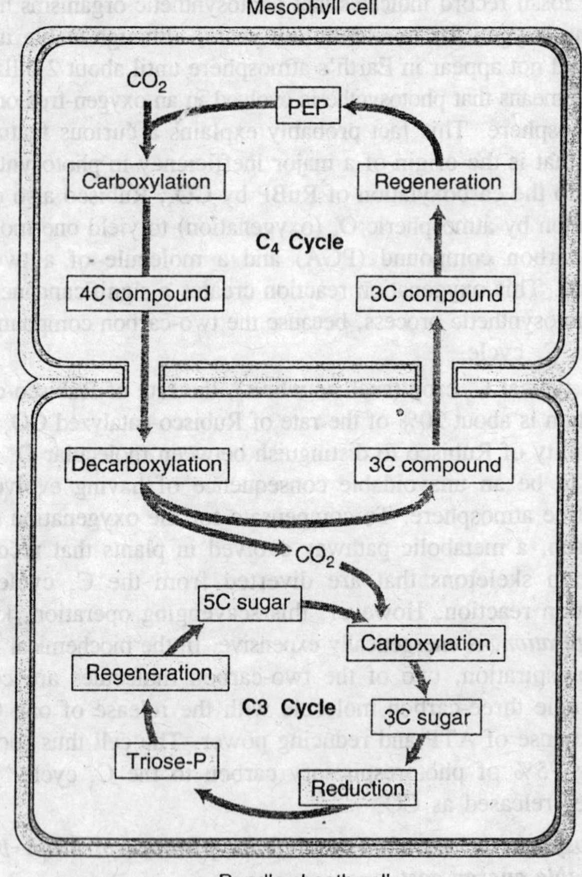

Fig. 6.5. The C_4 pathway involves two different cell types, mesophyll cells and bundle sheath cells.

vascular bundles of leaf. A key feature of C_4 photosynthesis is that the initial fixation of CO_2 takes place in mesophyll cells and involves the carboxylation of the three-carbon molecule phospho(enol)pyruvate (PEP) by an enzyme called *PEP carboxylase*.

Unlike Rubisco, PEP carboxylase does not bind O_2 in competition with CO_2, The four-carbon product of this reaction is transported to the bundle sheath cell and decarboxylated to release CO_2, The decarboxylation of the four carbon compounds significantly elevates the CO_2 concentration in the bundle sheath cell chloroplast where Rubisco and the other enzymes of the C_3 cycle are localized. The elevated concentration of CO_2 effectively competes with oxygen, virtually eliminating photorespiration. The three-carbon product of the decarboxylation reaction is transported back to the mesophyll cell so that the CO_2 acceptor PEP can be regenerated. Although the C_4 photosynthetic pathway effectively suppresses photorespiration, it is important to recognize that there are substantial energetic costs (from ATP).

In effect the C_4 cycle is a light-driven (that is, energy-driven) CO_2 pump, concentrating CO_2 in the bundle sheath cells, the site of the C_3 cycle in C_4 plants. Researchers estimate that only about 5% of all terrestrial higher plant species are C_4, indicating that the pathway with its higher energetic costs has an advantage in relatively few habitats. However, the much higher representation of C_4 species in hot/dry climates is well documented and probably reflects the higher water use efficiency associated with C_4 metabolism as well as the increasing affinity of Rubisco for O_2 with increasing temperature.

Photosynthetic Energy

Although photosynthesis is usually considered in the context of CO_2 reduction to form sugars, significant amounts of reducing power and ATP are used within the chloroplasts for other processes essential to plant growth and development. In many nonlegume herbaceous plants, most of the nitrate taken up by the roots is transported to the leaves, where it is converted in the cytoplasm to nitrite and imported into the chloroplast.

The reduction of nitrite to ammonia in the chloroplast consumes, per molecule, a third more reducing power than the reduction of CO_2 to carbohydrate. Nitrite reductase is located exclusively in the chloroplast and takes electrons directly from ferredoxin, thus competing with NADPH formation. The enzyme glutamine synthetase in the chloroplast transfers the newly formed NH_4^+ to glutamate forming glutamine at the expense of ATP. From glutamine, the reduced N can be channeled into a host of other amino acids in the chloroplast. The amino acids cysteine and methionine contain sulfur (S), and the reduction of sulfate to sulfide in the chloroplast requires, on a per mole basis, 50% more energy than does carbon assimilation. This multistep process also uses electrons that come directly from ferredoxin to produce sulfide, which is incorporated into cysteine. Although sulfate reduction is very energy intensive, the fact that the ratio of reduced sulfur to reduced nitrogen in plants is maintained within a narrow range near 1:20 implies not only an intricate coordination between these processes but also that the overall investment of photosynthetic energy is substantially greater for nitrate reduction.

Typically, chloroplasts expend less than 10% the amount of photosynthetically generated energy on nitrogen and sulfur reduction than is spent on carbon reduction. Many biosynthetic processes are unique to chloroplasts, and those that are not unique use enzymes encoded by unique genes. For example, a cell may contain two different but related enzymes for a specific step in amino acid biosynthesis: one enzyme located in the chloroplast and one in the cytoplasm. Each enzyme is encoded by a different gene, and the two enzymes are slightly different. Many herbicides specifically inhibit biochemical processes in chloroplasts because of the manner in which herbicides are identified. Researchers screen thousands of chemicals for their toxicity to plants and other organisms. Chemicals that are more toxic to plants than to other organisms are candidates for herbicide status. Because only plants have chloroplasts, it is not surprising that when researchers discover such chemicals' mode of action, they often find that the herbicides inhibit a biochemical process specific to the chloroplast.

Sucrose and other Photosynthate

Photosynthetic organs such as leaves are fully autotrophic and are considered to be source organs because they are sources of photosynthetic products such as sucrose. But plants also have heterotrophic tissues and organs that import the products of photosynthesis (photosynthate). Such *sink* organs fall into two categories (1) storage sinks such as developing tubers or seeds that accumulate carbohydrates, lipids, and proteins; and (2) metabolic sinks such as developing leaves or meristems that require the import of metabolites to sustain growth. Although seeds and tubers function as an important source of energy and metabolites during seedling growth or tuber sprouting, mature green

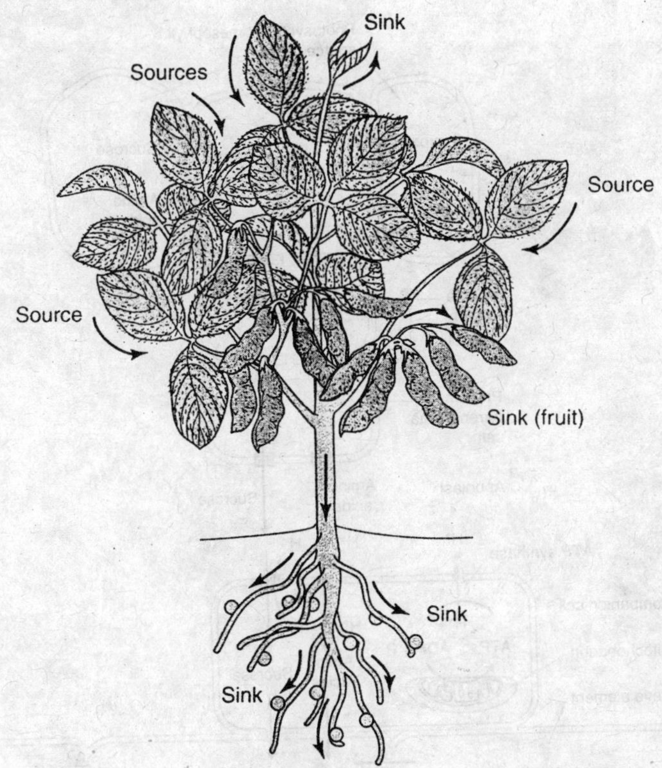

Fig. 6.6. Diagram of soybean plant showing patterns of photoassimilate distribution.

leaves are the fundamental source organs in the life cycle of plants. The function of a particular plant tissue as a source or a sink is dynamic, depending on its stage of development. Thus an expanding leaf changes from a sink to a source organ during maturation, but a mature leaf can later revert to a sink organ if it becomes heavily shaded by the leaves above. A potato tuber is a sink organ when it develops, but a source when the tuber sprouts. Such transitions are genetically programmed.

Long-distance transport from source leaves to sink organs requires the input of energy to drive mass flow

Long-distance transport of the products of photosynthesis is carried out by the phloem, a vascular tissue with two important cell types: sieve tubes and companion cells. Products of photosynthesis, such as sucrose or certain oligosaccharides and amino acids produced in the mesophyll cells of mature leaves, move by diffusion via plasmodesmata

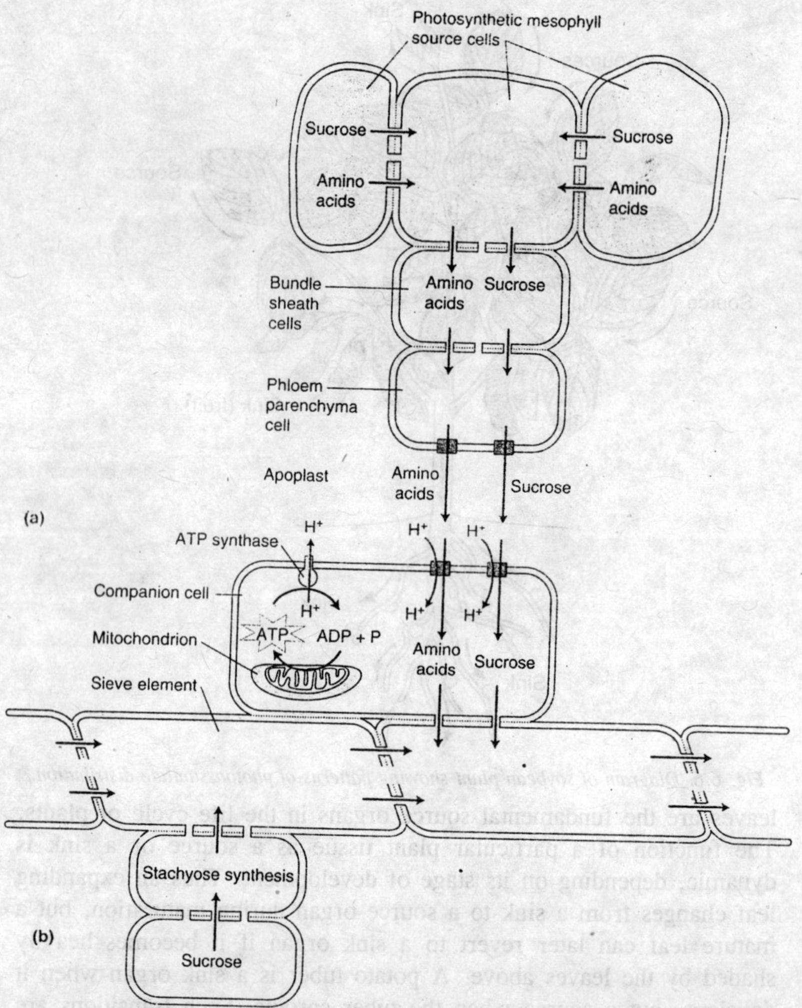

Fig. 6.7. Phloem loading. (a) Apoplastic phloem-loading pathway for sucrose and amino acids. (b) Symplastic phloem-loading pathway.

(channels that connect cells) toward the phloem, where loading into the sieve tubes for long-distance transport takes place. In most herbaceous crop plants, sucrose is exported from source cells near the phloem through the plasma membrane and into the cell walls. This space outside the cells is called the *apoplast*. Such export is from an area of high concentration in the cytoplasm to a much lower concentration in the apoplast and therefore does not require the input of energy. However, the subsequent loading of photoassimilates into the phloem is a different story, because the concentration of sucrose in the phloem sap of a source leaf is quite high.

Transporting sucrose from the leaf cell apoplast into the phloem against this high concentration gradient requires the input of significant energy resources in the form of ATP. In other species, such as squash, that transport raffinose oligosaccharides rather than sucrose, these compounds never enter the apoplast. Transport occurs via numerous plasmodesmata between the source mesophyll cells and the phloem. In these plants the concentration of photosynthate needed to drive mass flow in the phloem is achieved through the ATP-requiring biosynthesis of the oligosaccharides from sucrose. During periods of active photosynthesis, phloem transport velocities of 30-150 cm h^{-1} are typical. The direction of the mass flow is determined by site of consumption; up the aerial portion of a plant to a developing leaf, or underground to provide the energy roots need to function.

Sink strength regulates photosynthetic capacity of source leaves

To maximize growth, biomass production, and reproductive capacity, it is essential that photosynthetic production be coordinately balanced with mobilization, allocation, and use of photosynthate. A balance between the supply and demand for photosynthate in response to changing environmental conditions may be maintained in the short term by the modulation of metabolic factors and in the longer term by changes in gene expression that, for example, alter the capacity for photosynthate production. The rapid fine-tuning of photosynthetic carbon metabolism to match changes in the demand for photosynthate can be achieved through changes in enzyme activity.

Balance between production and use of triose phosphate is at the center of such metabolic control and operates to modulate carbon flux into sucrose. Thus when the sink demand for photosynthate declines, sucrose accumulates in mesophyll cells of source leaves, in turn elevating triose phosphate and diverting a larger portion of carbon flow from sucrose production to starch production. Sucrose plays important roles in plant metabolism and as a transportable form of energy in most plants. In addition, it serves a central regulatory role in maintaining the balance between the supply and the demand for photosynthate.

Many genes, whose products are involved at key points in photosynthetic metabolism, are either induced or repressed depending on the level of sucrose or other simple sugars in the source leaf mesophyll cytoplasm. For example, within hours of moving a plant from a condition of normal to elevated atmospheric CO_2, the genes of a number of proteins involved in photosynthesis (such as Rubisco) are repressed, as a direct response to increased soluble sugar levels in the cytoplasm. Signaling by sugar interconnects with signaling by hormones and stress, indicating that the signal transduction networks in plants are very complex.

Utilization of Carbon Dioxide

The photosynthetic assimilation of CO_2, an atmospheric gas, and its conversion into carbohydrates, takes place in the chloroplast-containing tissue inside the leaf, termed the *mesophyll*. Therefore, CO_2 must enter the leaves from the atmosphere. Furthermore, as with all biological reactions, photosynthesis takes place in water. This means that CO_2 from the atmosphere must dissolve into wet surfaces inside the leaf. The cost of exposing wet surfaces to the atmosphere to allow CO_2 to enter is the loss of water, yet water often limits plant production and survival on land. To understand how much water is lost to obtain CO_2 for photosynthesis and how the plant can regulate this loss, consider the physical processes that determine gas movement between surfaces and the atmosphere. The rate of water loss from a wet surface is determined by the law of diffusion, which states that the net flux of a gas (F_{gas}) is equal to the ratio of the concentration gradient (ΔC_{gas}) and the resistance *(r)* to transfer of that gas. The symbol F is used to indicate a flux, the symbol C to indicate concentration, and Δ to show the concentration difference between two points.

Fick's law of diffusion:

$$F_{gas} = \Delta C_{gas}/r$$

Using Fick's law of diffusion, you can predict the rate of loss of water from a wet surface. The air molecules immediately above a wet surface will be saturated with water vapour. At 25°C, air saturated with water vapour contains 23 grams of water vapour per cubic meter (g/m^{-3}). Such air is said to have a relative humidity (RH) of 100%. The air outside the leaf is almost always less saturated with water vapour and can be very dry. This means the water vapour concentration gradient between the inside of the leaf and the outside is usually quite steep. If the outside air is moisture laden, then the concentration gradient is less steep.

Molecular diffusion is the movement of molecules from a high to low concentration. In still air, the resistance to diffusion is proportional to distance and the rate of molecular diffusion. Wind accelerates movement, and decreases resistance. By assuming some reasonable values for the relative humidity (around 50%) and the resistance to diffusion, you can calculate that a square meter of surface would lose about 500 mg of water per second, or 1.8 kg (4 pounds) per hour. If a leaf really lost water at this rate, it would need to completely replace its water once every three minutes. How can plants expose wet surfaces to the atmosphere to obtain carbon dioxide and conserve their water? They do this by maintaining a barrier to water loss, the leaf epidermis.

A waterproof, waxy cuticle covers the outer surfaces of the epidermal cells. Within the epidermis there are stomates, small pores bordered by a pair of specialized epidermal cells, called *guard cells*. These cells alter their width to open or close the pore. When open, CO_2 can diffuse from the atmosphere to the wet surfaces of the photosynthetic cells of the mesophyll below. Typically, stomates open during the day, when photosynthesis requires CO_2, but close at night. Water vapour evaporates from the wet surfaces of the mesophyll cells surrounding the cavity below the stomate and escapes to the atmosphere through the open stomate. How much water is lost to gain CO_2? Let's return to the example of evaporation from a wet surface, but this time, consider the wet surface of the mesophyll below a stomate. For a well-watered plant in full sunlight, when the stomates are open stomatal resistance to water vapour diffusion is 10 times greater than from an open surface.

Water loss is therefore 10 times less, or about 50 mg per square meter per second. Furthermore, this is the minimum stomate resistance; at night the stomates will close and water loss will cease. How much water is actually lost for each milligram of CO_2 used in photosynthesis? Photosynthesis within the mesophyll removes CO_2, creating a lower concentration inside the leaf. CO_2 then diffuses down this gradient from the higher concentration outside the leaf. Typically the CO_2 concentration below the stomate pore of a C_3 leaf during daylight will be about 70% of that in the atmosphere, where it is about 360 parts per million of air.

Knowing the concentration gradient of CO_2, you can calculate that the rate of CO_2 diffusion over this same pathway is around 0.56 mg per square meter of leaf surface per second. Thus, the ratio of the mass of water lost per unit mass of CO_2 used is 89 in this example (50 mg m^{-2} s^{-1} divided by 0.56 mg m^{-2} s^{-1} = 89). Plants using the C_4 photosynthetic pathway fix CO_2 with PEP carboxylase, which is more efficient than Rubisco. As a result, the CO_2 concentration below the

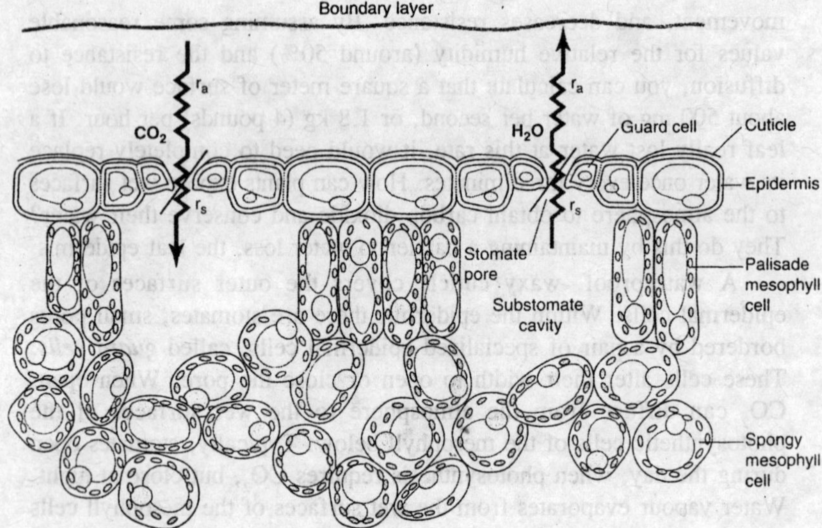

Fig. 6.8. Cross section of a leaf showing the path of diffusion of CO_2 and H_2O in and out of leaves.

pore inside the leaf is much lower in C_4 plants than in C_3 plants, and the diffusion gradient for CO_2 twice as steep. Thus C_4 plants can double the stomatal resistance and still take up CO_2 at the same rate. However, doubling the resistance halves water vapour loss, and C_4 plants only require about half the water needed by C_3 plants to assimilate the same amount of CO_2; that is, C_4 plants have greater water use efficiency than C_3 plants.

This advantage is reflected at the whole plant level: C_4 crops may produce twice as much biomass per unit of water transpired over their lifetime compared to C_3 crops growing in similar environments. The amount of water vapour that air can hold increases exponentially with temperature. The potential for evaporation of water similarly increases with temperature. At 40°C the mass of water required to assimilate 1 mg of CO_2 would rise from 89 mg at 25°C, in our example, to 300 mg at 45°C. This presents a particular problem to plants of hot semiarid regions, where daytime leaf surface temperatures may exceed 45°C and where both relative humidity and soil moisture are also typically very low. However, in these areas of the world nighttime temperatures may be quite low. One group of plants have evolved a way to take advantage of this marked difference in potential evaporation between night and day. These crassulacean acid metabolism (CAM) plants include the cacti. They open their stomates at night and use PEP carboxylase and respiratory energy to absorb CO_2 by making organic acids that are stored in cell vacuoles. In the daytime the stomates close to prevent water loss and the CO_2 is released from the acids to allow C_3 photosynthesis. If the average temperature at night is 10°C but 45°C during the day, the nocturnal opening achieves a water saving of 95%. This explains why CAM plants are the dominant perennial plants of the hot semideserts of the world.

Photosynthetic Efficiency for Photoprotection

All crops are inherently capable of photosynthetic efficiencies for carbon reduction that closely approach theoretical limits. With two photosystems operating in tandem, this high efficiency is possible only because the amount of light absorbed by the antennae serving the two photosystems is closely balanced. However, actual average CO_2 assimilation efficiency is much lower and the degree of inefficiency is often much greater than can be accounted for just by photo respiration, one of the major factors that limits efficiency.

Early in the day, the light levels are low and photosynthetic carbon reduction responds linearly to increasing light intensity; however, crop plants frequently encounter light intensities that exceed their photosynthetic capacity. Exactly what constitutes excess light for a

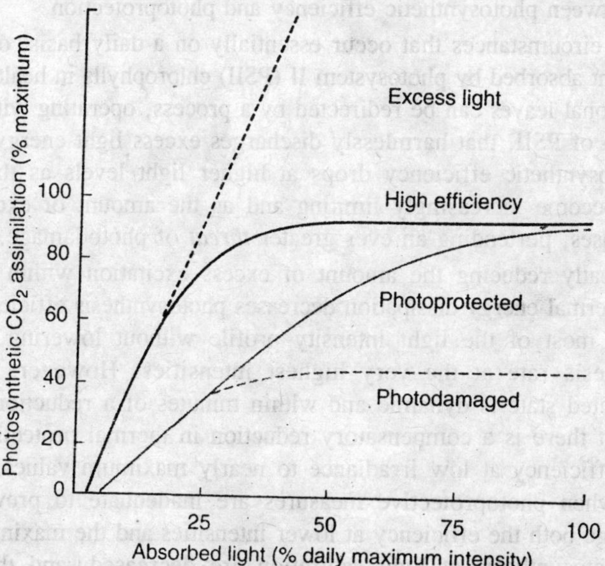

Fig. 6.9. A schematic of the light response of photosynthesis in a leaf in the high-efficiency state, in the photoprotected state, and the photodamaged state.

leaf depends on its instantaneous environmental conditions and can vary over quite a wide range of light levels. For example, irrigated field-grown sunflower is typical of C_3 crop plants, exhibiting maximum photosynthetic capacity during midmorning, with photosynthesis declining throughout the afternoon as stomates partially close in response to declining leaf water status. Thus, even under conditions that may not generally be considered stressful, partial stomate closure can substantially restrict CO_2 entry into leaves, causing even moderate light levels in the top layers of leaves of a crop to exceed the photosynthetic capacity for CO_2 reduction.

A dynamic process enabling leaves to regulate thermal dissipation of excess absorbed light trades photosynthetic efficiency for photoprotection

When environmental conditions prevent photosynthetic and photorespiratory carbon metabolism from using all the light absorbed by a leaf, the likelihood of biologically damaging oxidants forming through photosynthetic processes increases dramatically. Although some plants can reduce the amount of incident light they absorb, through strategic leaf and chloroplast movements, such strategies for rapidly reducing light absorption appear to play only a minor role in the challenge of coping with excess light. Instead, plants rely on a dynamic tradeoff between photosynthetic efficiency and photoprotection.

Under circumstances that occur essentially on a daily basis, over half the light absorbed by photosystem II (PSII) chlorophylls in healthy, fully functional leaves can be redirected by a process, operating within the antenna of PSII, that harmlessly discharges excess light energy as heat. Photosynthetic efficiency drops at higher light levels as these reactions become increasingly limiting and as the amount of excess light increases, portending an ever greater *threat* of photodamage.

By greatly reducing the amount of excess excitation within the antenna, thermal energy dissipation decreases photosynthesis efficiency throughout most of the light intensity profile without lowering the photosynthesis rate at the very highest intensities. However, the photoprotected state is dynamic and within minutes of a reduction in excess light there is a compensatory reduction in thermal protection, restoring efficiency at low irradiance to nearly maximum values. In contrast, when photoprotective measures are inadequate to prevent photodamage both the efficiency at lower intensities and the maximum rate of photosynthesis at light saturation are decreased, and these inhibitions do not rapidly recover as the excessive light condition recedes.

How effective is the tradeoff of efficiency for photoprotection?

Although photodamage has been documented in crops grown outside of their ancestral geographic range, the vast majority of plants in native habitats and even most crops under cultivation deal successfully with excess light, avoiding photo damage even under daunting environmental challenges. Clearly, photosynthesis is a balancing act in which plants trade photoprotection for photosynthetic efficiency. Apparently evolution has refined the photosynthetic apparatus by emphasizing high efficiency at limiting light with a failsafe that ensures that the plant can endure high intensities without accumulating photodamage. Under intense light other factors, such as maintenance of water status, take physiological precedence over maximizing photosynthesis. Although the tradeoff between efficiency and photoprotection is clear, it is less apparent how well the dynamic range of the tradeoff is suited for agricultural irradiances may prove an important factor in the in the ability of crop plant environments and productivity goals. Genetic variation higher varieties to maintain photosynthetic efficiency at somewhat search to improve photosynthetic productivity of crops.

Photosynthetic Efficiency and Crop Productivity

Plants frequently face periods of environmental stress sufficient to limit their growth and reproductive capacity. In natural populations, these periods represent a major force leading to genetic change, and those individuals with superior characteristics leave more progeny in succeeding generations. In agriculture similar forces are at work, but because of human intervention, successful genetic traits are not so closely linked to reproductive success, and so the meaningful selection criterion becomes production on a land area basis. However, because the plant's reproductive structures are often the sought-after agricultural product, there are clearly parallels between how environmental factors affect natural plant communities and how they affect crops. A comparison of average and record yields reveals the large impact of unfavourable environments on crop production.

The maximum production potential of a particular crop cultivar is called its *genetic potential*, and is set by the genetic makeup of the plant that can be expressed in optimum, nonlimiting growing environments. Although environmental constraints can never be entirely eliminated, a useful benchmark for estimating the genetic potential of a crop is the highest yield ever been attained under conditions of minimal unfavourable environmental conditions. A comparison of the world record yields versus the average yields for eight major crops of the temperate and tropical regions of the world show a stunning disparity. Even for rice, where the disparity is smallest, average yields are only 40% of record levels in temperate regions and below 30% in the tropics. These comparisons illustrate not only that the vast majority of agricultural environments greatly reduce yields but in addition that the largest shortfalls from the genetic potential of crops occur in the tropics and subtropics. But which environmental limitations are the most important? For eight major U.S crops, completely eliminating pests (insects, weeds, and disease) would bring the average collective yields up to about one third of the genetic potential of these crops, implying that abiotic (physical) stresses cause the largest portion of the yield suppression.

Abiotic factors that diminish yields are numerous, but the most significant ones can be categorized into two broad areas: unfavourable soils and unfavourable climates. Researchers estimate that 12% of the land surface in the United States and perhaps as little as 10% worldwide can provide a soil environment for plants that does not normally limit production. The impact of unfavourable climates is also large and widespread. Inadequate water, excessive water, and cold extract the largest toll in the United States, as judged from insurance payouts by the U.S. government to farmers for crop losses. Inadequate water availability is without question the single most important factor in limiting crop production throughout the world.

Abiotic stresses limit photosynthetic efficiency

Because plant growth is the result of many integrated and regulated physiological processes, it is seldom possible to assign limitations of plant growth to a single process. Certainly in the field a number of limiting processes control plant growth simultaneously and to varying extents. However, among the diversity of overlapping factors are likely to be some central physiological processes that are highly sensitive/responsive to the environment and are, therefore, dominant in determining plant response to stress. In many instances, the dominant process is photosynthesis. Plant growth as biomass production is in fact a measure of net photosynthesis integrated over time, so factors limiting plant growth are unavoidably the same factors that limit net photosynthesis. When soils are too dry to replace water loss from the leaves, plants respond by partially or fully closing their stomates, thereby lowering the rate of water loss and improving water use efficiency, but also lowering the rate at which CO_2 can enter the leaf.

Strong photoprotective measures, most importantly the thermal dissipation of energy within the light-harvesting antennas, trade off photosynthetic efficiency (productivity) for photoprotection. However, very severe or prolonged stresses can exceed the photoprotective capacity of plants and damage to the photosynthetic apparatus can occur. This creates a persistent decrease in photosynthetic efficiency (that is, the plant does not rapidly recover when the stress is removed). For example, when sunflower plants grown under well-watered conditions are suddenly deprived of water, photosynthesis decreases dramatically.

At atmospheric CO_2 levels, light saturated photosynthesis was inhibited 75% in this experiment. This decrease is not caused simply by closure of the stomates, and in such cases photosynthesis never fully recovers. Only new leaves that emerge after the stress recedes are again capable of high-efficiency photosynthesis. Certain plants have a significant capacity for adapting to stress. If such plants are exposed to a mild stress, they can better withstand a subsequent more severe stress. This phenomenon is called *acclimation*. Thus, sunflower plants grown under moderately dry conditions are able to acclimate so that they photosynthesize more rapidly under subsequent drought conditions. Understanding how plants cope with stressful environments will help breeders improve crop species more quickly. Drought stress very rapidly inhibits cell elongation, preventing growth.

Cell elongation can be sustained if the growing cells contain more soluble products of photosynthesis. An important aspect of acclimation to drought is the accumulation of photosynthate (soluble sugars and amino acids) in the regions of cell expansion. Thus the response of photosynthesis to drought is an important aspect of acclimation. Unfortunately, not all crop plants can measurably acclimate to all types of stresses. Thus although many wheat varieties can harden to withstand temperatures well below freezing, other important crop plants such as maize, soybean, and tomato have rather little ability to acclimate even to cool (above freezing) temperatures. These chilling-sensitive species were imported into temperate regions from warmer climates where they originated. Thus although genetic variability in osmotic regulation capabilities is providing powerful sources of drought tolerance, chilling tolerance in maize is not likely to be found within its ancestors' genetic repertoire. In this case, the need to understand the underlying biological mechanism of the lower temperature intolerance is even more crucial, because scientists need to introduce by genetic engineering the appropriate suite of genetic traits from sources outside the species.

Conversion of Solar Energy into Biomass

The efficiency with which plants convert intercepted solar energy into plant matter seems to be very low; just 1% in our example. This value was an average over the life of a wheat crop. Early in the season when crops are growing most rapidly, higher efficiencies are achieved for a few days. For C_3 crops, the highest efficiencies are about 3.5% and for C_4 about 4.3%. This section explores why even these record numbers are so low and whether photosynthesis in crops is really as inefficient as you might at first assume from such numbers. Only about 50% of solar energy can be used in crop photosynthesis, because the other half is in the near infrared part of the spectrum and is not energetic enough to drive photosynthesis. Leaves reflect some of the photosynthetically active light.

In C_3 plants, the minimum number of light photons required to fix one molecule of CO_2 is 8, regardless of wavelength within the photosynthetically active spectrum; that is, a red photon has the same effect as a violet photon. However, a violet photon contains about 70% more energy than a red photon. The additional energy of the violet photon is lost instantaneously on absorption as heat, representing an intrinsic photochemical inefficiency. Other pigments, such as the anthocyanins, may absorb light but cannot pass the energy on to photosynthetically active pigments. This type of light absorption is called *inactive absorption*.

In common with all syntheses, transfer of energy from the short-term energy stores of the chloroplast, ATP and NADPH, to the long-term energy store of most plants, carbohydrate, proceeds at the cost of energy. In C_3 plants, the efficiency of this conversion is about 35%. Because the C_4 pathway requires more ATP, carbohydrate synthesis in these plants has a lower efficiency. However, this difference is offset in C_3 plants by photorespiration, which converts a portion of this carbohydrate back to CO_2, Finally, mitochondrial respiration—necessary for synthesizing new tissues and maintaining existing tissues in all plants—uses about 40% of the remaining energy. All these losses encountered during the energy transformations of photosynthesis which illustrates that in C_3 plants a maximum of about 5% of the total incoming solar energy can in theory remain as biomass, and in C_4 plants, about 5.8%. So why do plants in the absence of pests, diseases, and environmental stress, fail even to achieve these modest efficiencies?

Even under optimal growing conditions it is impossible to remove all environmental limitations. For example, even when soil moisture is high, excessive evapotranspiration can outpace the vascular system's ability to deliver enough water for leaves to sustain fully open stomates. In addition, photosynthesis must operate over a wide range of solar energy input each day. It operates at maximum efficiency only at low light. In the next section we discuss how redesigning the arrangement of crop leaves improves the efficiency with which plants of some crops use photosynthetically active light within the canopy.

Improving Photosynthesis Performance

Earlier we showed that the major opportunity that remains for increasing maximum yields of major crops is increasing conversion efficiency (ε_i). Stressful growing environments can and do substantially decrease conversion efficiency. Does this represent an important opportunity to make improvements, and will this translate into higher crop yield? Another major cause of inefficiency is that leaf photosynthesis responds nonlinearly to increased solar energy. In C_3 crops, leaf photosynthesis is saturated at solar energy levels of about one quarter of maximum full sunlight, so any solar energy above this level is wasted. In this section we look at different potential approaches to overcoming these constraints and achieving a higher ε_i.

The distribution of light among leaves can be altered

A mature, healthy crop may have three or more layers of leaves; that is, above each square meter of soil may be 3 square meters of leaves. This ratio is described as a leaf area index of 3. If the leaves are roughly horizontal, the uppermost layer will intercept most of the direct light, about 10% may penetrate to the next layer, and 1 % to the layer below that. With the sun overhead, the photosynthetically active energy intercepted per unit leaf area by a horizontal leaf at the top of a plant canopy would be 900 J m^{-2} s^{-1}, or about three times the amount required to saturate photosynthesis. Thus at least two thirds of the energy intercepted by the upper leaves is wasted.

A better arrangement for an agricultural situation would be for the upper leaf layer to intercept a smaller fraction of the light, allowing more to reach the lower layers. This is achieved when the upper leaves are more vertical and lowermost leaves are horizontal, as in the example of Plant Y. For a leaf at a 75° angle with the horizontal, the amount of light energy intercepted per unit leaf area would be 300 J m^{-2} s^{-1}, just enough to saturate photosynthesis, but the remaining direct light (600 J m^{-2} s^{-1}) would penetrate to the lower layers of the canopy. By distributing the energy among leaves in this way, in full sunlight Plant Y would have over double the efficiency of solar energy

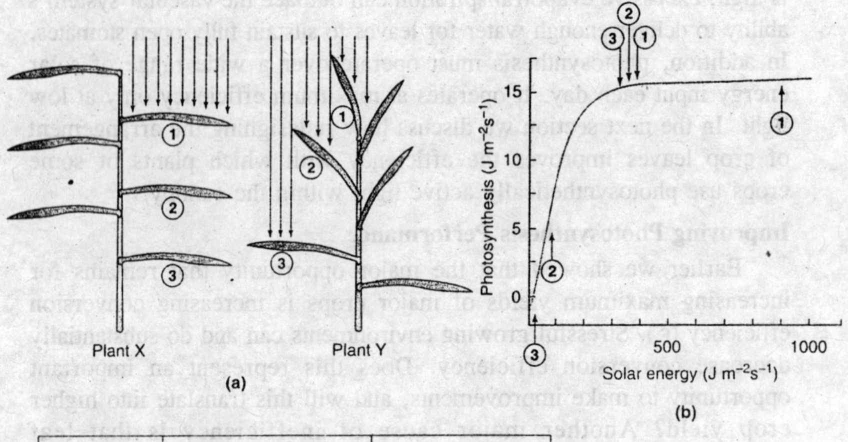

Plant Leaf layer	Solar energy J m^{-2} s^{-1} X	 Y	Photosynthesis J m^{-2} s^{-1} X	 Y
①	900	370	16	16
②	90	330	8	15
③	10	300	1	14
Total	**1,000**	**1,000**	**25**	**45**
Efficiency			*0.025*	*0.045*

(c)

Fig. 6.10. Light distribution in a crop canopy. (a) Plant X has horizontal leaves, such that the upper layer ① will intercept most on the incoming solar energy, shading the lower layers ② and ③ (b) Photosynthesis for a leaf. (c) From the graph of (b), the amount of solar energy and the photosynthesis for each leaf layer, and their totals, for the two plants are given.

use that Plant A has. Researchers have developed mathematical models to design optimum distributions of leaves for maximizing efficiency, which have been used as guides for selecting improved crops. This approach has been a major factor in improving rice productivity. Older varieties with more horizontal leaves such as Plant X have been replaced by newer varieties that have been bred to have more vertical leaves in the top layer, such as Plant Y.

Can rubisco be altered?

In considering how to redesign plant canopies, we noted that amounts of solar energy well below full sunlight saturate photosynthesis

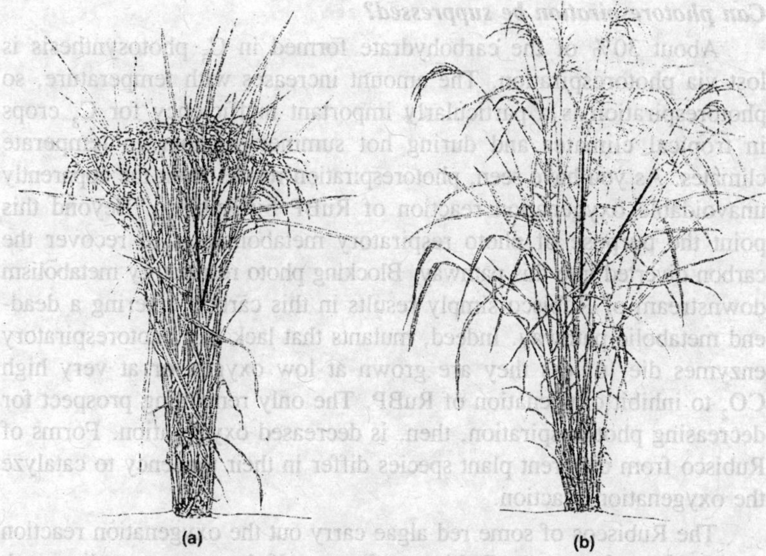

Fig. 6.11. Modern rice cultivars have been selected for vertical leaf orientation for the uppermost leaves (a), thereby improving light penetration into the canopy. This change in leaf orientation, compared to the more horizontal leaf position of older varieties (b), has been a major factor in improving the productivity of rice.

at the leaf level. Referring back for most of a sunny day solar energy exceeds the amount needed to saturate leaf photosynthesis. Are there other approaches to using this excess energy? The response of photosynthesis to solar energy is hyperbolic, rising rapidly with increases in light at low levels, but saturating at about 25% of full sunlight. Why does this saturation occur? Several analyses suggest that a major limitation is the amount of Rubisco in the leaf. So why not just increase the amount of Rubisco per unit of leaf area? Rubisco is already the most abundant protein in crop leaves. It is doubtful that leaves can profitably divert much more of their resources into making this one enzyme. An alternative would be to increase the efficiency of this enzymatic reaction. Build a better Rubisco! This would allow higher maximum light-saturated rates of photosynthesis without any additional investment into Rubisco. Researchers have shown that forms of Rubisco in some photosynthetic bacteria have catalytic rates four times faster than those from land plants. Modifying the catalytic site of crop Rubisco to that of the bacterial form is a possible approach to increasing efficiency of light use.

Can photorespiration be suppressed?

About 30% of the carbohydrate formed in C_3 photosynthesis is lost via photorespiration. The amount increases with temperature, so photorespiration is a particularly important inefficiency for C_3 crops in tropical climates and during hot summer weather in temperate climates. As you have seen, photorespiration results from the apparently unavoidable oxygenation reaction of RuBP by Rubisco. Beyond this point the purpose of photo respiratory metabolism is to recover the carbon diverted into this pathway. Blocking photo respiratory metabolism downstream of Rubisco simply results in this carbon entering a dead-end metabolic pathway. Indeed, mutants that lack any photorespiratory enzymes die, unless they are grown at low oxygen or at very high CO_2 to inhibit oxygenation of RuBP. The only remaining prospect for decreasing photorespiration, then, is decreased oxygenation. Forms of Rubisco from different plant species differ in their tendency to catalyze the oxygenation reaction.

The Rubiscos of some red algae carry out the oxygenation reaction only half as often as the Rubiscos of crops. If the genes encoding such algal Rubisco proteins could be transferred successfully to crops, or if crop Rubisco could be redesigned to mimic the red algal Rubisco, then crop photorespiration might be more than halved. Commercial growers of greenhouse crops take advantage of the properties of Rubisco to suppress photo respiration and obtain higher yields. In this closed environment, the CO_2 concentration is raised by injection of CO_2 to three or four times the outside concentration. This inhibits the oxygenation reaction of Rubisco, increasing photosynthetic efficiency and final yield. At the present time the global concentration of CO_2 is rising and this too may diminish photorespiration, but atmospheric change also includes many potentially negative effects for crops and natural ecosystems.

Global Interaction

Monitoring of atmospheric CO_2 concentrations at sites far away from any industrial sources of CO_2, such as the mountains of the Big Island in Hawaii and in Antarctica, has shown a steady increase since the late 1950s. Bubbles of air trapped in the ice of glaciers in Antarctica and Greenland can be dated and show that this increase began in the early 1800s, the start of the industrial era, and has accelerated dramatically over the last three decades. The increase is proportional to the amounts of CO_2 humans have released into the global atmosphere by burning fossil fuels and forests. In 1800, the concentration of CO_2

in the atmosphere was about 250 parts per million parts of air (ppm), in 2001 it reached 370 ppm, and by 2100 some researchers expect it to be between 600 and 700 ppm. If rising CO_2 were the only change occurring in the atmosphere, this might be expected to benefit crops. A doubling of CO_2 concentration would roughly halve photorespiratory losses in C_3 crops.

Photosynthesis would be further increased, because the present CO_2 concentration is insufficient to saturate Rubisco, with its surprisingly low affinity for CO_2. Finally, an increase in the atmospheric CO_2 concentration would allow plants to maintain the same photosynthetic rate, with a higher stomatal resistance, so less water would be lost for each CO_2 molecule assimilated by the plant. Elevating CO_2 concentration in greenhouses has been shown to increase yields of the major crops. However, enclosing crops such as wheat in a greenhouse affects growth, production, and appearance. Such plants may respond very differently to environmental treatments from plants in the field. But unlike nitrogen fertilizers or pesticides, CO_2 added in the field simply blows away. So how can researchers discover the effects of the future atmospheric CO_2 concentration on field crops, without using an enclosure? A new series of experiments being conducted at various locations around the world take advantage of natural air movements to enrich crops with CO_2 in the open air in a precise, controlled manner. These experiments use a new technology: Free-Air Carbon dioxide Enrichment (FACE).

FACE consists of rings of pipes that release CO_2 into the air. A computer continuously measures wind speed and direction, and the CO_2 concentration within the ring determines which pipes release CO_2 and how much they release. The CO_2 is released into the naturally moving air so that the concentration within the ring is elevated to a constant level. The advantage is that the crop is not enclosed and microclimate is unaltered, simulating the atmosphere of the future without otherwise altering the environment. The world's first such system, in Arizona, examined wheat crops over four years. Increase of CO_2 concentration to 550 ppm, the level possible for the second half of this century, resulted in a 28% increase in rates of photosynthesis; however, grain yield only increased about 10%.

One urgent task will be to select cultivars of crops that can translate the increased photosynthesis allowed by rising CO_2 into increased yield. But atmospheric change may have adverse effects on crop production. Rising atmospheric CO_2 concentration has several indirect negative effects on agriculture and may alter natural plant communities. Crops grown in elevated CO_2 throughout their life show decreased nitrogen and protein contents per unit mass, so increased quantity may be gained at the expense of nutritional value. Also, plants differ substantially in their ability to use the additional CO_2—some show large production increases, and others show none. In natural ecosystems, such differences may alter competitive balance, and lead to widespread extinctions. Because photosynthesis in C_4 plants is saturated at the present atmospheric CO_2 concentration, this group of plants is expected to lose out to C_3 plants where photosynthesis has the potential for substantial stimulation by increased CO_2.

Carbon dioxide is a potent "greenhouse gas," that is, a gas that traps the long-wave infrared radiation emitted by Earth's surface. This trapping warms the atmosphere in the same way the glass of a greenhouse warms the atmosphere that it encloses, hence the term "greenhouse effect." As a result, average global temperatures are expected to rise by 2-6 degrees C, and probably more, in this century. This may allow crop production at higher latitudes than at present, but is expected to depress yields in warm and tropical climates. It will also alter rainfall patterns, portending increased drought in some areas of the globe. Is photosynthesis providing some protection against atmospheric change?

Earth's atmosphere contains about 775 gigatons (Gt) of carbon (1 gigaton = 1 billion metric tons or 10^{15}g). Each year photosynthesis removes about 101.5 Gt of this carbon to the oceans and to land, and each year the respiration of plants, animals, and microorganisms releases about 100 Gt back into the atmosphere. This balance was stable for centuries, until humans started to release significant additional carbon to the atmosphere from fossil fuel use. Today fossil fuel combustion annually adds another 5.5 Gt and forest destruction 1.5 Gt, to the atmosphere. Because this anthropogenic emission of 7 Gt of carbon is about 1% of the total atmospheric concentration, the atmospheric concentration would be expected to rise at about 1% per year. However, the actual rise is about 0.5% per year, therefore half of the CO_2 that people add to the atmosphere is being absorbed somewhere—this somewhere has been termed the "missing sink".

Photosynthesis is apparently halving the rate of rise in CO_2 that would otherwise occur, and at present is protecting Earth from a more rapid rate of atmospheric and climate change. It is important to know where this photosynthesis occurs, to assess how long it may be sustained

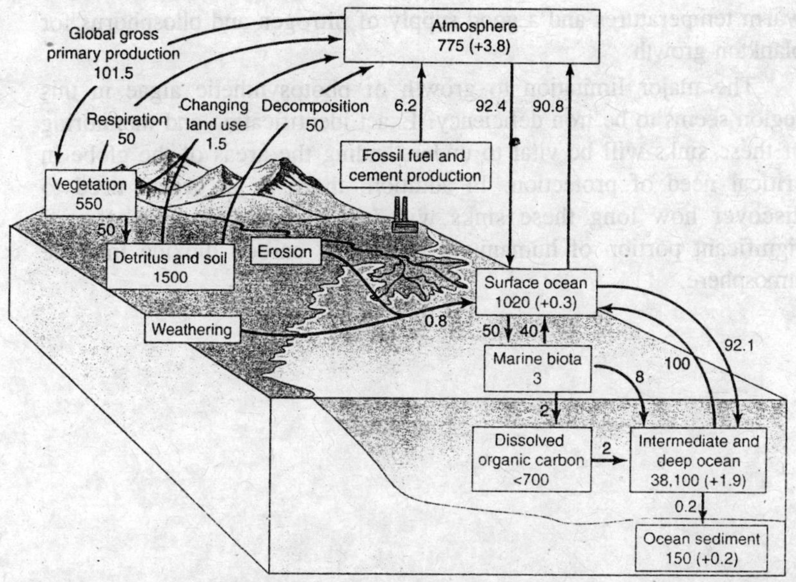

Fig. 6.12. The global carbon cycle, showing the reservoirs and annual fluxes average.

and to determine priorities for protecting the ecosystems that are providing this critical environmental benefit. How can people measure the photosynthesis and carbon dioxide exchange of landscapes, large tracts of ocean, and the globe? Two techniques are now becoming widely used within international networks. The first technique measures vertical wind speed and CO_2 concentration differences close to Earth's surface, to determine the amount of CO_2 being absorbed by surface vegetation. In the second technique, chlorophyll fluorescence-measuring devices lowered from research ships or from automated buoys in the ocean, are providing a detailed picture of the dynamics of photosynthesis in the oceans. This research into global photosynthesis is beginning to reveal where some of the additional carbon is and is not being consumed.

Originally scientists assumed that most of the additional carbon was deposited into the oceans. However, it is now clear that about half of the "missing sink" is actually on land. Areas of the temperate zone that people have allowed to revert from cropland to forest during the last few decades, appear to be a major sink for this additional carbon. Although tropical deforestation is a source of CO_2, remaining forest and regrowth secondary forest is another strong sink. In contrast, the tropical Pacific Ocean has very low photosynthetic rates despite warm temperatures and a good supply of nitrogen and phosphorus for plankton growth.

The major limitation to growth of photosynthetic algae in this region seems to be iron deficiency. Exact identification and monitoring of these sinks will be vital to understanding the areas of the globe in critical need of protection. In addition, there is an urgent need to discover how long these sinks will last and continue to offset a significant portion of humanity's release of carbon dioxide into the atmosphere.

7

Crop Diseases

Plants are marvelously productive under ideal growing conditions, but ideal growing conditions rarely occur. More typically, plants encounter abiotic stresses such as water shortage, lack of nutrients, or excess soil salinity, or biotic stresses in the form of plant pathogens, insect pests, or weeds. In this chapter we discuss (i) plant diseases caused by microbial pathogens, (ii) the intriguing biology of plant-pathogen interactions, and (iii) new biotechnology approaches that create plants with enhanced disease resistance.

Viral, Bacterial, and Fungal Infections

There are hundreds of thousands of different virus, bacteria, and fungus species in the world, and thousands of these are pathogens that infect plants. Anyone pathogen can severely depress the yield of a given crop. Pathogens may reduce yield by causing tissue lesions; by reducing leaf, root, or seed growth; or by clogging vascular tissues and causing wilt. Young seedlings can be overwhelmed by a pathogen and die soon after germination. Even in the absence of obvious symptoms, pathogens can cause a general metabolic drain that reduces plant productivity. Pathogens may also cause pre- or postharvest damage to the harvested product that can range from cosmetic blemishing to total decay. Given this arsenal of threatening microbes, you may think it a miracle that a crop can be produced at all. But the interaction of plants and pathogens is often very specific. Although some pathogens can infect many species of plants, many are highly specific as to the crop species and the part of the plant that they infect (root, fruit, leaves, stem, and so on). Furthermore, plants have specific defense mechanisms that keep infections under control. Like animals, after they have experienced an infection plants can even develop some immunity against some pathogens.

Plant breeders devote significant efforts to producing plant varieties with strong resistance to disease. Still, for any given crop the number of economically relevant pathogens is typically in the range of a few dozen species. As you read through this chapter, you will see that the different plant diseases illustrate many different biological phenomena. The microbes that do cause disease to a given crop can be devastating. From earliest recorded history to the present, every civilization carries stories of crop disease outbreaks that have caused famine, economic upheaval, mass migration, and death. For example, the Irish potato famine of 1845 and 1846 killed hundreds of thousands of poor Irish people. They could not afford to buy the more expensive wheat, so they died of malnutrition and starvation. This famine caused the first major wave of Irish emigration to the United States. The Bengal famine of 1943 was substantially attributable to an epidemic of brown spot disease on rice. Plant diseases can also alter the landscape dramatically, as has happened with the eradication of American Chestnut and American Elm trees across North America by single, destructive pathogen species. Less severe crop disease epidemics occur in most farming regions every few years, challenging the financial stability of farm families and farm communities.

Table 7.1. Rice diseases

Agent	*Plant organ attacked*	*Number of diseases*
Virus	Leaf	12
Bacteria	Leaf	4
Bacteria	Grain	3
Fungus	Leaf	11
Fungus	Stem, root	10
Fungus	Seedling	5
Fungus	Grain	10
Nematodes	Root	11

Disease Epidemics

What causes a disease epidemic? First, a virulent strain of pathogen must be present in sufficient numbers at the right place and time to start off the epidemic. Second, susceptible plant varieties must be widely present. Third and equally important, because pathogens are often sensitive to factors such as temperature, humidity, and wind,

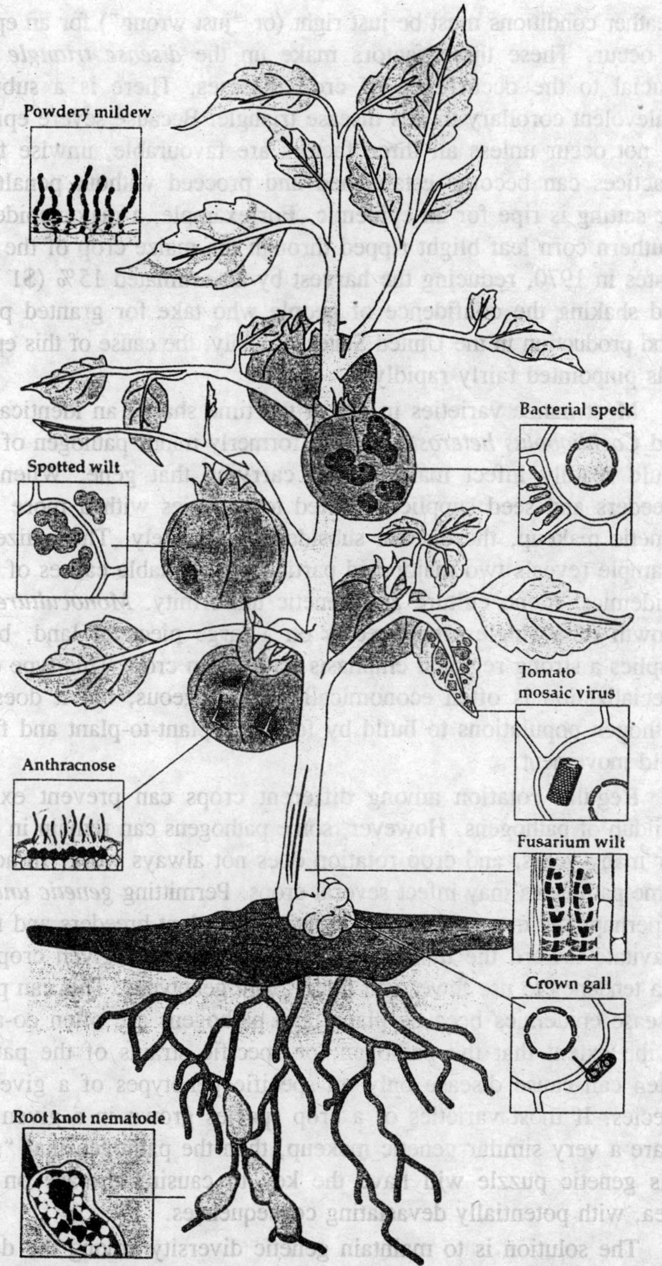

Fig. 7.1. Most microbes attack only a specific part of the plant and produce characteristics disease symptoms.

weather conditions must be just right (or "just wrong") for an epidemic to occur. These three factors make up the *disease triangle* that is crucial to the occurrence of crop diseases. There is a subtle and malevolent corollary to this disease triangle. Because severe epidemics do not occur unless all three factors are favourable, unwise farming practices can become established and proceed without penalty until the setting is ripe for an epidemic. For example, a major epidemic of southern corn leaf blight ripped through the maize crop of the United States in 1970, reducing the harvest by an estimated 15% ($1 billion) and shaking the confidence of people who take for granted plentiful food production in the United States. Luckily, the cause of this epidemic was pinpointed fairly rapidly.

Most maize varieties in use at the time shared an identical gene, and *Cochliobolus heterostrophus*, a formerly minor pathogen of maize, could readily infect maize plants carrying that gene. When maize breeders and seed suppliers shifted to varieties with a more diverse genetic makeup, the disease subsided immediately. The maize blight example reveals two major and partially preventable causes of disease epidemics: mono culture and genetic uniformity. *Monoculture* is the growth of a single crop species on a large piece of land, but also implies a strong regional emphasis on a given crop. This type of crop specialization is often economically advantageous, but it does allow pathogen populations to build by fostering plant-to-plant and field-to-field movement.

Regular rotation among different crops can prevent excessive buildup of pathogens. However, some pathogens can remain in the soil for many years, and crop rotation does not always work. In addition, some pathogens may infect several crops. Permitting *genetic uniformity* is perhaps the more dangerous practice. As plant breeders and farmers gravitate toward the most successful varieties of a given crop, there is a tendency to use fewer and fewer plant genotypes. This can promote disease epidemics because plants and pathogens are often co-adapted to the extent that the pathogen, or specific strains of the pathogen, often can cause disease only on specific genotypes of a given plant species. If most varieties of a crop species grown in a given region share a very similar genetic makeup, then the pathogen that "solves" this genetic puzzle will have the key to causing disease on a vast area, with potentially devastating consequences.

The solution is to maintain genetic diversity among the different popular varieties of a given crop species. In some cases disease-

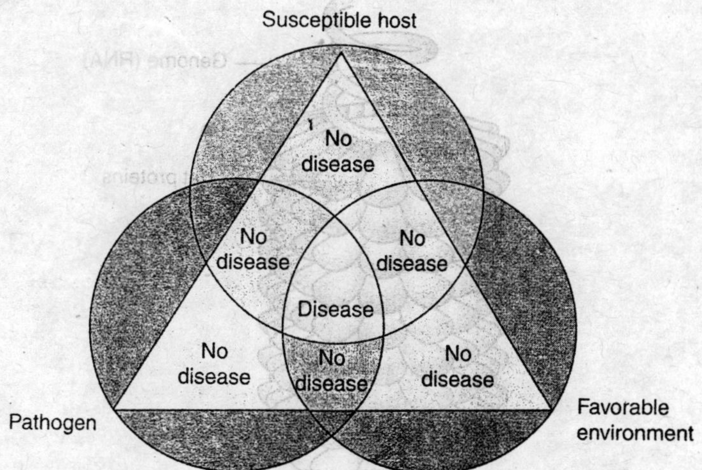

Fig. 7.2. The disease triangle.

favouring weather, virulent pathogens, and susceptible plants are common in a given region, and disease prevention tools such as fungicides or disease-resistant plant varieties are either too expensive or not available. Growers must then tolerate substantial annual losses, grow a different crop, or adopt cultural practices that greatly minimize the disease problem. This is true with rice blast disease across large regions in Asia.

In the late 1990s, the Chinese conducted a grand experiment in the Yunnan Province, in which thousands of farmers participated. To counteract the negative effects of genetic uniformity, each farm grew mixtures of different rice varieties with different levels of resistance to rice blast, the most serious rice disease. The level of rice blast infection was dramatically decreased, and in many areas the farmers were able to stop using fungicides. It is likely—although this was not examined—that other diseases were also reduced. This type of benefit has been observed in wheat mixtures as well. So why is this simple method not used more widely? First, the rice (or other crop) that is harvested is not of uniform quality, because many different lines are used; and second, mechanical harvesting is not possible, because the different varieties do not all mature at the same time.

Viruses and Viroids

Viruses and viroids are the smallest infectious agents that cause diseases. They are so small that people can only see them with an electron microscope. Viruses are not even cells, and plant virus particles

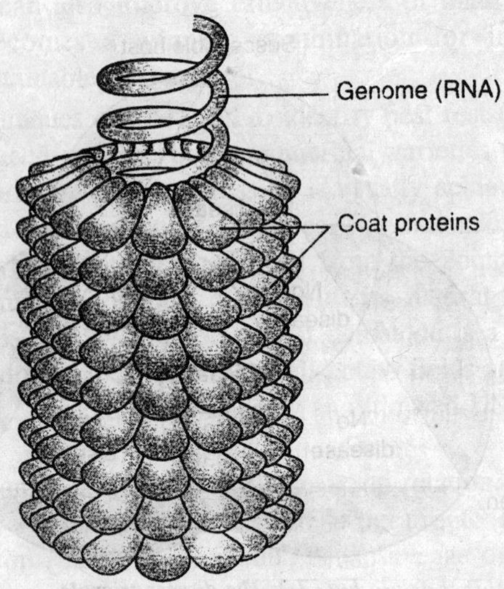

Fig. 7.3. Model of a virus particle showing the nucleic acid core surrounded by a protein coat.

often consist only of a nucleic acid (RNA or DNA) molecule wrapped up in a coat of protein. For example, the tobacco mosaic virus is a rod-shaped particle consisting of a coiled RNA molecule coated with spirally arranged coat protein molecules. The nucleic acid of most viruses has only enough genetic information for a few proteins. Viroids are even smaller, consisting of only a relatively small RNA molecule. Both viruses and viroids require the cellular machinery of their host to reproduce themselves.

Normally, viruses and viroids never quite kill their hosts, but because they divert the cellular functions of RNA and protein synthesis for their own ends (replication), they weaken the plant and diminish crop yields. Plant tissues can often be infected by one or more viruses and yet show no outward symptoms of this infection. However, many viruses cannot propagate in meristems. Plants regenerated from meristems are virus free, so horticultural industries use this type of clonal plant propagation to make virus-free plants. That viruses diminish the vigor of the plant is shown by the fact that the virus-free plants produced via meristem culture usually grow more vigorously than normal (virus-infected) plants.

Plant viruses don't harm people, and because plant viruses are everywhere, most people eat plant viruses on a regular basis. The nucleic acid molecule of a virus carries only a few genes. Its DNA

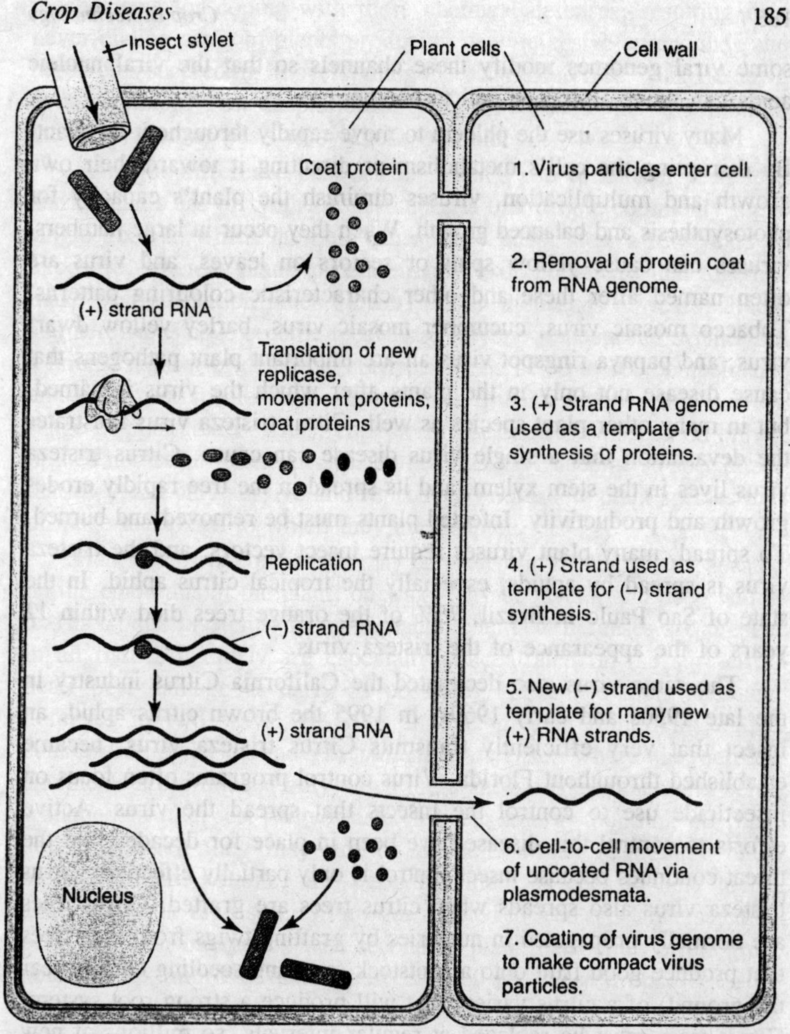

Fig. 7.4. Simplified life cycle of an RNA virus.

or RNA typically contains genes for the viral coat protein, for an enzyme that allows the nucleic acid molecule to replicate itself, and for a "movement" protein that lets the virus move from cell to cell. Viruses may enter the plant through a small wound, and they are often transmitted by insects such as aphids or leafhoppers. They then multiply within the infected plant cell. The plasmodesmata that connect plant cells exclude viruses, but the movement proteins encoded by some viral genomes modify these channels so that the viral nucleic acids can move from one cell to another.

Many viruses use the phloem to move rapidly throughout the plant. By disrupting the cell's metabolism or directing it toward their own growth and multiplication, viruses diminish the plant's capacity for photosynthesis and balanced growth. When they occur in large numbers, viruses can cause yellow spots or sectors on leaves, and virus are often named after these and other characteristic colouring patterns. Tobacco mosaic virus, cucumber mosaic virus, barley yellow dwarf virus, and papaya ringspot virus all are important plant pathogens that cause disease not only in the plants after which the virus is named, but in many other plant species as well. Citrus tristeza virus illustrates the devastation that a single virus disease can cause. Citrus tristeza virus lives in the stem xylem, and its spread in the tree rapidly erodes growth and productivity. Infected plants must be removed and burned. To spread, many plant viruses require insect vectors, and the tristeza virus is spread by aphids, especially the tropical citrus aphid. In the state of Sao Paulo in Brazil, 75% of the orange trees died within 12 years of the appearance of the tristeza virus.

The same virus also decimated the California Citrus industry in the late 1950s and early 1960s. In 1995 the brown citrus aphid, an insect that very efficiently transmits Citrus tristeza virus, became established throughout Florida. Virus control programs often focus on insecticide use to control the insects that spread the virus. Active efforts to control this disease have been in place for decades, but the threat continues because insect control is only partially effective. Citrus tristeza virus also spreads when citrus trees are grafted. Citrus trees are normally propagated in nurseries by grafting twigs from elite trees that produce good fruit onto a rootstock (a young seedling cut off near the ground) of a citrus variety that will produce a strong root system. Citrus trees must be replaced at regular intervals, so millions of new grafted trees are needed every year. It is important, therefore, to certify that the trees for grafting are free of virus and that the rootstocks used are virus resistant.

Plant Pathogenic Bacteria

Some bacteria play vital roles in crop production, such as those that fix nitrogen, transform nitrogen and sulfur compounds in the soil, or contribute to the decay of organic matter and the process of mineralization. However, many species of bacteria cause important diseases and contribute substantially to the annual 10-15% crop loss caused by diseases worldwide. Because they are microscopic in size, bacteria can enter the plant through natural openings or wounds. In some cases, they secrete enzymes that help dissolve the protective barrier of the epidermal cell wall and penetrate the plant in this way. Many bacterial diseases show up as spots on fruits, leaves, and stems. These spots, local infestations of the bacteria, diminish the photosynthetic capacity of leaves or disfigure fruits, thereby decreasing their commercial value.

Other bacteria cause soft rots by secreting enzymes that digest the cell walls of the plants, causing the plant tissues to become very soft or even liquid. Anyone who has left vegetables in the bottom drawer of the refrigerator for a month has witnessed the destructive action of *Erwinia* bacteria. Several other members of this genus cause postharvest losses of fruits and vegetables. Bacteria that invade the vascular conductive tissues of plants can clog such tissues so that water and minerals cannot get from the roots to the leaves. Such bacteria cause plants to wilt, and the diseases are accordingly referred to as *wilts*. These bacteria also secrete enzymes that dissolve cell walls of conductive tissues and the disease then spreads to other cells, quickly killing the plants.

The most economically important wilt disease is caused by *Ralstonia solanacearum*, a bacterial species that can infect plants from 44 different genera. The disease occurs especially in warmer regions and affects many crop plants, including bananas, tomatoes, and potatoes. Pierce's disease of grapes is caused by *Xylella fastidiosa*, a bacterium that clogs up the xylem of grapevines. The same bacterium causes a serious disease of citrus called citrus variegated chlorosis and infects apples, pears, and almonds. These bacteria are spread by a leafhopper (insect), the blue-green sharpshooter, that sticks its proboscis into the conductive tissue of the plant. Another insect vector with a new geographic range, the glassy winged sharpshooter, recently appeared in California. In the absence of effective control measures for this insect, the disease is threatening the California grape and wine industries. Because of the economic importance of the citrus industry to Brazil, scientists there chose *Xylella* as the first plant pathogen to have its entire genome sequenced. By knowing the genes that *Xylella* needs to complete its life cycle and pathogenic activities, we should be able to generate better strategies to combat the disease.

Bacterial pathogens can also cause blights and cankers. Blights are characterized by rapidly spreading infections that kill infected cells and tissues. Using fruit trees again as an example, a species of *Erwinia* causes fire blight of apples and pears. The disease, which can kill young trees in one season, devastated U.S pear orchards in the 1930s and still kills numerous trees annually. Citrus canker, a disease caused by a *Xanthomonas* bacterium, devastated the citrus industry in the 1920s. In 1984 it allegedly reappeared in some of Florida's largest nurseries, and millions of citrus seedlings and young trees were immediately destroyed because the agricultural authorities were afraid there might be a repeat of the earlier devastation.

A new outbreak of citrus canker is again spreading through Florida as this book goes to press. There are no chemical sprays to effectively contain such outbreaks, and the only way to keep the disease from spreading at present is to destroy (burn) all trees suspected of being infected. The genus *Agrobacterium* includes soil-dwelling bacteria that can cause plant tumors (crown gall disease) or other outgrowths (hairy root disease). *Agrobacterium* transfers a segment of its DNA from a plasmid to the genome of the infected plant cell, causing the plant to express genes encoding enzymes that create unique food compounds usable by *Agrobacterium* but not by other microbes or organisms. Other transferred genes specify the synthesis of the hormones auxin and cytokinin that cause these "food factory" plant cells to proliferate, forming tumors. Plant biotechnologists are now using this fascinating capability of *Agrobacterium* to transfer DNA to the host plant (discovered in the mid-1970s) to introduce desirable genes into plants.

Pathogenic Fungi

Fungi and oomycetes are a group of microscopic organisms that include molds, mushrooms, mildews, and yeasts. Most of the approximatey 100,000 known species of fungi are strictly *saprophytic*, living on dead organic matter that they help decompose. A small minority—about 100 different species—cause diseases of humans and animals, but more than 8,000 species cause plant diseases, some of which are extremely damaging. There are also fungi that benefit plants. Mycorrhizal fungi live in close association with plant roots and help in the acquisition of phosphate and other key minerals.

Certain endophyte fungi live within the plant and produce substances that protect the plant against grazing animals or bacteria. Fungal pathogens are of concern not only because they cause yield reductions. Some fungi produce chemicals that are toxic or carcinogenic to people. When these fungi infect the part of the plant that is consumed, the health of humans and animals can be adversely affected. Some fungi that infect ears of maize produce such mycotoxins. People generally will not eat such infected ears when they purchase sweet corn, but infected ears may contaminate batches of maize used in processed foods. The vegetative body of a fungus, called a *mycelium*, consists of a mass of long filaments or hyphae. Fungi propagate by means of spores that may be formed asexually (vegetatively) or as a result of a sexual mating process between two individuals.

The sexual/asexual distinction is important for plant disease control because asexual reproduction generates additional copies of essentially the same individual, whereas sexual reproduction generates offspring with new gene combinations that may have different disease-causing abilities. Spores are microscopic and easily carried away by water, wind, people, and animals. Pathogens such as the rust fungi are particularly problematic because they reproduce rapidly by sexual means and produce wind-borne spores that can spread over wide geographic regions. Other properties of fungal pathogens are important determinants of how a disease persists and is propagated. Almost all plant-pathogenic fungi spend part of their lives on their host plants and part in the soil or on plant debris on the soil. But some need to spend part of their lives on dead host tissues in order to complete their life cycles. Others live continuously on their hosts and only their spores may land on the soil, where they remain inactive until a new host appears.

Still others are "obligate parasites" that cannot live in the absence of plant host tissue. Other fungi have a saprophytic stage and can live on any organic matter in the soil. These latter pathogens, which often have a wide host range, can survive in the soil for years, even in the absence of their hosts. How do fungi enter the plant and cause disease? Like bacteria, they can enter plants through natural openings or wounds, or by "forcing" their way into the plant. Fungi secrete enzymes that break down the large macromolecules of the cell wall, and if the plant's defense is not quick or strong enough, the fungal hyphae quickly grow and spread from cell to cell. Once inside, they can affect the plant in many ways. Some produce toxins that alter the permeability of the cell membrane, destroying its ability to regulate what goes in and out of the cell. Many fungi secrete slime that accumulates in the vascular tissues, preventing transport of water and nutrients from roots to the shoot, causing the plant to wilt and die. Like similar diseases caused by bacteria, these diseases are called *wilts*.

Other fungi produce plant hormones, and as a result the plant loses control over its own developmental processes. Some soil-borne fungi attack seedlings as soon as they germinate, killing the root system. Seedlings that look healthy one day are gone the next, and many seedlings die even before they emerge from the soil. This type of disease is known as "damping off" Many fungi simply cause necrotic (dead) spots on leaves, stems, fruits, and seeds of the plants they attack. As with viruses and bacteria, they decrease the vigor of the plant, render the seeds and fruits less fit for human consumption, or decrease the market value. The rust diseases caused by different Basidiomycetes are among the most destructive diseases known to humanity. They have caused numerous famines and economic depressions and still reduce the world grain harvest by an estimated 10% per year.

Various subspecies of *Puccinia graminis* infect different cereals such as wheat, oats, corn (maize), and barley, and produce long, narrow rust -coloured blisters on the leaves and stems—hence the name *rust*. Heavy infections greatly diminish plant growth and seed size. The only practical control of wheat stem rust (*Puccinia graminis* var. *tritici*) is through breeding varieties resistant to infection by the pathogen. Researchers have identified resistance genes in the wild relatives of wheat and then introduced these genes into domesticated wheat by crossing and backcrossing. The genotypes of this rust fungus continually change because the fungus mates and produces new progeny, so new pathogen "races" regularly appear that are not daunted by the resistance genes being used.

Some wheat breeders continue to work with race-specific resistance genes, seeking more complex and durable resistance gene combinations. Other wheat breeders focus on forms of rust resistance that are less effective, but which are durable because they act against any race of the pathogen. For plant breeders who battle cereal rust diseases, the job is never finished.

Rice blast, a disease that was discussed previously is another example of a very serious fungal disease. The causal fungus produces a toxin that kills plant cells wherever it grows. Rice blast was recorded as a disease as far back as the Chinese Ming dynasty in 1637 and is widespread in Asia. The blast fungus thrives on plants that have been fertilized with nitrogen, and the high-yielding varieties introduced as part of the Green Revolution are therefore especially vulnerable. These rice varieties require nitrogen fertilization to reach their high yields, but rice blast generally reduces by 20% the yield increase that nitrogen fertilizers induce in rice.

Chemical Strategies for Diseases

An important advance in agriculture was achieved in the 1880s with the discovery of "Bordeaux mix." This blend of copper sulfate and lime was originally applied to maturing grapes to deter thieves, but was subsequently found to suppress the very problematic grape downy mildew disease. This was one of the first *fungicides*. Over the last hundred years, numerous other compounds with antifungal or antibacterial activity have been discovered. These compounds have gone a long way toward allowing a cheap and predictable supply of food and other plant products. Unfortunately, the expense and toxicity of these compounds often complicates their use. Fungicides may be applied as sprays or as dusts on plants in the field, or as a seed coating prior to sowing. Many of the older broad-spectrum fungicides must be present before infection, because contact with the growing pathogen is important and the fungicide does not move beyond the plant surface. These compounds often must be applied many times during a growing season. Some newer fungicides can move systemically within the host plant, allowing "curative" action against existing infections.

Many newer fungicides are also less broadly toxic, affecting a narrow set of target organisms. The development of fungicides parallels the development of pesticides, from broadly toxic to toxic against many fungi to specifically toxic to a few target species. Fungicides can be expensive to use, so they are often impractical for grain crops. However, they are widely used on vegetable, fruit, and flower crops. For example, 95% of all grapes and potatoes grown in the United States are treated with fungicides, but fungicides are not used on most maize, rice, or wheat. In these grain crops, disease-resistant varieties keep fungal diseases at bay.

Antibacterial compounds include copper or sulfur sprays, and also well-known antibiotics such as streptomycin or tetracycline. Because of their expense, these compounds also are used almost exclusively on fruit and vegetable crops. An additional problem arises with repeated use of some antibacterial compounds and fungicides: *pathogen resistance*. For example, some bacteria carry genes that allow the bacteria to degrade, export, or otherwise resist specific antibiotic compounds. When an antibiotic is used for many years continuously, strong selection pressure is applied, permitting the initially small population of resistant bacteria to become dominant in the pathogen population of that area. The antibiotic is then no longer effective and other contral strategies are needed. Bacterial movement on contaminated seeds, seedlings for transplant, or even shoes or farm equipment can then spread these antibiotic-resistant strains to other farms or regions. The human or broader environmental toxicity of fungicides and antibacterial treatments remains a significant concern. A number of compounds that were formerly used are no longer acceptable.

Fungicides are normally "registered for use" by government agencies for certain crops, and then only under certain use restrictions (for example, no application within a month of harvest). Agricultural chemical companies devote very substantial resources to the synthesis and testing of new compounds. They search for chemicals that will have efficacy (the ability to kill a pathogen or slow its growth not only in the laboratory but also under realistic field conditions). But these compounds, designed to be toxic to pathogens, must also have acceptably low toxicity to humans and other non-target organisms such as fish or birds. Also, they may have low persistence so that they do not remain present in their toxic form. Even after a compound with efficacy has been identified, it can cost millions of dollars to carry out toxicity testing and certification. Companies take on these costs, however, because an acceptable compound can return many millions in profit if the marketplace accepts it. Increasingly, and often with good reason, society at large is rejecting the use of many broadly toxic fungicides and other pesticides. As with so many other issues in society, it is necessary to balance the benefits of pesticide use with the associated costs.

There is widespread agreement that the most noxious pesticides should be banned, but where do people draw the line? As methods to detect toxicity become more refined, should we ban highly useful compounds that pose a very small but detectable health risk? If researchers can determine that larger doses of a compound are toxic, can we tolerate the presence of extremely small amounts of the compound on food? Is the use of toxic compounds more acceptable in developing nations that are more pressed to produce food? Experience shows that when toxic pesticides are banned, the manufacturers protest vociferously, but alternative ways of dealing with the pests or diseases soon emerge. Clearly, chemical strategies for disease control are extremely important, but also problematic.

Fungicides and antibiotics help control destructive plant pathogens and have been extremely valuable in allowing reliable, large-scale production of food crops. But these pesticides can be expensive to use. In addition, they do not always work particularly well, because pathogens may become resistant to a given compound, and the compounds may be toxic to non-target organisms. Much attention has therefore returned to an older method of disease control, also imperfect, but easier for the grower to use. That method is the planting of crop varieties with inherent, genetically determined disease resistance.

Defense Genes

Plants have coexisted with pathogens for millions of years, and as a result they have evolved many different defense mechanisms. Defenses that are always in place include thick cell walls that are difficult to penetrate and a waxy cuticle (leaf/stem surface) that tends to dry out rapidly, providing less support for the growth of fungi and bacteria. In some cases farmers may be able to avoid a pathogen, for example, by planting after the cool, moist soil conditions that favour certain pathogenic fungi have largely passed. In addition to physical barriers that block infection, plants contain a diverse array of antimicrobial compounds.

The coevolution of plants with pathogens and insects has resulted in the production by plants of an incredible array of secondary metabolites that include phenolics, tannins, glycosides, flavonoids, and many other chemical families. Some of these compounds attract pollinators or promote seed dispersal, but others function mainly as defense compounds. Antimicrobial peptides or enzymes are also present in plants. It is important to note that some antimicrobial compounds are always present, but the synthesis of many others is induced only after the plant has been infected with the pathogen. Infection-induced antimicrobial compounds are also called *phytoalexins*. The response of a plant to microbial infection is multifaceted and involves:

1. Increased expression of a large number of genes in cells at the site of infection.
2. Activation of pre-existing enzymes.
3. Strengthening and cross-linking of the cell walls.
4. Secretion of phenolics into the cell walls.
5. Generation of signaling molecules that will move locally or systemically to activate defenses in other plant cells.
6. In some cases, the hypersensitive response, a beneficial plant cell death response.

The genes that are expressed more actively when a plant is infected are sometimes termed *defense genes*. A better name, also in use, is *pathogenesis-related genes* (PR genes). This name is better because

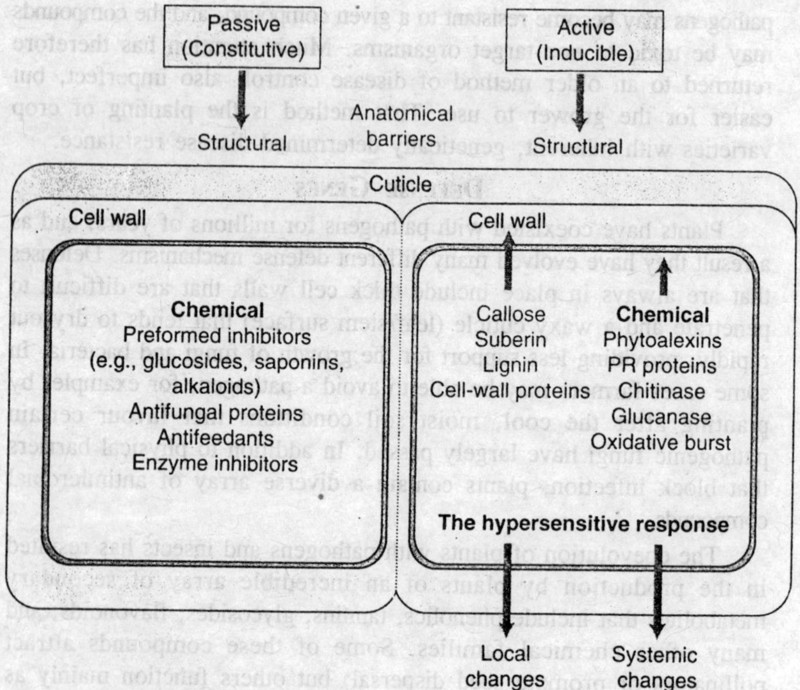

Fig. 7.5. Different types of plant defense, based on existing anatomical or biochemical features of plant cells (left) or active changes induced after a challenge by pathogens (right).

genes expressed upon infection by a pathogen do not necessarily contribute to defense against this pathogen. But many of the most prominent PR genes do in fact encode proteins with known antimicrobial properties that help build the defenses of the plant. These include chitinases and glucanases that can degrade the cell walls of invading pathogens, antimicrobial pep tides that are not enzymes but rather are directly toxic to microbes, and enzymes that control the pathogen-induced biosynthesis of antimicrobial compounds.

For other pathogenesis-related genes, scientists do not understand how if they contribute to plant defense. Why are antimicrobial responses activated only after infection? Why don't plants just make these compounds all the time? It turns out that many antimicrobial responses are costly to the plant: They consume valuable energy and mineral resources; they may also be mildly or severely toxic to the plant. Scientists have generated plants in the laboratory that express antimicrobial responses all the time, but these plants are often stunted and produce less seed. Clearly, it is advantageous for the plant to be able to quickly sense a pathogen invasion and to respond with a massive defense effort rather than to express this response continuously.

Successful Pathogens

How does the plant know a pathogen is present? Recognition often involves the binding of a molecule coming from the pathogen to a molecule of the plant. This is similar to human use of antibodies, which allow specific cells of the human immune system recognize a foreign substance. The plant genes that encode these recognition proteins are called resistance genes. The details of the process are still under study, but scientists believe that each resistance gene encodes a protein that recognizes a specific pathogen compound and then activates host defense responses. The protein encoded by a resistance gene may recognize a virus coat protein, or a bacterial virulence factor that is secreted into the host, or perhaps a fungal protein that is present on the pathogen surface.

Early detection of the pathogen is followed by a strong and rapid induction of defense responses in the cells immediately surrounding the infection. Plants that have undergone a strong, resistance gene-mediated defense response often form numerous necrotic spots that are evidence that the plant has "committed suicide" on a microscopic

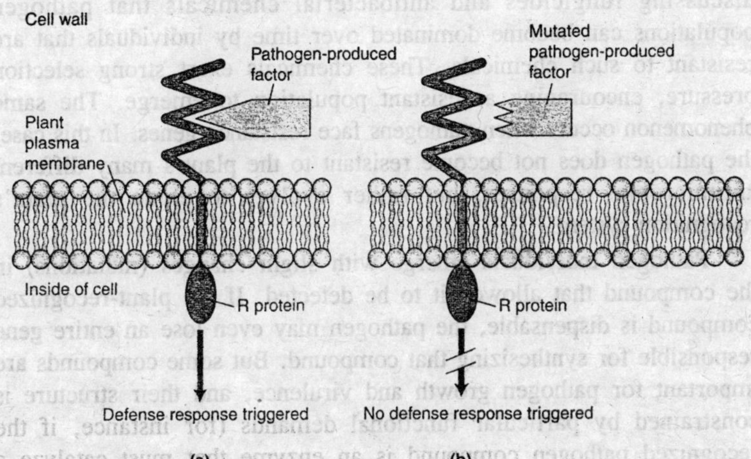

Fig. 7.6. Recognition of a factor from the pathogen by the resistance gene (R-gene) protein. (a) The pathogen factor binds to the R protein and activates the defense response inside the cell. (b) The pathogen factor has mutated and is not recognized by the R protein.

scale. The plant actively sacrifices infected cells as part of a larger defense response that contains the pathogen in a small walled-off region of dead cells. This rapid *hypersensitive response* programmed cell death process may be adaptive not only by walling off the pathogen but also because it releases antimicrobial compounds, releases signaling molecules that elicit defenses in other host cells, and kills off a host cell that might otherwise support the growth of a virus or a biotrophic fungus. The infected cell dies, but the plant is saved if the pathogen is contained. Different resistance gene products control defense activation by detecting extremely different pathogens: viruses, bacteria, fungi, or even nematodes or insects. When the first plant disease resistance genes were characterized in 1993-1995, an interesting finding emerged. The proteins encoded by these genes share similar protein structures.

The mechanisms for pathogen recognition are highly conserved across many different plant species and diseases. What is so interesting about that? It implies that the different specifics of pathogen recognition probably evolved from a small number of progenitor resistance genes. New resistance genes with new pathogen recognition capabilities arise over time. This evolution has been crucial for the ongoing battle of plants to keep pathogens at bay. It may seem odd that a pathogen would produce compounds that allow the plant to sense the pathogen's presence and activate barriers against it. We noted earlier when discussing fungicides and antibacterial chemicals that pathogen populations can become dominated over time by individuals that are resistant to such chemicals. These chemicals exert strong selection pressure, encouraging a resistant population to emerge. The same phenomenon occurs when pathogens face resistance genes. In this case, the pathogen does not become resistant to the plant's many different antimicrobial responses, but rather evolves to elude the plant's recognition system.

Pathogen individuals emerge with slight changes (mutations) in the compound that allowed it to be detected. If the plant-recognized compound is dispensable, the pathogen may even lose an entire gene responsible for synthesizing that compound. But some compounds are important for pathogen growth and virulence, and their structure is constrained by particular functional demands (for instance, if the recognized pathogen compound is an enzyme that must catalyze a specific biochemical reaction). Plant resistance based on recognition of a conserved and essential pathogen structure is much more likely to be long-lived ("durable") over many years in the field. If pathogens can evolve to elude the plant's defenses, they can still respond. Recent studies have shown convincingly that they do. A given plant carries a few hundred different resistance genes—as many as 0.5% to 1% of all the genes in the plant. And although most genes in an organism are highly conserved across many generations, resistance genes change over surprisingly short time periods.

Crop Protection

Plant breeders must keep ahead of pathogen evolution if they are to keep farmers supplied with disease-resistant plants. In some cases, pathogens seem to evolve slowly and resistance genes in the crop "last" for decades. It is only a small challenge for plant breeders to maintain those genes in the best varieties of a particular crop. But other pathogens evolve rapidly to escape plant resistance. If appropriately effective resistance genes are not available in the breeding lines that are in use, plant breeders must extend their search and access other plant sources. In traditional plant breeding, those plants must be interfertile (sexually compatible) with the crop variety—they must be of the same or very closely related species. Wild relatives, also called wild accessions, serve as an important source of new disease resistance traits. Introducing resistance genes into elite cultivars involves extensive backcrossing to eliminate all the undesirable genes from the wild plant.

An improvement on the use of wild accessions comes when breeders can use older varieties or landraces of the crop in question. These landraces, which often come from local farming cultures where the varieties were saved from generation to generation, serve as a valuable reservoir for disease resistance traits. As world commerce brings increasing uniformity across wide areas, many local farmers have switched to the latest new varieties and the older varieties have been lost. But academic breeding programs, companies, and governments now recognize that the older varieties are a precious genetic resource, and germ plasm collections (seed banks) that maintain at least some of these older varieties exist for virtually all crop species.

Some pathogens, such as the rust fungi of wheat mentioned earlier in the chapter, evolve at alarming speeds. In these cases, single resistance genes effective against the current pathogen population have often lost effectiveness in the first year or two after a new crop variety is released, or even before it is released. In light of such failures, many plant breeders have shunned these strong resistance genes in favour of less effective but more broadly and durably active resistance traits. These forms of resistance are often controlled by multiple genes that each make a small contribution to resistance. Breeding for this type of resistance can be very difficult, but it may be preferable if strong resistance genes prove ineffective.

Antimicrobial Compounds

With the availability of recombinant DNA genetic engineering methods, a number of new approaches to plant disease control have become very promising. In one of the simplest approaches, a gene for a single antimicrobial compound or protein is expressed in the plant. The gene may encode a chitinase, for example, that degrades the chitin-containing cell walls of many fungi. This antimicrobial compound approach has been tried, but success has been limited because no single protein that is highly effective at stopping pathogen growth without damaging the plant has been identified.

There are indications that resistance can be improved by expressing two or more such antimicrobial proteins in the same plant. The approach is very appealing in principle, and successes are likely to emerge as more compounds are tested in the future. A number of other creative approaches are being used to engineer plants with improved disease resistance. For example, the pool of resistance genes available to plant breeders is widened substantially if genetic engineers can move resistance genes between plant species that are not sexually compatible. This type of transfer has succeeded, for instance, between tomato and

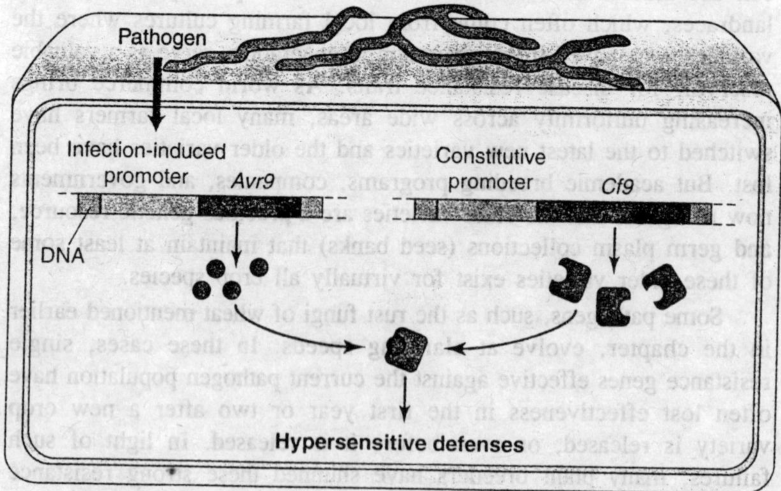

Fig. 7.7. New approach to protecting plants against infection using gene technology.

pepper and between tobacco and tomato. When a tobacco gene that confers resistance to tobacco mosaic virus was incorporated into tomato, the tomato plants became resistant to the same virus. But this may be just the beginning. It is becoming feasible to design or select new resistance genes in the laboratory in response to specific disease organisms, and to then place these genes in crop plants.

The advantage of working with resistance genes or other defense-activating proteins is that they act as the "master switch" that turns on the plant's own multifaceted defense responses. The feasibility of a related approach was recently demonstrated, in which an unmodified plant resistance gene is used to provide widely targeted disease resistance.

Genetic Engineering

Strategies for genetically engineering resistance against virus diseases have been successful enough to be released in commercially available crop species. If researchers introduce genes of a virus, such as the gene that encodes the virus coat protein, into the plant genome under control of a general promoter so that the protein is always made, virus development is often disrupted. There is a very low level of infection, but virus multiplication is severely inhibited. Scientists do not yet understand how or why this works. The presence of too much coat protein in the plant cells may make it difficult for the virus to ever "undress" (remove the protein coat from the nucleic acid molecule) and start replicating in the cell, because the balance is shifted toward an excess of coat protein. As an alternative, a natural antivirus mechanism of plants can be put to work. If an introduced gene (such as a virus gene) causes production of mRNA or other RNA intermediates that can form double-stranded (base-paired) forms of RNA, the plant interprets this gene as foreign and shuts down the expression of both the host and the virus copies of that gene. This process is known as *gene silencing*.

Virus-derived resistance has given very promising results and one commercial success. Papaya ringspot virus threatened the Hawaiian papaya crop with extinction; the industry was saved by the introduction of genetically engineered papaya plants that express the coat protein of this virus in their cells. In the past the industry controlled this disease by using pesticides to kill the insects that transmit the virus. In Hawaii, the insects had become resistant to the pesticides, making this type of control impossible. The GM (genetically modified) papayas are now marketed as "pesticide free." These results are particularly important because, although fungicides and antibacterial pesticides are available, no effective chemical control strategies that are directly antiviral have been invented for plants. Genetically determined disease resistance, plant cultivation methods, and insect control remain the only viable option for controlling virus diseases.

Plant Immune System

As noted, plants, just like humans, defend themselves by activating defense responses when infected, and resistance genes in the plant allow early recognition of the pathogen and rapid induction of a strong defense response. Infected plants also undergo a systemic induction of elevated defensiveness that lets them meet the next infection with a stronger defense response. A plant can be infected on one leaf or region and then, over the next day or two, can activate defenses all over the plant. This process, known as *systemic acquired resistance*, does not require the pathogen-specific resistance genes discussed earlier. It can be effective against pathogens to which the plant is not initially resistant. One leaf may be infected and badly damaged due to a poor plant defense response, but later infections by the same pathogen strain on other leaves cause much less disease damage because the plant meets them with a stronger defense response. Salicylic acid (a chemical closely related to aspirin, which is acetyl salicylic acid) is a key

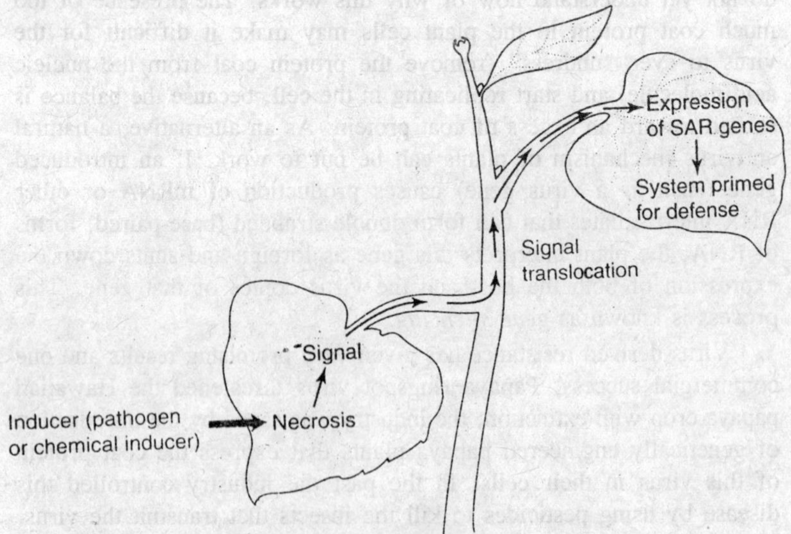

Fig. 7.8. Model for the transmission of the signal that brings about systemic acquired resistance (SAR).

mediator of the cellular responses that activate systemic acquired resistance. For example, salicylic acid causes the plant protein NPRI, which normally dwells in the cytoplasm, to move into the nucleus and interact with transcription factors that activate defense gene expression.

Intriguingly, salicylic acid does not turn on high expression of all antimicrobial defense genes throughout the plant. Instead, it potentiates the plant, placing it in a state of alert or watchfulness so that the plant turns on defenses quickly when the next infection occurs. Although systemic acquired resistance is normally activated by pathogen infections, simple spraying of salicylic acid onto plants activates a similar response. This is hard to do in a practically beneficial way because too much salicylic acid is toxic to the plant, but scientists have identified compounds that are structurally similar to salicylic acid and that last longer and have less potential for damaging the plant. These represent a new concept in chemical disease control: The compounds are not directly toxic to microbial pathogens or other organisms, but instead activate plant defense responses that constrain the pathogen.

An alternative approach is being tested that does not use sprays, but instead involves transgenic plants that express the NPRI protein at high levels. As with the sprays, this elevated NPRI expression potentiates the plant, promoting a more effective defense response against pathogens that might otherwise cause serious disease.

8

GM Crops

In this chapter we will look briefly at some of the broader issues and concerns surrounding the use of GM crops. Some of these are scientific; others are economic, social, ethical and moral. Complete volumes could, and should, be written about each issue and its influence on plant biotechnology. It is only by careful consideration of all the issues that any meaningful dialogue can be entered into and decisions made. Some of these issues may appear more relevant than others, but it is worth remembering that very few decisions are taken in isolation, and decisions made in one part of the world can influence, for better or worse, the whole world.

Current State of Transgenic Crops

It seems appropriate to spend a little time reflecting on the current status of GM plants in broad terms. Despite public opposition and political difficulties in some regions of the world, particularly in Europe, the area of land cultivated with transgenic crops continues to increase. Figures from the International Service for the Acquisition of Agri-biotech Applications (ISAAA) clearly show the importance of transgenic crops in both developed and developing countries. It is estimated that in 2001 over 52 million hectares, or 130 million acres, of land was planted with transgenic crops (an increase of 8.4 million hectares, or 19%, on figures for the year 2000), compared with only 1.7 million hectares in 1996. The USA has the greatest amount of land dedicated to transgenic crops, but several developing countries grow a significant area. Over 5.5 million farmers (compared with 3.5 million in 2000) grew transgenic crops, many of these being in developing countries.

Who has Benefited from these First-generation GM crops?

An analysis of the figures shows that a large area is planted with GM crops. However, just two traits, herbicide resistance and insect resistance (singly or in combination), and four crops, grown predominantly in North America, account for the vast majority of this area. The agribiotech multinationals, like Monsanto and Syngenta, drove the development of these crops. The main beneficiary, other than the agribiotech multinationals, appears to have been the farmer, who has generally enjoyed increased yields and revenues from growing GM crops, despite a sometimes considerable premium on GM seed. Monsanto puts this increase in revenue at $US16.46 per acre from *Bt* GM maize in areas where the European corn borer is a problem. This increased revenue is, in many cases, associated with a decrease in the use of herbicides and/or insecticides (although the figures are disputed). This reduction in the use of potentially harmful chemicals was trumpeted as one of the real benefits of these first-generation GM crops. However, it is interesting to note that surveys in the USA have shown that environmental benefits are not a driver for the adoption of GM technology by farmers.

In many cases, but not all, there does seem to be a real reduction in the use of herbicides or pesticides associated with growing GM crops. For example, the introduction of herbicide-tolerant cotton resulted in increased yields but no decrease in herbicide usage, whereas the adoption of herbicide-resistant soybeans led to significant reductions in herbicide use and only small increases in yield. The herbicides used on GM crops are also considered to be more environmentally friendly than are some of the alternatives. There does therefore appear to be a real environmental benefit associated with these GM crops. Other environmental benefits from growing GM crops might also be apparent. For instance, more environmentally friendly cultivation methods (non-tilling) can be used that reduce soil erosion.

Biodiversity may also be increased as weeds can be allowed to grow postemergence if herbicide-resistant crops are being grown. However, agricultural management has to respond to the use of these crops if negative effects on the environment are not to materialize. Some farmers in Canada have encountered problems with multiply herbicide-resistant volunteer canola for instance, and the altered herbicide application regime used with some GM crops can also bring about its own problems (increased spread from later sprayings, when the crop is taller, for instance). What is noticeable about these first-generation crops, though, is the absence of engineered traits that are not primarily for agronomic benefit.

The agribiotech multinationals invested a large amount in the development of these crops with -the expectation that the investment would eventually be repaid and produce a profit. The technology was relatively simple and there was obvious commercial gain to be had. However, other traits that could considerably improve the general well being of farmers and consumers in both the developing and developed worlds are conspicuous by their absence. This commercially driven development has implications for the future of GM crops in general. Distrust of the large multinationals, who control both the GM seed and the herbicide for instance, contributes significantly to the distrust of GM technology, particularly in Europe.

The perception, at least, is that GM crops have failed to live up to the considerable claims that were made for them. Many people therefore view future developments in GM crops, ones that could have significant benefits for the people of the world and the environment, with suspicion. They are seen by many as being unnecessary (at the very least) and commercially driven.

What will Drive the Development of the Future Generations of GM crops?

Given the considerable opposition to GM crops in some quarters, it is interesting to briefly consider whether GM crops will continue to develop in the same way. Commercial concerns will undoubtedly continue to drive the development of some GM crops, but what is being seen now is the development of GM crops with traits other than agronomic. We have already looked at the example of 'Golden Rice', which is a pioneering example of a GM crop developed for the developing world and without the backing of a commercial company. Like many of the new developments in GM crops it was primarily developed in academia with backing from public funding and charitable institutions. Later in this chapter we will look at some future developments in GM crops that will have potential benefits for the environment and the people of the developing world. However, until the benefits of these 'new generation' GM crops become apparent, the technology faces some considerable hurdles to widespread acceptance.

CONCERNS ABOUT GM CROPS

There is a considerable degree of public concern in many parts of the world about the real and perceived issues associated with GM crops. Some of these are based on fears about the potential impact of the technology on food safety, human health and the environment. At the other end of the spectrum, there is moral or ethical opposition to GM crops on the basis that they are wrong in principle. These public concerns can have a massive impact on plant biotechnology and its adoption. In this section we will look at Some of the issues of public concern about GM crops, and examine what plant biotechnologists are doing to allay those public fears.

Antibiotic Resistance Genes

Previous chapters have looked at the use of antibiotic resistance genes as selectable markers for transformed plants. Plant transformation is a very low-frequency event and some means of selecting the transformed cells 'from the untransformed cells is required. The use of antibiotics as selective agents was already well established as a fundamental tool of molecular biology and cloning in other organisms, particularly in *Escherichia coli*. Their use in plant biotechnology was therefore logical. However, the use of antibiotic marker genes has proved to be one of the hurdles to the widespread acceptance of GM crops. In 1996, Novartis sought approval for a maize variety that carried an ampicillin resistance gene. The UK Advisory Committee on Novel Foods and Processes (ACNFP) blocked this approval for a considerable length of time, but eventually the maize line was approved for cultivation in France. However, concerns about GM crop safety and lack of market acceptance meant that this GM maize was never widely cultivated.

This episode was in some ways unusual, as ampicillin resistance genes are not widely found in the current generation of GM crops. It did, however, help to set the tone for much of the discussion about the safety of antibiotic resistance genes in GM crops. Ampicillin is an antibiotic of the penicillin family that is widely used to treat a variety of infections in humans. The presence of a resistance gene in a genetically modified organism (GMO) released into the environment was perhaps bound to raise fears about creating antibiotic-resistant bacteria, and particularly human pathogens. However, there are several scientific arguments that indicate that this scenario is unlikely and here we will look at just a few:

1. The antibiotic resistance genes used in creating GM crops were originally isolated from bacteria. Resistance to antibiotics, which can confer a selective advantage, occurs naturally in bacteria, and is often carried on plasmids that can be readily transferred

from one strain to another. The development of antibiotic-resistant strains of bacteria is therefore a real problem in hospitals, where antibiotics are used routinely and create a selective pressure for the development of resistance. Against this background, the transfer of antibiotic resistance genes from plants to bacteria (for which there is no known mechanism) will not significantly alter the pool of antibiotic resistance genes in the environment. The transfer of intact, functional, antibiotic resistance genes to gut flora from ingested plant material is also highly unlikely, though not impossible. There is certainly little evidence for the transfer of DNA from food to gut flora, and it is difficult to see why a transgene would behave differently from the DNA in which it is integrated. Even if resistance genes were transferred, no real selective advantage would accrue and the resistant bacterium would be unlikely to survive.

2. Many of the antibiotic resistance genes commonly found in GM crops (such as *nptII*) confer resistance to antibiotics that are not used to treat disease in humans, their use having been superseded by less toxic and/or more effective alternatives.

Despite these explanations of the relative safety of antibiotic resistance genes in GM crops, plant biotechnologists are seeking ways to obviate the need for their use or presence in GM crops. Various alternative selectable marker genes have been developed and technologies that remove selectable markers, so called 'clean-gene' technology, are also being developed.

Herbicide Resistance and 'Super-weeds'

It is possible to use selectable marker genes that confer resistance to herbicides as an alternative to antibiotic resistance genes. However, one problem associated with the use of herbicide resistance markers is the potential for the creation of 'super-weeds'. Gene transfer of herbicide resistance genes, predominantly via cross-pollination, to weedy relatives of GM crops could create such 'super-weeds'. This problem is unlikely to occur if the herbicide resistance gene was only used as a selectable 'marker during regeneration from tissue culture. However, the creation of GM crops engineered specifically to express herbicide resistance, as a desirable agronomic trait to simplify crop production, is more problematic. Gene transfer and the creation of herbicide-resistant weeds is not a theoretical consideration.

The usefulness of some herbicides has already been reduced by the presence of naturally occurring herbicide-resistant weedy relatives of several major crops. The transfer of herbicide resistance genes from GM crops could exacerbate this problem. Practically every major crop species has weedy relatives that could be cross-pollinated by a GM crop. The transfer of herbicide resistance to weeds is already an issue, though at present it is more a question for agricultural management than widespread public concern.

The transfer of herbicide resistance genes to weedy relatives may result in the weed becoming resistant to one or more herbicides, but it will still be susceptible to other herbicides. This may, of course, be of major concern to the farmer (or others), as the herbicide of choice may be rendered ineffective against some weeds. However herbicide resistance genes confer no selective advantage on weeds that are not subject to treatment with the herbicide, and therefore the trait is unlikely to spread throughout the population. Plant biotechnologists can also decrease the use of herbicide resistance genes as selectable marker genes by again either employing alternatives or by using clean-gene technology. A particularly attractive alternative is to engineer the chloroplast genome, which is discussed later in this chapter. Chloroplasts are, in most, but not all cases, inherited maternally, so negating the chance of gene transfer by cross-pollination.

Gene Containment

Preventing the transfer of foreign genes from GM crops to other plants is a wider environmental issue, a concern that does not only apply to herbicide resistance genes. A great variety of foreign genes are being introduced into GM crops, but the environmental impact of these genes is currently difficult to predict. In view of these concerns and to prevent what has been described as 'genetic pollution', several strategies for preventing foreign-gene transfer are being developed. Gene transfer usually occurs through pollen, although GM crops, if a wild relative pollinated them, could also serve as a female parent for hybrid seeds.

The dispersal of seeds from GM crops amongst weedy relatives could also produce mixed populations, with introgression of, for example, a herbicide resistance gene resulting in herbicide-resistant weeds. The potential for foreign genes from GM crops to be transferred to weedy relatives depends on a great many variables. If gene transfer is through pollen, the amount produced, the longevity, the dispersal mechanism and distance as well as the proximity of weedy relatives that are sexually compatible with the GM crop all influence the frequency of transfer. Dispersal by seeds obviously depends on whether the GM crop is allowed to set seed, whether the seeds are dispersed amongst or close to weedy relatives and the viability and longevity of those seeds.

Techniques for gene containment

A wide variety of techniques are capable of preventing or reducing gene transfer. Some of the techniques are at present, theoretical or have only been demonstrated in laboratory situations. It is also interesting to note that at least one of the techniques, so-called 'terminator technology', has itself been the cause of much opposition to GM crops as seed can not be collected and saved by the farmer for planting in subsequent years.

Big Business

The threat of terminator technology being introduced led to protests in many parts of the world, and was seen as yet another example of big business imposing its wishes on farmers and consumers alike. It is informative to look at the example of terminator technology as it illustrates some interesting points about the plant biotechnology industry and public perception. Terminator technology has not been introduced commercially, although most of the big plant-biotechnology firms are thought to have the technology.

Terminator technology has the potential to make GM crops safer by reducing gene transfer to weedy relatives. The publicity has, however, focused on its potential use to prevent farmers from saving a portion of their grain for use as seed in subsequent years, thus tying them to an expensive source of fresh seed each year. Concerns that terminator technology had already been introduced led to GM cotton crops being burnt in India amid wide-scale unrest. For terminator technology to become a problem for farmers in developing countries (and even in the USA, where 20-30% of soybean farmers reuse seed) they would already have to be growing seed, with terminator technology engineered in, from a commercial company like Monsanto. Otherwise seed can continue to be kept as normal. There is concern, though, that future development will mean that the highest yielding varieties (which governments or banks will promote) will come with technology protection schemes (TPS) attached. Farmers will, therefore, be either forced to pay for seed every year or continue to grow the low-yield varieties.

Apomixis

In May 1998, in what may well come to be heralded as the beginning of a new era for GM crops, leading researchers in apomixis declared that the technology needs to remain available to developing countries. The Bellagio Apomixis Declaration recognized that apomixis offers too great a benefit for it to be controlled by a small number of concerns (which inevitably would be western-based biotechnology companies protecting their investment with patents).

The needs of commercial concerns (that will undoubtedly playa major role in technology development) are not ignored, but the emphasis is put on novel approaches to patenting, licensing and technology development to ensure 'openaccess' to the technology. So what is apomixis, and why is it such an important technology? In plants, seeds are usually produced as a result of fertilization in sexual reproduction. In some plants, however, seeds can be produced without fertilization, a process known as 'apomixis'.

Natural apomixis occurs in a limited number of plant species (about 400 in 40 families), but is not found in any major crop plant. Plants from apomictic seeds are genetically identical to the mother plant and are therefore clones. Any desirable features of the mother plant are retained in subsequent generations. Apomixis, if it could be introduced into cereal crops like rice, wheat and maize, would provide major advantages to both plant breeding/biotechnology companies and farmers, particularly those in the developing world. Much of the seed sown by farmers is high-performance hybrid seed, produced from parents that have different, desirable characteristics.

Hybrid seed is expensive, though, because of the continuous effort needed to produce seed from the parents, as hybrid plants do not breed true and the desirable characteristics are lost. If apomixis were introduced into the hybrid, it could be cultivated indefinitely, as the offspring would be genetically identical and therefore retain the desirable traits. This is known as 'fixing heterosis' (where the desirable characteristics of both parents are combined in one plant). Plant biotechnology companies are interested in genetically engineering apomixis into crop plants because it can be used to considerably speed up the development' of new varieties.

The advantage to farmers is that seed, even from high-performance hybrids, can be saved and reused, without the loss of desirable characteristics. This frees the farmer from having to buy expensive seed every year, and would make expensive, high-performance, hybrids available to farmers in the developing world. The advantage of apomixis, the freeing of farmers from dependence on expensive seed is, of course, one of the potential hurdles to its development, as seed companies cannot protect their investment. Several ways round this problem have been suggested, such as coupling apomixis and terminator technology, so that farmers who could afford to buy new seed each year are supplied with seed with the terminator technology activated. Seed for farmers in the developing world would not have the terminator technology activated and seed could therefore be reused. The problem, according to some from the plant biotechnology industry, may not be all that great, as farmers in the developed world (the natural market for hybrid seed) prefer to grow the latest hybrids anyway, and would not save and reuse on a large scale.

Regulation of GM Crops and Products

We have looked at a few of the issues surrounding GM crops and seen that some of the concerns are about the safety of such crops, both to human health and the environment. It is perhaps appropriate to look at some of the regulations that apply to all genetically modified organisms (GMOs). The future of GM crops is, to a large extent, determined by the regulatory framework that is applied to their growth and processing. As a minimum expectation, regulations should ensure that the GM crops and products derived from them are safe (both to the environment and to human/animal health). However regulations pertaining to GM crops, particularly in the European Union (EU), cover additional aspects and reflect a range of public concerns. These additional regulations cover aspects such as the labelling of GM foods and GM -derived products so that consumers can exercise choice over whether to eat GM ingredients. Regulations are also designed to ensure that the public is reassured and confident that any GM crops or products are safe and that they are not being consumed without their knowledge.

European Union (EU)

Brief introduction to EU regulatory processes

In EU countries, national laws implement various Council Directives and Regulations governing the regulation of GM crops and products. These include the key legislative elements Directive 90/220/EEC on the deliberate release into the environment of genetically modified organisms (which is being replaced by Directive 2001/18/EC in late 2002) and Regulation (EC) 258/97 concerning novel foods and novel food ingredients. Directive 90/220/EEC (and subsequent amendments) requires an extensive environmental risk assessment (ERA) to be carried out before any authorization to place the GM crop on the market under the Directive can be given. The company responsible for the GM crop (the notifier) submits a technical dossier and a summary (the notification) to the competent authority of a Member State.

The technical dossier is examined, with particular reference to the ERA, to ensure compliance with the Directive. If the outcome is favourable then the process progresses, with the dossier being forwarded to the European Commission. The competent authorities of all Member

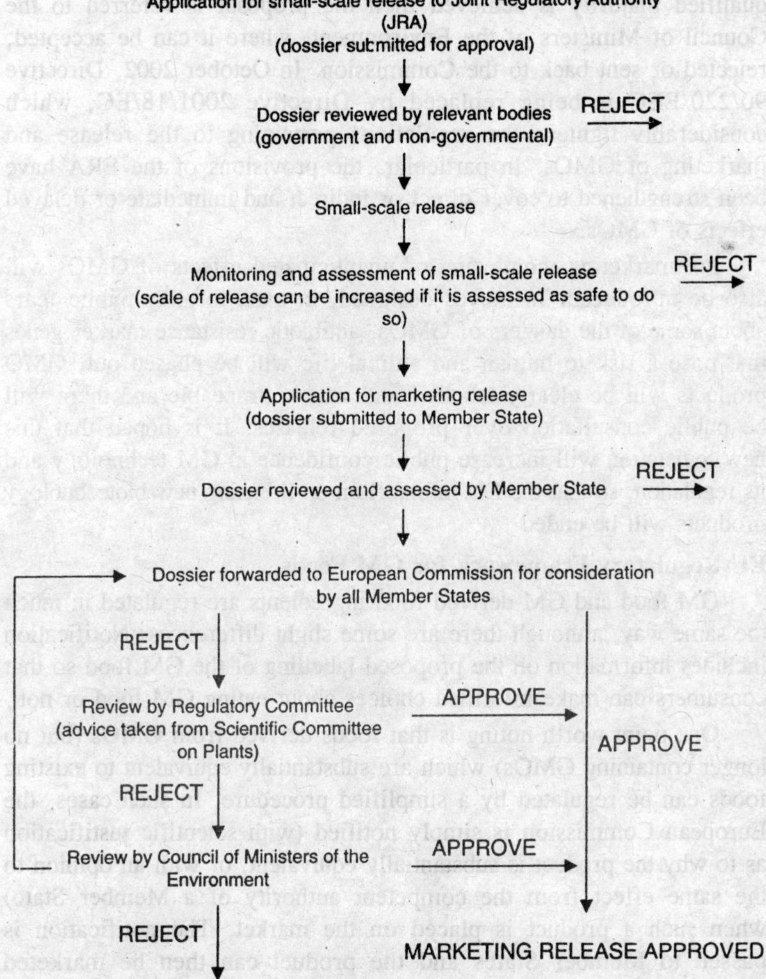

Fig. 8.1. A flow chart summarizing the regulatory process for the release of GM crops into the environment in the EU.

States evaluate the notification and raise any objections. If no objections are received, the product can be placed on the market under the consent of the Member State where the original notification was made. If objections are raised, the proposal is subjected to further scrutiny.

A Commission proposal is submitted to the Regulatory Committee of Member States and is subjected to a vote. This proposal takes advice from the Scientific Committee on Plants. If a qualified majority of Member States support the proposal, then it is accepted. If no qualified majority is achieved then the proposal is referred to the Council of Ministers of the Environment, where it can be accepted, rejected or sent back to the Commission. In October 2002, Directive 90/220/EEC is being replaced by Directive 2001/18/EC, which considerably tightens the regulations pertaining to the release and marketing of GMOs. In particular, the provisions of the ERA have been strengthened to cover direct or indirect and immediate or delayed effects of GMOs.

Postmarketing monitoring of unanticipated effects of GMOs will also be introduced. In moves designed, in part, to allay public fears about some of the dangers of GMOs, antibiotic resistance marker genes that pose a risk to human and animal life will be phased out, GMO products will be clearly labelled and will be traceable and there will be public consultation over proposed releases. It is hoped that this new legislation will increase public confidence in GM technology and its regulation, so that the EU moratorium on licensing new biotechnology products will be ended.

EU Regulatory Framework for GM Foods

GM food and GM-derived food ingredients are regulated in much the same way, although there are some slight differences. Notification includes information on the proposed labelling of the GM food so that consumers can make informed choices about eating GM food or not.

One point worth noting is that foods derived from GMOs (but no longer containing GMOs) which are substantially equivalent to existing foods can be regulated by a simplified procedure. In such cases, the European Commission is simply notified (with scientific justification as to why the product is substantially equivalent, or with an opinion to the same effect from the competent authority of a Member State) when such a product is placed on the market. The notification is passed to Member States and the product can then be marketed throughout the EU.

Concerns about GM food safety

Whatever the scientific limitations of the work carried out by Pusztai and his colleagues, the revelations certainly contributed to the general anti-GM food feeling in the UK. Several scientists have supported Pusztai's findings, and, as was pointed out the area of GM-food safety clearly needs further investigation. One of the contributory factors to the public reaction to the Pusztai report was undoubtedly the loss of public confidence in food safety following the bovine spongiform encephalopathy (BSE) outbreak.

The public feeling, particularly in the UK, against GM foods was badly underestimated by some, and the use of the full weight of the UK academic hierarchy to investigate Pusztai's findings only served to create the impression that Pusztai was being 'silenced' by government, academia and big business. Thus, a bad situation for plant biotechnology was made even worse.

It is important to remember there is no evidence that GM foods are any less safe than non-GM foods. However, the issue of GM food safety has now become much broader. How do plant biotechnologists assess the safety of GM foods, and how does this data get fed into a meaningful debate involving all interested parties? Does the public trust either the information or, indeed, the scientists? If GM foods are to accepted in Europe, it is incumbent upon plant biotechnologists and governments to engage in a real dialogue with the public, and not simply criticize or ignore public attitudes in a post-BSE Europe.

Part of this dialogue must deal with the very real scientific issues surrounding concepts like substantial equivalence. What sort of compositional analysis, if any, is sufficient, or more than sufficient, to prove substantial equivalence? What alternative methods (like transcriptomics and metabolomics) could be used in analyzing GM foods? Is substantial equivalence a scientifically valid and, more importantly, publicly acceptable concept? Part of the dialogue also needs to emphasize the real benefits (e.g. improved nutritional or health values) that GM crops and food could bring to the consumer, rather than to the agribiotech multinationals.

USA

In the USA, regulations pertaining to GM crops are implemented by three federal agencies. The US Department of Agriculture (USDA) and, in particular, the Animal and Plant Health Inspection Service (APHIS) under the authority of the Federal Plant Pest Act determines whether a transgenic crop is likely to become an agricultural or environmental pest. The USDA regulates the transport, import and testing of transgenic plants. In many cases, field tests of GM crops can be conducted if APHIS are simply notified and six criteria regarding safety of the GM crop are met and if the applicant meets certain standards to ensure biological confinement. If these criteria and standards are not met, a more involved process is required to obtain a permit for release from APHIS. The field tests and the subsequent reviews by federal authorities can take many years to ensure the safety of the GM crop to human health and the environment.

GM crops that are to be commercialized can be granted 'non-regulated' status by APHIS upon petition. This process requires that extensive data dealing with the nature of the modification and its effects on the crop and the wider ecosystem are presented to APHIS. If non-regulated status is granted, the product and any offspring can be released in the USA without APHIS review. APHIS can prevent the continued use of a GM crop if it is thought to present a risk of becoming a pest. However, the USDA regulatory process has been criticized recently (in a series of National Academy of Sciences reports) for not being rigorous, inclusive and transparent.

The Environmental Protection Agency (EPA) ensures the safety of pesticides, including those produced biologically. Under the authority of the Federal Insecticide, fungicide, and Rodenticide Act, the EPA regulates the testing, distribution, marketing and use of plants that have been genetically engineered to be pest-resistant. The EPA also sets tolerances for herbicide residues on herbicide-resistant crops. The EPA issues permits for field-testing and registrations for the commercialization of GM crops that are pest-resistant.

The Food and Drug Administration (FDA), under authority of the Federal Food, Drug and Cosmetic Act, regulates foods and animal feed derived from transgenic plants. Genetically engineered foods have to meet the same safety standards as all other foods. However, many GM crops do not need premarket approval from the FDA because they do not contain substances that are significantly different (they are deemed to be substantially equivalent) from those already present in other foods.

One or more of the federal agencies may regulate a particular GM plant, depending on the nature of the modification and its intended use. Thus, herbicide-tolerant crops would be regulated by the USDA (to ensure the GM crop would not become a pest), the EPA (the new use of a herbicide) and the FDA (to ensure the crop was safe to eat).

If no food or feed use is intended, the FDA need not be involved in the regulatory process.

As well as federal regulation, some states require additional regulation of GM crops. The acceptance and regulation of GM crops has been a source of considerable tension between the EU and the USA. The EU moratorium has, for instance, cost the USA over $200 million annually in corn sales to EU Member States. The new regulations and the lifting of the moratorium should open up the EU market to (properly assessed) GM imports from the USA. However, the USA claims that some aspects of EU regulation put imports from the USA at an unfair disadvantage.

Future Developments in the Science of Plant Biotechnology

In previous chapters we have looked at some of the developments taking place in plant biotechnology. Some of these developments are essentially market-driven, whilst some are aimed at either making the technology safer and more acceptable (clean-gene technology, or chloroplast transformation) or bringing the benefits of plant biotechnology to people in developing countries. These developments cannot take place in isolation, because social, political, economic, ethical and moral pressures all influence the development of plant biotechnology. In this last section we will look at just a few of the developments taking place in plant biotechnology.

'Greener' Genetic Engineering

Although many of the products of plant biotechnology on the market today can offer real environmental and health benefits (from reduced pesticide use or provitamin A rice, for instance), considerable opposition to GM crops remains. However, new approaches to GM crop production are becoming apparent and might help to make the technology more acceptable. These new approaches, which make use of data from genome sequencing projects together with our increased understanding of the fundamental biology underlying agronomically important traits, rely on plant genes to produce the desired effects. Gene transfer from widely differing organisms and the creation of 'Frankenfoods' is avoided, and the emphasis shifts towards modification of plant growth, development and physiology to bring about the desired changes.

This new approach could bring benefits to the developed and developing world alike. Plant genes that control or influence agronomically important traits such as time of flowering, dwarfing, bolting and disease resistance have been identified and manipulated to produce agronomic benefits. These benefits could be applied to a range of crops, benefiting producers and consumers from a range of economic backgrounds.

For instance, rice takes only a little more than 6 months to produce a crop in some areas. Introduction of a gene from *Arabidopsis*, called '*LEAFY*' (which promotes flowering), into rice accelerates flowering just enough to allow two crops per year, with only a small reduction in yield per crop. Similar approaches have been taken to speed up the reproductive development of citrus trees. The constitutive expression of either '*LEAFY*', or another *Arabidopsis* gene known as '*APETALA*', shortens the juvenile phase of citrus trees, so that they crop as early as their first year, as opposed to having a delay of more than 6 years.

The development of dwarf strains of cereals like wheat and rice led to the 'Green Revolution' of the 1960s and 1970s, which transformed countries like India into net wheat exporters. So great an achievement was the development of these dwarf strains that Dr N. E. Borlaug was awarded the Noble Peace Prize for his pioneering work in this area. Dwarf strains are generally higher yielding as less of the photo assimilate goes into growth and more goes into the grain. They also have the advantage of being less prone to damage from wind and rain. Dwarf strains can, however, take a considerable time to produce by conventional breeding, and in many cases breeding in dwarfing characteristics has led to the loss of other, desirable characteristics. Scientists are now beginning to manipulate the mechanism (which involves the plant hormone gibberellin) responsible for stem growth directly, using GM technology. Both a dominant-mutant form of the *Arabidopsis GAI* (gibberellin insensitive—a gibberellin derepressible repressor of plant growth) gene and the wild-type gene, when introduced into rice, confer reduced responses to gibberellins and result in dwarf rice plants. New, dwarf cereal strains could, therefore, be produced relatively quickly using this approach. The ability to directly manipulate 'elite' varieties also means that the other desirable characteristics might not be lost.

These examples are interesting, not only because of the hope they hold out for increasing crop yield and productivity, but because they also highlight how information and resources from fundamental investigations conducted in model plant species, like *Arabidopsis*, can directly benefit crop plant biotechnology. These examples also illustrate how plant biotechnology is beginning to address a criticism sometimes levelled against it, that it does not allocate resources to projects aimed at increasing plant yield that would benefit farmers and consumers in developing countries. Other approaches to the problem of plant growth and yield, such as improving the photosynthetic efficiency of plants, are also being adopted.

Transgenic plants are also being developed that have an improved nutritional status and/or health value. Some of these, like provitamin A rice, are quite specifically aimed at the developing world, others offer potential benefits to people in both the developing and developed worlds. The new generation of transgenic crops also includes plants with obvious environmental benefits. Transgenic plants can be used, for instance, to detect and treat heavy metal contamination. Transgenic plants can also be used to remove toxic explosives that contaminate some land. Transgenic plants can be used as factories to produce biodegradable, renewable and CO_2-neutral alternatives to oil-based products, for instance, which, as oil becomes a rarer and more expensive resource, will allow the dwindling resources to be conserved and only used in applications where no alternative is feasible.

No plant biotechnologist claims that plant biotechnology and, more specifically GM crops, is the answer to all the world's problems. There are undoubtedly other issues, such as wealth distribution, that require urgent attention. What these examples show, however, is that plant biotechnology can make a difference to life expectancy and the quality of life, and the environment, in both the developed and developing worlds. Some of these developments would simply not be achievable through conventional plant breeding. Plant biotechnology should be seen for what it is: not a panacea, but part of an integrated package of developments aimed at achieving the goal of sustainable growth and development, and food security, throughout the world.

9

Crop Evolution

Agriculture is one of the major technological innovations of humankind. However, the appearance of agriculture was more than a mere technical advance. Since its inception some 10,000 years ago, agriculture has permitted civilizations to develop and converted wild landscapes into fields and pastures. The change from hunting-gathering to agriculture was so great that the term "Neolithic Revolution" is used to describe it. The Neolithic or New Stone Age was a phase in human evolution characterized by the development of more advanced stone tools. About the same time, humankind became settled down and started living in villages. Sedentarism, in turn, led to a reduction in birth spacing, which eventually caused higher population growth. Societies changed from an egalitarian to a hierarchical mode of organization. Most importantly, crops and farm animals were domesticated. Domestication is a change in the genetic makeup and, hence, morphological appearance of plants and behaviour of animals, such that they fit the needs of the farmer and consumer.

Procuring Food by Hunting

Human origins can be traced back 4 million years when apelike ancestors of humans adopted an upright posture. Walking on its hind legs allowed *Australopithecus afarensis* to free its hands to manipulate objects. Nevertheless, subsequent evolution was still slow, It was not until 2.5 million years ago that *Homo habilis* appeared, with an enlarged body and brain. It is also around that time that the first stone tools appeared. The *Homo erectus* descendants of *Homo habilis* lived from 2 million years ago to 400,000 years ago. The brain size of *Homo erectus* was larger than that of its predecessor, and body height was comparable to that of modern humans. Gathering and scavenging were the main food procurement activities.

Homo erectus was the first hominid species to master fire, some 500,000 years ago, and to migrate out of Africa into Asia and across a land bridge into Indonesia. *Homo neanderthalensis* was the immediate predecessor of *Homo sapiens* in Europe, the Near East, and northern Africa. They lived from approximately 250,000 to 30,000 years ago. Their body and brain were larger than those of modern humans. They lived in small bands practicing cooperative hunting of large game animals, which may have been made possible by their language capability. They may have been interested in music, as shown by the discovery of a flute in some of their remains. They cared for their sick and buried their dead.

Homo sapiens is not a direct descendant of *Homo neanderthalensis*, but rather a more gracile, nimble, and clever relative of the latter. This species originated in Africa some 50,000 years ago and also migrated to other continents. In Europe, they displaced or killed their Neanderthal predecessors. Their tools of stone, bone, and antlers were of standard size and shape, indicating a greater mastery of tool-making techniques. They also made multipiece tools such as harpoons, spear-throwers, and bows and arrows, and knew how to make ropes, which they used to construct nets, lines, and snares. They owned sewn clothing, adorned themselves with jewels, and buried their dead in elaborate

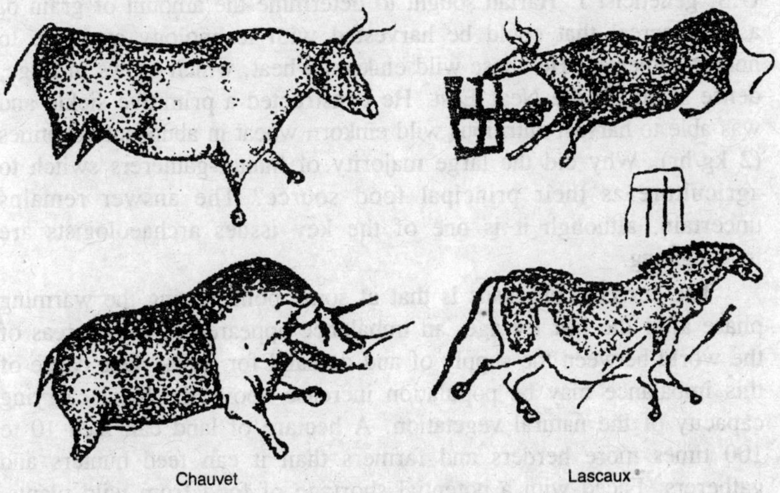

Fig. 9.1. Cave art by hunter-gatherers.

rituals, indicating a belief in an afterlife. They are well known for their magnificent cave paintings and drawings in southern France and northern Spain, demonstrating both artistic inclinations and a keen sense of observation of their environment and particularly of their food sources.

Approximately 15,000 years ago, hunter-gatherers started focusing their attention on a broader range of food items, including small game, waterfowl, fish, and wild grains. In addition, they developed special tools such as grinding tools and storage pits to deal with these foods. Whether this change took place by the hunter-gatherers' choice, or out of necessity because of dwindling numbers of large game animals, is still a matter of speculation. Additional evidence on the food procurement practices of hunter-gatherers comes from the study of contemporary hunter-gatherers. They have an intimate knowledge of the animals they hunt and plants they gather. Their existence is far from precarious, as they can obtain a diverse and abundant diet, with relatively little work. For example, the !Kung San of the Kalahari Desert in southern Africa eat 23 of the 85 edible plant species and 17 of the 55 edible animal species they know. They spend two and a half days per week procuring food and the rest of the time as leisure.

Similar observations have been made for other groups. For example, the Cahuilla native Americans of southern California used over 170 species, and their yearly harvest of oak acorns—their staple food—was copious enough to feed each family. In a classic experiment, U.S. geneticist J. Harlan sought to determine the amount of grain of a wild cereal that could be harvested with technology available to hunter-gatherers. He chose wild einkorn wheat, which grows in large, dense stands in the Near East. He constructed a primitive sickle and was able to harvest nutritious wild einkorn wheat in abundant quantities (2 kg/hr). Why did the large majority of hunter-gatherers switch to agriculture as their principal food source? The answer remains uncertain, although it is one of the key issues archaeologists are investigating.

The current consensus is that at some point during the warming phase after the last ice age, an imbalance appeared in some areas of the world between the supply of and demand for foods. One cause of this imbalance may be population increase above the local carrying capacity of the natural vegetation. A hectare of land can feed 10 to 100 times more herders and farmers than it can feed hunters and gatherers. Faced with a potential shortage of food from wild plants, gatherers may have turned to more intensive forms of plant management, such as seeding new stands of wild grains. Another reason may be climate change. For example, in the Near East the warming trend after the last ice age was interrupted by a short-lived cold spell. During this cold spell, the gatherers may have intensified their management of the vegetation. Whatever the reasons may be, there was a gradual transition from hunting and gathering to herding and farming. Herders-farmers also continued to rely on natural vegetation to supplement their food.

Beginning of Agriculture

A large part of human knowledge about the origin of agriculture comes from studies on the origin of individual crops. Two types of information are particularly relevant in this regard, as proposed in 1885 by Alphonse de Candolle, a Swiss botanist who initiated the studies on crop origins. The geographic distribution of wild relatives of a crop provides a general idea where a crop may have originated. Careful botanic explorations are necessary to determine the precise distribution of wild progenitors. It is within this area that a crop was probably first cultivated—provided, of course, that the contemporary distribution of wild crop relatives matches that of these relatives at the time of domestication.

Additional genetic studies involving crosses between the crop and presumptive wild ancestors and a comparison of their DNA can then identify in more detail a specific region of domestication. Archaeological studies provide a wealth of information on the transition from hunting-gathering to agriculture. For example, they can tell us more about the type of society in which agriculture was practiced: Were people living sedentarily, in settlements? What was the size of these settlements? Did people herd animals as well as raise crops? Were the plants raised still wild, and what was their age at the time of domestication? In wheat, as in many other cereals, a major difference between the wild progenitor and domesticated descendants lies with seed dispersal. Wild plants spontaneously shed their seeds at maturity, assuring their dispersal.

In contrast, to minimize yield losses early farmers have generally selected domesticated plants to hold on to their seeds. For cereals in particular, archaeologists can distinguish the two dispersal modes by specific morphological traits of the seed. Analyses of archaeological plant remains can also determine the approximate age of the materials. This is done through carbon-14 dating. As a result of these analyses,

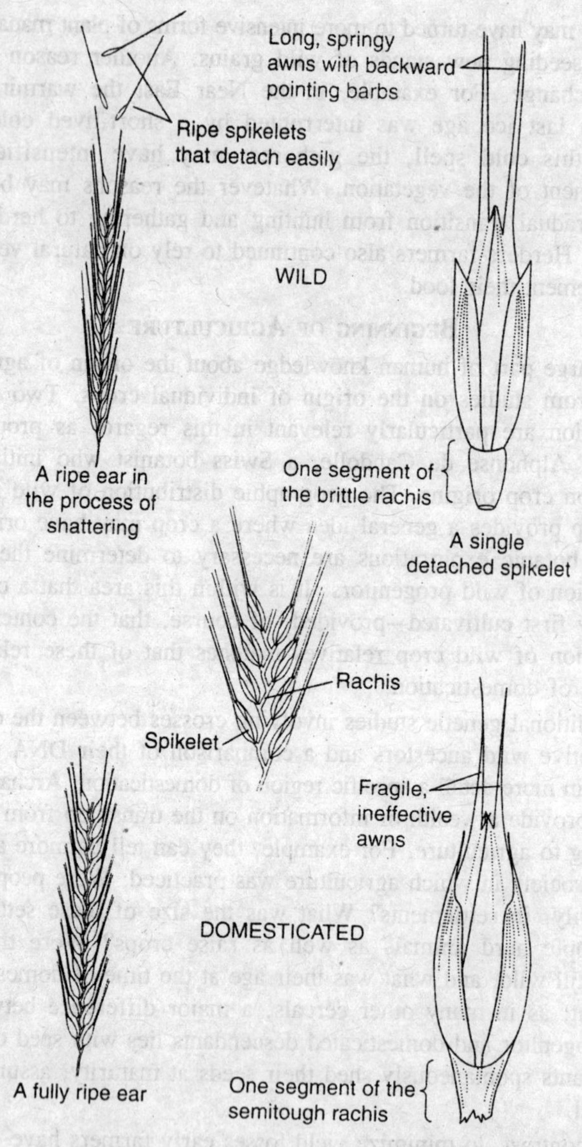

Fig. 9.2. Comparison of wild and domesticated grains of wheat.

people have a much better idea of the age of the earliest remains of crop plants and in some cases their wild relatives. A striking observation from this table is that agriculture originated at similar times (approximately 10,000 years ago) in widely different regions of the world, probably because of a climate change acting on a global scale. This climatic event was the end of the last ice age. The warming trend that followed the last ice age increased rainfall and led to changes in the geographic distribution of vegetation.

In Mediterranean and monsoonal climates, where most crops originated, the warmer postglacial period led to wetter rainy seasons but also to longer, drier summers. These climatic changes favoured annual plants that could complete their life cycle within the rainy season and could survive the dry season as seeds. Where did our crops originate? As proposed in the 1920s and 1930s by Nicolai Vavilov, a Russian geneticist, there are a limited number of areas in the world where crops originated. These areas are generally located in tropical or subtropical regions, at middle elevation (approximately 1,000-2,000 m) in areas with varied topographies, including river valleys, hills and mountainsides, and plateaus. These regions often have a climate with distinct wet versus dry seasons, either Mediterranean climates (wet winter) or savanna or monsoon climates (wet summer).

Table 9.1. Examples of the age of earliest crop plant remains

Location	*Crop*	*Age (years before present)*
Mesoamerica	Squash	10,000
Mesoamerica	Maize	6,300
Central America	Cassava, Dioscorea yam, arrowroot, maize	7,000-5,000
Fertile Crescent	Einkorn wheat	9,200-8,500
Fertile Crescent	Lentil	9,500-9,000
Fertile Crescent	Flax	9,200-8,500
China	Rice	9,000-8,000

Six major independent agricultural centers of origin have been identified. These are (1) Mesoamerica (the southern half of Mexico and the northern half of Central America); (2) the Andes of South America, including its foothills region on the eastern slope; (3) the "Near East" (Southwest Asia); (4) the Sahel region and the Ethiopian highlands in Africa; (5) China; and (6) Southeast Asia. In addition, there are a few minor centers of domestication such as North America and Central Asia. In each of these regions, people domesticated different sets of plants. However, in most regions they domesticated crops with similar uses. For example, they domesticated cereals in most of the regions, as well as crops of the legume family.

Cereals and legumes complement each other nutritionally. Whereas cereals are deficient in lysine, an essential amino acid for the human body, legumes are deficient in methionine, cysteine, and tryptophan, three other essential amino acids. When eaten together, however, the two types of crops complement each other. Legumes also complement cereals agronomically, as they provide additional nitrogen obtained by symbiotic nitrogen fixation to cereal crops. Most plants were domesticated in a single region. However, people domesticated several crops in more than one place.

Vicarious domestications are domestications in widely different places of similar plants, whether they belong to the same or different species. Examples are common bean (Mesoamerica and the Andes), cotton (Africa, India, the Andes, and Mesoamerica), and rice (China and Africa). When multiple domestications take place, people often select the same trait in different regions. Farmers have selected the similar seed colour or colour patterns in beans of the Andean and Mesoamerican areas. After domestication, the fate of each crop differed. Some became extinct or nearly so, because they were displaced by other crops introduced from other areas. This was the case for some crops domesticated in North America such as goosefoot or lambs-quarter (Chenopodium) and marsh elder (*Iva annua*).

The cultivation of lambs-quarter, originally domesticated by Native Americans in eastern North America, was discontinued in the eighteenth century. It is now extinct. Other crops have survived to this day. However, the extent of their distribution differs very much from crop to crop. For some crops, the distribution remains limited to their original domestication center. Others have been dispersed to a limited degree outside their domestication center, and still others have acquired major production zones outside the domestication center. Finally, most major crops are now very widely cultivated. Most production for these crops takes place outside the respective domestication centers. The hunting-gathering societies gradually receded and were replaced by farming and pastoral societies. Societies governed by immediate food rewards for their efforts were replaced by delayed-reward societies.

Harvesting a crop (the delayed reward) is preceded by several operations including land clearing, plowing, and planting, all of which can be quite labour intensive. The need for additional labour as well as the reduction in birth spacing allowed by a sedentary lifestyle led to larger villages. To maintain cohesion in these larger groups, people performed religious rituals and festivals. Control of resources such as irrigation water created inequalities in these early agricultural societies. Eventually, more structured and hierarchical societies appeared, culminating in the creation of city-states led by a ruler. Examples of the earliest cities are Mari and Ebla in Mesopotamia, the area between the Tigris and Euphrates streams, which is surrounded by the Fertile Crescent. In this area, well-known early civilizations eventually came into being, including Sumer, Akkad, and Babylon.

Agriculture played a major role in the development of these civilizations by generating a surplus of food, which allowed a part of society to pursue activities other than food procurement. These activities included specialized crafts, religion, the military, trade (including that of crop products), and administration. Thus the initiation of agriculture had a profound effect not only in biological terms through the domestication of plants and animals, but it also caused a profound change in human societies, a change that has lasted through today.

Wheat Domestication

Wheat is one of the most important food plants in the world, principally as a source of calories. It is grown on an area of 215 million hectares on five continents. Total production in 2001 was 600 million metric tons. The word *wheat* is a generic term for a number of related grain crops belonging to the genus *Triticum*. Among the most important of these crops are bread wheat (*Triticum aestivum*) and pasta wheat (*Triticum durum*). An additional species—currently cultivated only as an animal feed in mountainous agricultural areas of Turkey, Italy, and Spain—is einkorn wheat (*Triticum monococcum*). This species was, however, the original domesticated wheat species. For millennia, it constituted the main staple crop in the Near East and surrounding areas. For example, the last meal of the man whose frozen 5,000-year-old remains were recently discovered in the Alps consisted mainly of einkorn.

Most wild relatives of wheat, which belong to the genera *Triticum* and *Aegilops*, are distributed in the Near East, a region encompassing modern-day Turkey, Lebanon, Israel, Jordan, Syria, Iraq, and western Iran. They are concentrated in the mountainous regions surrounding the alluvial plains of the Tigris and Euphrates, the region called Mesopotamia, on the west, north, and south. Because of its shape and role in the origin of agriculture, this region has been called the Fertile

Crescent. The genomes of these *Triticum* species underwent two major changes during their evolution. First, they went through a normal process of evolution and chromosome divergence leading to different versions of the basic seven-chromosome set of wheat species.

Geneticists have labeled these different sets with uppercase letters: A, B, D, and so on. Second, the *Triticum* species form a polyploid series that includes diploid species (two sets of seven chromosomes), tetraploid species (four sets of seven chromosomes), and a hexaploid species (six sets of seven chromosomes). These tetraploid and hexaploid species arose from two rounds of crosses between species with different basic chromosome sets. Normally, crosses between these species yield sterile hybrids. However, occasional doubling of the chromosomes, either in the gametes or in the progeny, led to a fertile polyploid progeny. There are two evolutionary lineages in domesticated wheat.

In the einkorn lineage, a wild, diploid species with an AA genome from the Fertile Crescent was the progenitor of the only domesticated wheat species, called *einkorn wheat*. It is a hardy species, although not high-yielding compared to its cousin, emmer wheat. It was therefore grown in more remote, often mountainous, areas. It was also an evolutionary dead-end because it did not lead to polyploid descendants as did the pasta and bread wheat lineage. In this lineage, two diploid wild species with an AA and a BB genome, respectively, crossed to

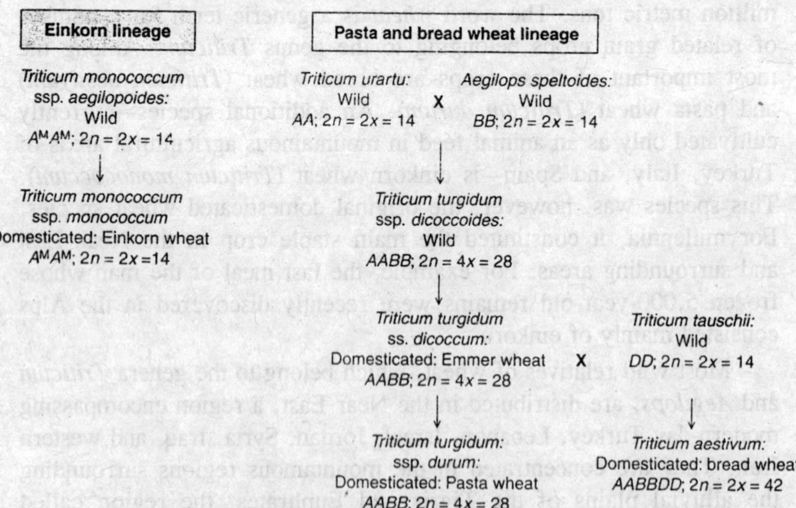

Fig. 9.3. Origin of the three different cultivated wheat species: einkorn wheat, pasta wheat, and bread wheat.

give a wild tetraploid progeny, wild emmer wheat or *Triticum dicoccoides*, with anAABB genome.

Domesticated emmer wheat (*Triticum dicoccum*) appeared some 9,000 years ago in the Fertile Crescent. The grains of this wheat were covered with tightly adhering leaflike structures called *glumes*, hence this primitive wheat is called "hulled-grain wheat" Because these glumes hinder processing the grain after harvest, farmers selected a variant with thin glumes that could easily be separated by threshing after harvest; hence the name "free-threshing" or "naked-grain wheat," also called "pasta wheat" (*Triticum durum*) because its grains are used to make pastas, such as spaghetti or macaroni. Emmer wheat became quite popular and spread within the Fertile Crescent and beyond. When it reached the region south of the Caspian Sea, it crossed with another wild diploid species with a DD genome (*Triticum tauschii*) to yield a hexaploid species with an AABBDD genome, *Triticum aestivum* or "bread wheat."

Triticum tauschii is adapted to the more continental climate of Central Asia, with both colder winters and hotter summers. Bread wheat therefore gained a broader adaptation beyond the Mediterranean climate favoured by its forebear, emmer wheat. This accounts in part for the widespread cultivation of bread wheat around the world. Another reason may be its bread-making qualities. In contrast with tetraploid wheats, bread wheat grains contain sticky proteins that trap the CO_2 gas that originates from yeast fermentation, resulting in leavened bread. In contrast, bread made from tetraploid wheat, such as pita bread from the Middle East, is always unleavened or flat.

Domestication of Rice

The rice most people know is actually of Asian origin and is called *Oryza sativa*. It plays an important role in human nutrition, because half of humankind relies on it for daily intake of calories. World production is around 600 million metric tons on an area of 150 million hectares. China and India are by far the largest producers. The highest per capita level of consumption takes place in Bangladesh, Cambodia, Indonesia, Laos, Myanmar (Burma), Thailand, and Vietnam. In Asia, rice has two wild relatives, the annual *Oryza nivara* and the perennial *Oryza rufipogon*. These two wild species have widely overlapping distributions in subtropical Asia from the Indus valley across Southeast Asia and into China, the Philippines, New Guinea, and northern Australia. In these regions, wild rice grows in flooded sites owing to its unusual ability to transport oxygen from emerged leaves to flooded roots. Most cultivated rice is derived from *Oryza nivara*. Scientists generally think that the first domesticated varieties were paddy or lowland rices or varieties growing in inundated fields, a habitat similar to that of their wild ancestors.

"Paddy" is a Malay word for threshed, unhulled rice and, by extension, for the shallowly flooded field in which rice is generally grown. From these evolved the rain-fed or upland rices that can be grown without irrigation in rainy climates. Deep-water or floating rices, which can have stems of up to 5 meters long, are generally grown in southern Asia and contain more genes from *Oryza rufipogon*. The actual region where Asian rice was domesticated is still uncertain, partly because the two wild relatives are widely distributed in tropical and subtropical Asia, and partly because the archaeological record is sketchy in those areas: The hot and humid climate does not favour long-term conservation of plant remains. Archaeologists have identified the oldest remains of rice in the Yangtze and Yellow River valleys in China. These remains are some 8,000 years old.

A possible clue to the actual domestication pattern in rice comes from the observation that there are two major types of rice cultivars. The *japonica* types, also called *sinica* or *keng* (sticky), have short grains and are generally adapted to more temperate climates. The *indica* types, also called *hsien* (not sticky), have long grains and are generally better suited for more tropical climates. These two groups may have actually resulted from separate domestications in different areas of the distribution of the wild relatives. Alternatively, the two groups could have resulted from a single domestication followed by divergence into indica and japonica types. Both groups contain paddy and upland varieties, but the deep-water rices are only found in the indica group. From its center of origin in Southeast Asia, rice was dispersed over the entire world and eventually into the New World and Africa. In Africa, however, a local species, *Oryza glaberrima*, is grown mainly in western Africa. Farmers domesticated it from a local wild relative, *Oryza barthii*. West Africans grow it especially in remote areas, because it is more resistant to adverse conditions. Recently, because it is lower yielding and tends to drop its seeds, it has been losing ground to its Asian cousin. These multiple domestications of related rice species in different parts of the world are examples of vicarious domestication.

Domestication of Maize and Beans

Maize (from the Arawak Indian name *maize*, also called *corn* in the United States) is a food crop that is grown as such in many countries of the world. More recently, it has been grown for other uses such as animal feed and for industrial uses (such as alcohol production) as well. World production in 2001 was about 600 million metric tons. The major producer was, by far, the United States, followed by Brazil, France, Mexico, India, and Italy. Where did a crop with such a worldwide distribution arise? Several types of evidence, including botanical, archaeological, and folk oral history, point to Mexico and Central America as the center of origin for this crop. Maize was—and still is—the staff of life for people in Mexico and Latin America (as well as Africa and Asia). It is the main source of calories, but has other uses as well, for example, as forage and roofing material.

There is an extensive vocabulary in native languages of Mexico designating maize or its products. The likely progenitor of maize, called *teosinte*, grows in dispersed populations on the western slope of Mexico and Central America, where their distribution parallels that of the ancient Mesoamerican civilizations. They grow in regions receiving summer rains, often close to cultivated maize fields. Their flowering time tends to parallel that of maize, hence crosses between maize and teosinte are possible. Farmers cannot easily rid their maize fields of teosinte until the plants flower, when crossing may already have occurred. Teosinte and its hybrids with maize can therefore become weeds inside maize fields.

A comparison of teosinte and maize DNA shows that teosinte is the closest relative of maize. Domestication of maize probably took place in the mountainous regions of western Mexico. Maize is often cultivated with other crops such as beans and squash. Pole beans use maize as a support for growth, and squash plants cover the ground between maize and bean plants. This traditional way of growing these plants may have predated domestication. In several regions of Mexico, wild bean vines climb on teosinte plants. Therefore, the first farmers not only may have domesticated maize, beans, and squash, but may also have understood how to grow them from the way they grow in natural vegetation. Similar multiple-crop cultivation systems are still widely used in Africa. Although maize, beans, and squash have all been domesticated in Mesoamerica, there are differences as well among the domestication patterns of these crops.

Maize was domesticated only once, in Mesoamerica. Beans, in contrast, were domesticated at least twice. In addition to the Mesoamerican domestication, which eventually gave rise to bean types such as the pinto, pink, and black beans, another domestication in the southern Andes led to kidney beans and green beans. Different species of squash were domesticated in different locations in Mexico and Central and South America. These are further examples of vicarious domestications. Nowadays, maize, beans, and squash are grown worldwide, their distribution spreading since 1492 after the conquest of the Americas. The crop exchanges that followed the voyages led by Christopher Columbus (hence called the Columbian exchange) led to widespread dispersal of crops between the "Old World" (Eurasia and Africa) and the "New World" (the Americas). Old World crops such as wheat and rice became widely grown in the Western Hemisphere, and conversely crops from the Western Hemisphere such as maize, potato, and tomato became a significant part of Old World agriculture.

Genes in Domestication

The cultivated agricultural environment in which crop plants find themselves differs substantially from the natural environment in which the wild relatives of modern crops grew and still grow. The selection pressures the plants experience are very different in these two situations. Here we discuss just a few properties of plants.

Seed Shattering

In a natural environment, plants are responsible for their own propagation. They do so by shedding their seeds spontaneously at maturity, by various mechanisms depending on the plant species. In wild grasses, the rachis (the stem) of the inflorescence becomes brittle at maturity. This lets the seed drop to the soil around the plant. In wild legumes, fibers in the pod wall contract at maturity, causing the pod to open suddenly and like a slingshot propel the seeds around the plant. Imagine that a farmer arriving at a field ready for harvest discovers that half the harvest has already been scattered by the plant's seed dispersal mechanism! Thus a major change during domestication has been the disappearance of natural seed dispersal mechanisms. Domesticated plants keep their seeds on the plant.

Seed Dormancy

Wild plants are subject to a very variable environment compared to cultivated environments, particularly with regard to rainfall. To germinate, seeds need sufficient moisture, usually supplied by rainfall. Plants have developed a mechanism to deal with the erratic timing and amount of rainfall. Seed dormancy prevents premature germination of seeds in a moist environment. By delaying germination, dormancy

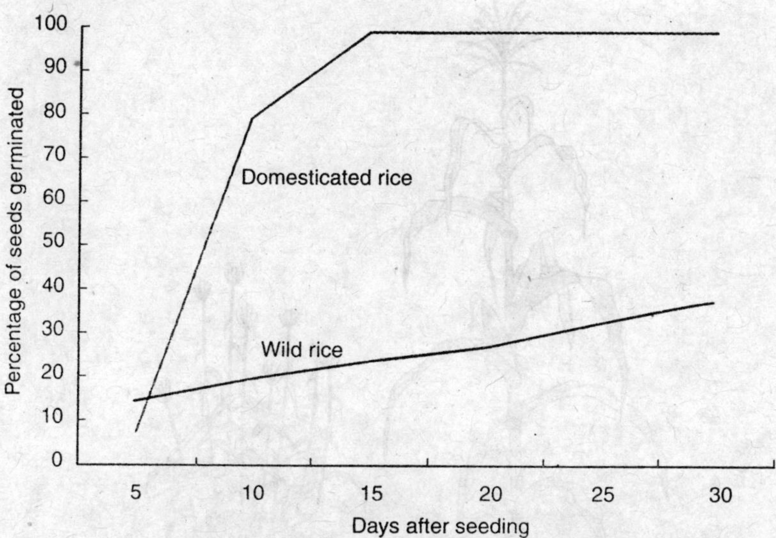

Fig. 9.4. Germination of domesticated rice and wild rice.

gives seeds a chance to germinate in a moister, more favourable environment, whether in the year immediately after seed dispersal or in later years. Dormancy lets wild plants accumulate a seed bank in the soil that allows such species to survive temporary, unfavourable conditions. A major goal of farming is to provide a more predictable and less competitive environment for crop plants (to the extent possible given the vagaries of year-to-year weather variability). Farmers plant their seeds in soils that have accumulated sufficient moisture following the first rains. To achieve a dense and regular stand of plants, most seeds must germinate more or less simultaneously and within a few days after planting. Thus domesticated plants have lost seed dormancy. Their seeds germinate readily when located in moist, sufficiently warm soils.

Growth Habit and Harvest Index

Wild plants also confront strong competition for light, water, and nutrients from other plants, whether they belong to the same or a different species. To stave off this competition, at least in part, plants have developed special growth habits. The viney, climbing growth habit of some plants lets them grow on top of other plants and thus access light. Other plants have a very branched growth habit that allows them to shade out neighbouring plants. Regardless of the growth habit, vegetative development (stems and leaves) is as important as actual

Fig. 9.5. Growth habit of maize and teosinte.

seed production for survival of the species. Wild plants are therefore said to have a low harvest index, defined as the weight of the harvested part (usually seeds) divided by the sum of the total biomass, including the harvested and nonharvested aboveground parts of a plant. Typically, the harvest index for wild plants is around 0.2 to 0.3. In other words, a plant that weighs 1 kilogram (after drying) carries 200-300 grams of seed. Modern varieties have a harvest index around 0.4 to 0.5. To avoid excessive competition among domesticated plants, these usually have a more compact growth habit with fewer and shorter branches.

Many crops with a wild relative growing as a vine have acquired an upright, nonclimbing growth habit, more like a bush. For example, the ancestor of the common bean was a vine, and wild relatives of the common bean still have a twining growth habit. Most commercial production of common beans is from "bush bean." But pole beans are still widely grown in gardens all over the world. Highly branched plants have seen a reduction in the number and length of their branches, as illustrated by maize and its wild relative teosinte.

Adaptation to Photoperiod

The timing of flowering and seed production is also of great importance for wild plants. Many plants live in an environment that favours growth for only part of the year. This is the warm season in temperate regions or the humid season in subtropical areas. It is therefore important to schedule flowering so that seed development does not unduly extend into the unfavourable season. In addition, the dispersal mechanism of many seed plants relies on the drying phase associated with maturation to operate. This scheduling is achieved by timing flowering as a function of the length of the night. Some plants—called *short-day plants*—will flower when the night exceeds a certain number of hours. These are usually plants of tropical origin, where the length of day and night are more similar. Other plants—called *long-day plants*—will flower only in short nights of 8-10 hours, occurring in temperate, higher-latitude locations.

Gigantism and Diversity of Harvested Organs

The harvested organs of crop plants range from fruits such as apples or strawberries; seeds such as those of beans, soybean, rice, or maize; to leaves such as those of lettuce or spinach. Also, as Charles Darwin pointed out, the harvested parts of crop plants have acquired much broader diversity in shape and colour, making them less camouflaged than their wild counterparts, as shown by the seeds or fruits of maize, beans, and squash. A part of these changes is driven by selection for novelty by farmers and consumers alike.

Resistance to Diseases and Pests and the Presence of Toxic Compounds

Wild plants are subject to the constant attacks of diseases and pests. Unlike animals, they cannot escape by moving away from their would-be attackers. Plants have, therefore, established a whole set of defense mechanisms that camouflage or make them unpalatable to attackers. These include spines, hairs, toxic subtances, and usually tan to dark seed colours. Examples of toxic compounds are morphine in poppies, digitalis from foxglove, cyanic-add-producing compounds in cassava and lima bean, and glucosinolates in some plants of the cabbage family. The presence of these compounds has not prevented humans from domesticating some of these plants. People have adapted by selecting varieties with a reduced level or complete absence of the compound. Alternatively, they have developed processing techniques to remove the compounds altogether. Wild cassava contains high levels of chemicals that produce hydrogen cyanide.

Domesticated varieties of this crop can be divided into two groups: sweet varieties that contain very low levels of these compounds and can be eaten after minimal processing such as boiling and drying, and bitter varieties that contain much higher levels of these compounds. The latter can only be consumed after more extensive detoxification. The traits distinguishing crops and their wild ancestors are quite similar in all crops; together, these traits are called the *domestication syndrome*. The change from the wild to the domesticated version of these traits—for example, from seed dispersal in the wild to seed retention in the domesticated state—is the result of a selection pressure resulting from human cultivation. This selection took place either unconsciously or consciously on spontaneous mutations in genes controlling the domestication syndrome in populations of wild relatives as they were being cultivated by the first farmers. Because of positive selection, the frequency of these mutations gradually increased until it reached 100%, at which point the crop could be considered fully domesticated.

Domestication is therefore a selection process that results in genetic changes in plants and confers adaptation to a cultivated environment and acceptance by farmers and consumers. Research in maize, bean, rice, tomato, and pearl millet has shown that the inheritance of these traits is usually simple and that only a few genes are involved despite the large phenotypic differences between crops and their wild relatives. These genes generally have a major effect on the phenotype, and their expression tends to be relatively independent of environmental influences. The genes appear to be concentrated in a few places on the chromosomes (loci) in the genome of domesticated plants. This relatively simple inheritance suggests that domestication could have happened fairly quickly over a time span of a few decades to a few centuries.

The actual time it took to domesticate plants was probably longer, up to a millennium. It is likely that in the initial years of the transition to agriculture, plants were cultivated only occasionally. This decreased the selection pressure on plants and hence the speed of conversion to a domesticated type. In addition, a limiting fracture may well have been the combination of different domestication traits into a single plant through crosses within the cultivated fields. Nevertheless, the end result was a number of crops (and farm animals) that depended on humans for survival. Without planting and harvesting by humans, fully domesticated crop plants would most likely face extinction. Conversely, humans nowadays would be hard pressed to derive enough food by hunting and gathering. Only crops and farm animals can assure a sufficient production of food for humanity, currently and in the foreseeable future.

Genetic Diversity

The preservation of biodiversity in the world's natural ecosystems has become of major concern to humankind, and this concern extends to the preservation of crop biodiversity. The genetic diversity of most modern crops is often limited compared to their wild ancestors. This reduction in diversity during crop evolution is by no means recent but began with the initial cultivation of plants. Selecting certain plants with desirable traits such as absence of seed dispersal, a more compact growth habit, and features increasing attractiveness to consumers reduced genetic diversity during the domestication process itself. After domestication, crops were disseminated from their center of origin into other regions of the world. For many crops, these new cultivation areas had a very different environment from that of the domestication center.

Most modern crops originated in tropical or subtropical areas but are now grown also in markedly temperate areas. This dispersal generally involved small samples of seeds or planting material, further reducing the genetic diversity of crops as they spread around the globe. In this new environment, farmers subjected crops to a second round of selection that adapted them to these new conditions.

Landraces

Crops grown in the original centers of domestication or elsewhere, especially in developing countries, did not result from scientific plant breeding until recently, hence they are called *landraces* or local varieties. Farmers recognize individual landraces based on their morphological traits, such as seed colour or size or tuber shape. These landraces have usually been grown for many centuries by local communities and may therefore have been selected for adaptation to local conditions or tastes. They are generally believed to be better adapted to unfavourable or stress conditions such as infertile soils or drought conditions than cultivars resulting from breeding programs. In addition, farmers have selected landraces for specific uses. For example, farmers in the neighbourhood of Mexico City grow a maize variety that produces husks (surrounding the ear of the maize plant), which can be sold as a food-wrapping material on the markets. Indians of the Bolivian highlands use special potatoes (*Solanum x juzepczukii:* triploid: $2n = 3x = 36$, S. x *curtilobum:* $2n = 5x = 60$) to prepare a freeze-dried food called *chunos*.

The preparation takes special advantage of the particular environmental conditions prevailing where they live, namely the dry mountain air in that part of the Andes and the often freezing nights. Thus, landraces are characteristic of the marginal environments where they originated. Such environments are often environmentally, economically, and culturally heterogeneous. This heterogeneity is reflected in the genetic heterogeneity of the landraces. The heterogeneity of the landraces may be one reason why farmers continue to grow them even after the initiation of breeding programs to develop new cultivars.

Genetic heterogeneity buffers the effect of the spread of diseases and the year-to-year variability and even unpredictability of the weather. Their adaptation to local unfavourable conditions and to special consumer uses also explains their survival to this day. Nevertheless, as communications become easier, rapid changes in economic and social conditions of farmers and progressive reduction in cultural diversity threaten the continued existence of landraces. Proposals have recently been made to reward farmers for their conservation of landraces. However, the practical implementation of this proposal is a major stumbling block. In addition to their use by local farmers, landraces are also used by plant breeders as a source of interesting traits they can introduce into modern varieties. Examples of such traits are the tolerance of Central American bean landraces to soils low in phosphorus, the high protein content of two Ethiopian barley landraces, and the drought tolerance (and thicker, deeper roots) in upland (rain-fed) rice from Africa, South America, Bangladesh, and Ethiopia.

Loss of Genetic Diversity

A third bottleneck developed during the 20th century. Scientific plant breeding, for reasons of practicality, focused on an ever-decreasing number of so-called elite varieties. These elite materials are ones shown in previous years to have superior characteristics such as higher yields, disease and pest resistance, and high quality traits such as bread-making ability in wheat and long and strong fibers in cotton. This progressive reduction in genetic diversity is well illustrated by the situation in common bean. As indicated earlier, common bean was domesticated at least twice. All common bean cultivars therefore derive from one of two distinct lineages. One lineage originated in the wild beans growing in Mesoamerica, most likely Mexico, and the other from the wild bean populations in the southern Andes, somewhere between southern Peru and northwestern Argentina.

The genetic diversity in both lineages was reduced, first by domestication and other events such as crop failures after domestication. In a second step, dissemination of beans to other regions of the world such as North America and Europe as well as the development of new bean cultivars further reduced genetic diversity. Because both the industry and the public require uniformity, a limited number of varieties in each crop now occupy a significant proportion of the land devoted to that crop.

Although high genetic diversity is not a precondition for high yields, reduced genetic diversity in crops carries some risks that have been dramatically exposed in recent history but that may have existed since domestication. The best known incident is the Irish potato famine, which started in 1845 and lasted for several years. Similar epidemics affecting genetically uniform crops have occurred in the past, such as the leaf rust epidemic of coffee on the island of Sri Lanka in the early 1870s and the corn leaf blight that struck the United States in 1970. The coffee plantations on Sri Lanka, destroyed by leaf rust, were replaced by tea plantations, which may be one reason why the English became enthusiastic tea drinkers. The 1970 epidemic of Southern corn leaf blight prompted plant breeders to increase the genetic diversity of their varieties.

Table 9.2. The diversity of our crops is limited because of cultivation of a few varieties over a large area

Crop	*Total Number of Varieties*	*Major Varieties* Number	*Major Varieties* Acreage(%)
Beans, dry	25	2	60
Beans, snap	70	3	76
Cotton	50	3	76
Maize	197	6	71
Peanut	15	9	95
Soybean	62	6	56
Wheat	269	9	50

Genetic Erosion

The loss of crop plant biodiversity, also called *genetic erosion*, affects all components of the genetic resources of crop plants: the wild progenitors, landraces, obsolete cultivars, advanced breeding lines, and elite cultivars. For example, wild progenitors suffer from a loss of habitat; landraces are lost, replaced by modern varieties and by market demands for specific types that are only a subset of what a farmer grows, and modern varieties have a narrow genetic base because plant-breeding programs have limited resources. Genetic erosion has existed since the origins of agriculture and throughout the different phases of crop evolution. Recently, however, the pace of loss has accelerated.

To a large extent, the ultimate cause of this recent genetic erosion is the explosive human population growth in the last decades. This growth has fueled the need for ever higher levels of food production in response to increased demand. As a consequence, deforestation has spread wider than ever before to make room for additional fields and pastures. Excessive grazing by domestic animals also destroys the natural habitats of wild relatives of crops. The introduction of higher-yielding varieties displaces existing farmers' varieties, and until recently, plant-breeding programs focused on an ever-shrinking genetic base to develop new cultivars. The international community's response to genetic erosion of crops has been the creation of both off-site (*ex situ*) and on-site (*in situ*) conservation programs. *Ex situ* programs typically include gene banks, which include cold storage facilities, where the seeds of hundreds of thousands of different lines are kept, and a kind of botanic garden where live specimens of trees (for example, fruit trees such as apple or avocado) are grown and propagated.

In addition, thousands of varieties of certain plants such as cassava and potato that cannot be stored below freezing or in botanic gardens are kept in sterile culture flasks and propagated regularly. *In situ* programs seek to maintain landraces or their wild relatives where they grow originally, either on the farm or in native vegetation, An example of an *in situ* program is the Biosphere Reserve of the Sierra de Manantlan in western Mexico. In this reserve, maize and beans are grown in close proximity to their respective wild relatives.

Crop Hybridization

Plants have a full range of reproductive systems from cleistogamy (self-fertilization within a closed flower bud) to obligate outcrossing, depending on the species. Even cleistogamous and other highly self-fertilized plants do cross at least occasionally with other plants, depending on environmental conditions and genetic factors. For example, spontaneous mutations can cause pollen sterility, which sharply increases outcrossing because all the progeny result from fertilization with pollen from other, male-fertile plants. Large populations of insect pollinators also increase outcrossing. On the whole, plants are much more promiscuous than animals. Like their wild counterparts, crops are also subject to outcrossing. This is especially true when crops and their wild relatives belong to the same species. A species is defined as the sum total of all individuals that can interbreed, whether actually, as in outcrossing species, or potentially, as in self-fertilized species. Plant breeders have introduced the concept of a gene pool to describe the ease with which these crosses can take place.

The first gene pool (GP1) contains the crop and its wild progenitor. Sexual crosses within this gene pool, whether among cultivars or between cultivars and their wild progenitors, are easy to make and yield viable and fertile progenies. The boundaries of this gene pool correspond to the concept of biological species. The second gene pool (GP2) contains species that can be crossed with GPI but with a lower success rate. Progenies resulting from this cross have somewhat impaired viability and fertility. The third gene pool (GP3) contains species that can only be crossed with difficulty with GP1. Often special techniques such as tissue culture and embryo rescue must be used to secure progeny for the cross.

The progeny have limited viability and fertility. The fourth gene pool (GP4) contains species that cannot be crossed with GP1. These contain plant species as well as species of other kingdoms. Although genes cannot be transferred by the traditional sexual crosses, they can be introduced by laboratory techniques such as Agrobacterium-mediated transformation. With these techniques, GP4 becomes an additional source of genetic diversity for improving crops. In addition, information from species in GP4, especially model species such as *Arabidopsis thaliana* and rice, can help breeders identify genes to improve crops.

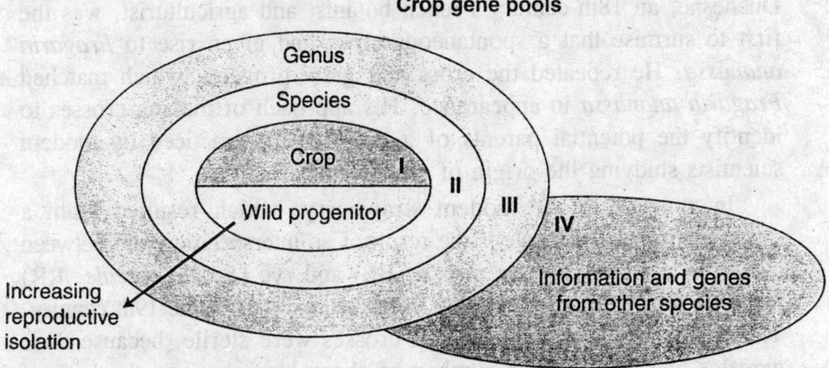

Fig. 9.6. The gene pool concept as used by plant breeders.

Hybridizations between and within species are important in shaping the genetic diversity of plants in general and of crops in particular. Hybridizations in crops take place between varieties, between varieties and their wild progenitors, and with other species altogether, whether they are wild or domesticated. Although crops generally tend to adopt a more pronounced selfing reproductive system during and after domestication, outcrossing remains occasionally important and can produce bursts of additional genetic diversity. In selfing crops, even low levels of hybridization can lead to genetic changes over several generations. Hybridizations can also lead to new crops. We have already discussed the origins of wheat and the role that interspecific hybridization played in the evolution of hexaploid wheat. Another example is found in the bean family where the year bean (*Phaseolus polyanthus*) resulted from a cross between an ancestor of common bean (*Phaseolus vulgaris*) as a female parent and runner bean (*Phaseolus coccineus*) as a male parent followed by backcrossing to the runner bean parent. Other, more recent examples of hybridization leading to new crops include strawberry and triticale.

The modem strawberry varieties sold in most stores and markets, with their large fruits, belong to the species *Fragaria ananassa*, which is an octoploid species with 56 chromosomes. This species is the result of a cross between two other octoploid species, *Fragaria chiloensis* (a species growing on the Pacific coast of North America) and *Fragaria virginiana* (a species growing in the eastern half of North America). Surprisingly, the cross between these two species did not take place in North America but in 18th-century France, where the two parental species had been introduced as agricultural curiosities. Antoine Duchesne, an 18th-century French botanist and agriculturist, was the first to surmise that a spontaneous cross had given rise to *Fragaria ananassa*. He repeated the cross and grew progeny, which matched *Fragaria ananassa* in appearance. His approach of making crosses to identify the potential parents of a crop is still practiced by modem scientists studying the origin of crop plants.

In contrast to the modem strawberry, which resulted from a spontaneous cross, triticale is the result of an interspecific cross between durum wheat (*Triticum durum*, AABB) and rye (*Secale cereale*, RR) made by plant geneticists starting in the second half of the 19th century. The ABR hybrids from the initial crosses were sterile, because their gametes had an irregular number of chromosomes given the lack of pairing of chromosomes during meiosis. However, doubling of the chromosome number of the hybrids led to fertile progeny (AABBRR). Breeders capitalized on these fertile hybrids to make further crosses and selections that have yielded today's triticale varieties. These combine the higher yield of wheat with the vigor, winter hardiness, adaptation to acid soils, and higher seed protein content of rye. Most triticale production takes place in Eastern Europe. The crop is used primarily as animal feed (seeds) and forage (young plants).

When a crop and its wild progenitor can be easily crossed with each other, hybrid populations between the two can often be identified in the crop's center of origin. These hybrid swarms can be of varying sizes and longevities. Some are short-lived, whereas others persist for several years. Normally, they are not expected to last indefinitely, because their hybrid status renders them ill adapted to either the cultivated or the natural environment. However, backcrossing the hybrids to either parent can increase adaptation to either environment. For example, hybrids can acquire the same time to maturity and the same morphology as the domesticated parent yet still conserve the ability to disperse seeds. Before harvest, it is very difficult to distinguish the hybrid from the crop, especially before flowering, because there are few distinguishing characteristics in leaves and stems. During harvest, seeds of the hybrid mature but drop to the ground, where they remain until the next growing cycle. They then germinate and grow as a weed in whatever crops are grown in the next cycle. It is by this mechanism of hybridization and backcrossing that some of the most troublesome weeds have appeared. Examples are shattercane, a weedy sorghum, and red rice, a weedy rice. Both are distributed not only in the crop center of origin (Africa and Asia, respectively) but through the seed trade have been distributed to other continents as well, including North America.

Polyploids and Crops

Many wild plants are diploid—just like all animals—and contain two copies of each gene and of each chromosome in their cells. However, plants seem to be more flexible than animals in their genomic content: There are also many polyploid species. These contain more than two genomes in each cell: species containing four or six genomes are called *tetraploid* or *hexaploid*, respectively. Polyploids probably arise from spontaneous chromosome doubling in the diploid parent plant or from the union of "unreduced" gametes, which, through anomalies in the meiotic process, contain twice the usual number of chromosomes. There are two basic polyploid types. In autopolyploids, the same basic set of chromosomes is repeated. As a consequence, chromosomes of one set can pair with comparable chromosomes of the other sets. Examples of autopolyploids are potato and alfalfa.

Allopolyploids have sets of chromosomes that differ somewhat from each other and the chromosomes of one set usually do not pair with the chromosomes of the other sets. Wheat, cotton, tobacco, peanut, canola, coffee, and plantain (a starchy type of banana) are examples of allopolyploid species. Following the hybridization event that gives rise to the allopolyploid, the different chromosome sets gradually diverge. For ancient polyploidization events, the divergence is so pronounced that it becomes difficult to recognize duplicated chromosome sets. Only recently has evidence for their existence been obtained from DNA information. Examples of ancient allotetraploids are maize and soybean.

Polyploidy is a frequent phenomenon in plants, caused in part by the promiscuity of plants, already mentioned. Whereas hybridization may be a necessary condition for the formation of allopolyploids, it is by no means a sufficient condition for their survival, and other reasons have been proposed. For crops particularly, the larger size of polyploids compared to diploids is an advantage, because selection during domestication has generally favoured larger harvested organs such as seeds. The combination of different genomes, each conferring adaptation to a specific environment, may have broadened the adaptation of polyploids and ensured their successful dispersal outside of their immediate centers of domestication.

An additional advantage of polyploids resides in the existence of multiple copies of the same gene. Each of these copies can then diverge and acquire slightly different functions. Also, new interactions can appear between genes belonging to the different genomes, which in turn lead to the appearance of novel, unprecedented traits or the enhanced expression of traits existing before the polyploidization event. An example of the former is the ability to make leavened bread from bread wheat (*Triticum aestivum*, an allohexaploid with 2n = 6x = 42 chromosomes; AABBDD genome) but not from its progenitors, the domesticated allotetraploid wheat *Triticum durum* (2n = 4x = 28; AABB genome) and the wild diploid species *Triticum tauschii* (2n = 2x = 14; DD genome). Somehow interactions between genes of the A, B, and D genomes led to biochemical changes in the seed proteins of bread wheat that are responsible for dough rising prior to baking.

Diploid and tetraploid wheats are strictly used for making unleavened bread. Cotton is an example of increased expression of traits. Most cotton production comes from two allotetraploid domesticated species, *Gossypium hirsutum* or upland cotton and *Gossypium barbadense* or Sea Island or pima cotton. Both species originated in the Western Hemisphere. Upland cotton may have been domesticated in the Yucatan peninsula of Mexico, and pima cotton was domesticated in South America. The two species have 2n = 2x = 52 chromosomes and result from the cross between a maternal A genome and a paternal D genome ancestor some 1 to 2 million years ago, well before the advent of modern humans and the initiation of agriculture. Strangely enough, the A genome is actually native to the Old World and must somehow have been transported to the New World.

Two related, diploid A genome species in the Old World, *Gossypium arboreum* in Asia and *Gossypium herbaceum* in Africa, produce fibers that are spinnable, that is, twine sufficiently so that individual fibers can be combined to make thread that can be woven into ropes or cloth. The diploid species have fiber yields that are 20-30% lower than their tetraploid relatives, which explains why they are no longer widely grown except in certain parts of the world such as India. The donor of the D genome has seeds covered with fuzzy fibers that cannot be spun. Yet genetic analyses show that the D genome in tetraploids provided genes that increase fiber production and quality. Again, scientists suggest that interactions between genes of the A and D genomes explain this observation.

Various species belonging to the genus *Brassica*, which includes cabbages, turnip, and rapeseed, form an interesting network of genetic relationships, which are illustrated by the Triangle of U, named after the Japanese agronomist N. U, who deciphered these relationships. The tips of the triangle contain three diploid species with differentiated genomes, A, B, and C. These include *Brassica campestris* (turnip, Chinese cabbage, and the oilseed crop turnip rape), *Brassica nigra* (black mustard), and *Brassica oleracea* (different cabbages, kale, Brussels sprout, broccoli, and cauliflower). Along the three sides are the respective allopolyploids with the combination of genomes of the diploid species. Most important among these is *Brassica napus*, which includes rapeseed or canola, and is grown in cooler climates and is an important source of oil.

The various examples presented here clearly illustrate the important role polyploidization has played in the development of some of the world's most important crops. Polyploidization can lead to an increase in size of harvested organs, it can broaden the adaptation of crops to a wider set of environmental conditions, it can lead to the formation of new crops, and it can result in novel gene interactions that result in new traits.

Crop Genomes

Until a few years ago, determining the nucleotide sequence of a piece of DNA was a cumbersome activity. Recent technical progress, however, has made it possible to sequence large stretches of DNA, including entire eucaryotic genomes, in a short period of time. The first complete genome sequences of two plant species, the small weed *Arabidopsis thaliana*, and two varieties (Indica and Japonica) of rice were obtained in 2001 and 2002, respectively. The DNA sequences of the genomes of other crops will undoubtedly be deciphered in the near future. These sequences constitute an additional genetic resource because they will provide information on how plant in general and crops in particular function. They will tell scientists how specific landraces or cultivars or their wild relatives resist diseases, tolerate insects, survive drought spells, or have certain nutritional characteristics. In turn, this information will help plant breeders develop better crop cultivars. A comparison of DNA sequences between wild progenitors and their domesticated descendants will also allow a more accurate determination of the domestication history of the crop.

One of the major differences between maize and its progenitor teosinte is the branching of the plant. Maize has a single stem, ending in a tassel on which the male flowers are located. A markedly shortened side branch ends in an ear where female flowers and therefore seeds are located. Teosinte, in contrast, has numerous side branches with tassels at their tips and ears as side branches. This major difference in appearance between maize and teosinte is controlled—surprisingly—by a single gene, *tb-1* or *teosinte branched 1*. A comparison of the sequences of *tb-1* and the DNA region immediately upstream of the gene has yielded interesting information: It confirms that domestication has induced a genetic bottleneck in the species. However, this bottleneck is only apparent in the upstream, regulatory region and not in the gene itself. This observation suggests that selection operating during domestication affected primarily the regulatory region. Whether this pattern of selection can be generalized to other domestication genes or genes in general during evolution, remains to be determined. The *tb-1* sequences of maize are very similar to each other and fell on the same branch of a genealogical tree. In addition, the *tb-1* sequences of teosinte that resemble most closely those of maize are those of populations from the Balsas River basin in western Mexico, confirming earlier evidence. It is remarkable that an area at a relatively short distance away in western Mexico has been proposed as domestication area for common bean. This suggests that early farmers may have domesticated the two crops (and other native crops) together.

Further information from DNA sequences of other maize genes suggests that domestication could have involved a small population of individuals and could have been completed over a short period of time spanning anywhere between 300 and 1,000 years. However, no archaeological remains are yet available to independently verify this conclusion.

As explained earlier, crop plants typically have lost the ability to disperse their seeds. However, at plant maturity many crops still display a partial level of seed dispersal, which can obviously lead to losses at harvest time. A better understanding of the genes involved in seed dispersal will help breeders develop cultivars that lack this trait. Recently, researchers have identified two genes in *Arabidopsis thaliana* that cause explosive dehiscence (opening) of the fruit and, hence, dispersal of the seeds contained in it. The fruit of *Arabidopsis* resembles a pea pod and consists of two halves that are joined at the edges, referred to as valves. Two valves, one on each edge of the fruit half, are joined by a dehiscence zone. It is along this zone that the two halves separate after the fruit matures and dries. This explosive separation causes seed dispersal. The two genes identified in *Arabidopsis* are called *SHP1* and *SHP2* (*SHP* for "shatterproof").

In their active state, these genes are responsible for the development of fruit valve margins. The presence of a single active gene is sufficient to cause seed dispersal. Loss of function of the two genes blocks development of the dehiscence zone and deposition of lignin next to this area. As a consequence, the fruits remain closed and the seeds cannot be dispersed. *Arabidopsis* is a close relative of cabbage. In cabbage species that are grown for their oil-rich seeds, such as canola or rapeseed, premature opening of the fruits reduces crop yield. Learning the *SHP* gene sequences is a first step for canola breeders to isolate similar sequences in canola and develop improved cultivars without fruit shattering.

This last example shows how domestication of plants is an ongoing process. Although plants (and animals) were first domesticated some 10,000 years ago, plant breeders and farmers continue selecting and developing new varieties that satisfy public demand for higher-yielding crops with stronger resistance to prevailing diseases and pests, and with improved nutrition, taste, and other useful traits. Our ability to find individual genes and use these for crop improvement by relying on plant breeding, including genetic engineering, has raised the question about the ownership of genes.

10

SAFETY EVALUATION

Throughout history, plant breeders have sought to be genetically modify food crops to improve yield and increase resistance to disease and plant pests. Initially these improvements were achieved by selecting seed from superior plants and reproducing these with continual selection and breeding. Traditional breeding methods have increased corn and wheat yields by approximately 100% over the last half century. However, traditional plant breeding methods are slow and unpredictable. To introduce a desired gene or set of genes by conventional breeding methods require a sexual cross between parental lines followed by repeated backcrossing between the hybrid offspring and one of the parents until progeny with the desired characteristics are obtained. Genes are only accessible from plants that can be sexually crossed, and many genes besides the desired gene(s) will be transferred. Biotechnology provides an opportunity to overcome some of the limitations of traditional breeding by enabling plant geneticists to identify and clone specific genes encoding desirable traits, such as protection against insect pests, and to introduce these genes selectively into already useful varieties of plants.

Sexual compatibility is no longer a limiting factor for transfer of desired traits. The transformation process is faster and more efficient because successfully transformed plants can readily be identified. Numerous traits are being assessed for their potential to yield products with the ability to (1) protect plants against various insect, fungal, and viral pests and plant pathogens; (2) provide selectivity to preferred herbicides; (3) improve agronomic performance such as crop yields; (4) increase nutritional value of food for humans and farm animals; (5) reduce naturally occurring toxicants, antinutrients, or allergens; (6)

modify the ripening process to improve the flavor of fruits an vegetables; (7) use plants as factories to make environmentally friendly biodegradable polymers for packaging materials; and (8) use plants to produce pharmaceutical products more cost effectively, and so on. Since the initial reports of the first genetically modified plants 15 years ago, nearly all agronomically important crops have been genetically modified.

By the end of 1995, more than 70 different crop species had been transformed. At least 56 different crops have been planted in at least 34 countries and grown in more than 15,000 individual field sites. According to the U.S. Animal and plant Health inspection Service (APHIS), the most frequently field-tested traits are insect protection, virus protection, fungal resistance, herbicide tolerance, and food quality enhancements. The number of transformed crops and field trials will continue to expand as more new crop varieties are developed through biotechnology for eventual entry to the marketplace. At least 34 genetically modified plants have successfully completed regulatory review by the appropriate regulatory agencies. Genetically modified plants were grown commercially in 1996 on approximately 7 million acres in various world areas with approximately 30 million acres planted in 1997. Biotechnology provides plant breeders the opportunity to develop new varieties of food crops more efficiently and with greater potential benefit than has been possible with conventional breeding practices.

As with any technological innovation, there must be assurance that the technology will deliver food "as safe as" that developed through traditional breeding programs. This chapter addresses the safety assessment strategies for new varieties of food products developed through biotechnology. The approaches are consistent with the guidance developed by various international organizations such as OECD (1993, 1996, 1997), FAO/WHO (1996), and WHO (1991, 1995). An example of the application of these strategies to assess the safety of a genetically modified food plant. To understand how safety assessment approaches have evolved, it is instructive to review briefly how new traits are introduced into food crops through conventional breeding compared with biotechnology.

Introduction of Genes into Food Crops: Conventional Breeding Compared With Biotechnology

Conventional Breeding

Genetic modification of plants has been practiced for hundreds of years with considerable success by plant breeders. Plant breeding has

Table 10.1. Examples of plant biotechnology products that have successfully completed regulatory review in at least one country

Company	*Genetic Trait*
AgrEvo Canada, Inc.	Glufosinate-tolerant canola
	Glufosinate-tolerant corn
	Glufosinate-tolerant soybean
Agritope, Inc.	Modified fruit-ripening tomato
Asgrow Seed Co.	Virus-resistant squash I
	Virus-resistant squash II
Bejo-Baden	Male sterility/glufosinate-tolerant chicory
Calgene, Inc.	Flavr Savr™ tomato
	Bromoxynil-tolerant cotton
	Laurate canola
China	Virus-resistant tomato
Ciba Seeds	Insect-protected corn
Cornell U./U. of Hawaii	Virus-resistant papaya
DeKalb Genetics Corp.	Glufosinate-tolerant corn
	Insect-protected corn
DNA Plant Technology	Improved ripening tomato
DuPont	Sulfonylurea-tolerant cotton
	High-oleic-acid-soybean
Florigene	Carnations with increased vasa life
	Carnations with modified flower colour
Monsanto	Improved ripening tomato
	Insect-protected potato
	Insect-protected cotton
	Glyphosate-tolerant cotton
	Glyphosate-tolerant canola
	Insect-protected corn
	Glyphosate-tolerant corn
Mycogen	Insect-protected corn
Northrup King	Insect-protected corn
Plant Genetic Systems	Male sterile oilseed rape
	Male sterility/glufosinate-tolerant corn
University of Saskatchewan	Sulfonylurea-tolerant flax
Zeneca/Petoseed	Improved ripening tomato

become a very sophisticated branch of applied genetics. Breeders have developed elegant procedures for crossing plants to introduce and maintain desirable traits such as increased yield and resistance to disease. With conventional breeding, cultivars highly adapted to cultivation can be improved by crossing them with closely related highly adapted cultivars to combine the desired features of the parents. If the parents are highly adapted to cultivation, their progeny tend to be highly adapted also. However, if a desirable traits is not available in highly adapted cultivars, the breeder will cross-adapt cultivars with cultivars from different geographic areas, more primitive varieties, or wild species.

The greater the diversity in genetic material of the parents, the greater the chance that undesirable characteristics (genes) will be introduced into the progeny. The progeny may be an "off-type", which means it has less desirable agronomic properties, such as stunted growth or poor yield, than parent cultivars. Eliminating plants with undesirable traits while retaining plants with desired features may require many backcrosses carried out over several generations. The greater the difference in genetic content of the parents, the more difficult sexual crossing can be. In vitro procedures such as embryo culture and protoplast fusion have made it possible to cross parents that may not be sexually compatible.

Biotechnology

Biotechnology methods do not replace conventional breeding practices but can facilitate the introduction of desirable traits into the plant genome more efficiently and with greater precision. Unlike traditional breeding in which thousands of genes maybe introduced into progeny form their parents, only one or a few genes are typically introduced using biotechnological techniques. The source of the desired trait the breeder wishes to introduce into new crop varieties is no longer limited to those from sexually compatible species. This greatly expands the opportunity to introduce new traits to improve crop varieties.

The techniques used to introduce new traits into food crops vary depending on the kind of plant being transformed. Seed plants have been divided into two subclasses; monocotyledonous plants (monocots), whose seeds have a single cotyledon (meaning "seed leaf"), and dicotyledonous plants (dicots) or those with two cotyledons. Dicots (broadleaf plants like tomato, cotton, and soybean) were the first seed plants to have genes introduced via biotechnology. The first genetically modified plants were produced in 1982 using *Agrobacterium tumefaciens*, a bacterium that can transfer a portion of its own DNA (T-DNA) into the genome of plants. The disease-causing sequences from the

Agrobacterium T-DNA have been deleted followed by insertion of DNA that encodes for a desired trait such as insect protection. Regulatory signals are added to enable the gene to function optimally in the plant. Border sequences in the plasmid delineate the desired genes that will typically be inserted in the plant genome.

The T-DNA is incorporated into a plasmid derived from *Escherichia coli*, which is transferred to *A. tumefaciens* via a conjugation process. *A. tumefaciens* containing the engineered plasmid is incubated with selected tissue from the host plant, and the desired genes are stably inserted into plant chromosome. The inserted genes can then be transferred to new plant varieties using traditional breeding methods. Monocots, which are represented by agronomically important crops such as wheat, rice, and corn, were not initially amenable to transformation with *A. tumefaciens*. Recently, more aggressive strains of *A. tumefaciens* with broader host ranges have been developed that have been used to introduce genes into some cereal crops. In addition, techniques such as protoplast transformation and particle bombardment have been used to transfer DNA directly into cells where it is stably inserted into the plant genome. These techniques have been particularly valuable in monocot plants for which the *Agrobacterium* transformation method was not effective or efficient.

In particle bombardment, very small metal beads (1-μ diameter) are coated with DAN and shot form a "gun" into the target monocot cell. Some of the plant cells will incorporate the desired gene(s) into their genome. Whether *Agrobacterium* transformation or particle bombardment is used to introduce genes into plat cells, only a small percentage of eligible plant cells will be successfully transformed. To identify the transformed cells in culture, genes for selectable markers have been included with the genetic information inserted into plant cells. The marker genes provide resistance to antibiotics, herbicides, and other substances added to the cell culture to inhibit the growth of nontransformed cells.

Plant cells that survive have been successfully transformed and can therefore be identified and regenerated into whole plants. As with conventional breeding, off-types are typically discarded. Progenies with normal agronomic properties that express the desired phenotype or trait are backcrossed with commercial crop varieties to generate seed bearing the new trait. The chances of success in identifying a progeny with the desired trait are much higher with recombinant DNA techniques because the breeder knows that the trait was successfully inserted into the plant genome. With conventional breeding, many generations of backcrossing may be required to identify a progeny with the desired trait.

History of Safety Assessment for New Crop Varieties Developed by Conventional Breeding

During this century, our food supply has steadily improved in quality, variety, nutritional values, safety, and economy through the use of conventional breeding techniques to improve food crops. The history of safe use of new varieties of food crops developed by classical breeding techniques is based on several factors: (1) confidence and experience with the procedures used to generate new crop varieties; (2) knowledge of the composition of the food crop, including important nutrients and toxicants, if present; and (3) observation of the agronomic properties of new crop varieties that have been developed during this century via traditional breeding, only a very limited number of new varieties have presented safety concerns. The more well-known examples are (i) increased psoralens in certain varieties of celery that caused photo-dermatitis in food handlers; (ii) increased glycoalkaloid content in the Lenape variety of potatoes (glycoalkaloids can cause gastrointestinal discomfort); and (iii) increased cucurbatin levels in vegetable squashes that leave a bitter taste. In these three examples, the food crops contained endogenous toxicants affording protection against plant pests. The levels of these endogenous toxicants were inadvertently increased in these new varieties. These examples have caused plant breeders to monitor new varieties of food crops that naturally contain potentially harmful toxicants or antinutrients more carefully to be certain that levels of these substances are within acceptable limits. For example, the United States and Canada have set acceptable limits for glycoalkaloid levels in potatoes that new potato varieties must meet.

Regulatory Oversight of New Crop Varieties

Conventional Breeding

In the United States, there are no premarket regulatory requirements governing the introduction of new crop varieties developed through conventional breeding (with the exception that the variety must not exceed standards set for the level of certain natural toxicants such as glycoalkaloids in potatoes). Food safety is assessed through postmarketing measures under the Food, Drug, and Cosmetic Act. This practice is based on the long history of safe introduction of new crop varieties developed through conventional breeding. As discussed earlier, plant breeders monitor the quality of new plant varieties before they are introduced into commerce. In Europe and for some crops in Canada, new varieties of food crops must be registered with the government. This is not done for safety reasons but more as an assurance to the farmer that the new variety will perform at least as well as commercial varieties in the field. In the United States, the market place determines the performance acceptability of new crop varieties.

Biotechnology

Because biotechnology is relatively new, there is considerably more regulatory oversight for the introduction of new crop varieties developed through this approach. In the United States, the regulatory authority to ensure the safety of food and feed products derived from plant biotechnology resides within the Food And Drug Administration (FDA). The U.S. Department of Agriculture (USDA) has the authority to ensure that genetically modified plants will not become plant pests. The Environmental Protection Agency (EPA) has the authority to evaluate the safety of plants that have been genetically modified for protection against plant pests such as insects, fungi, bacteria, and viruses.

The EPA also regulates herbicides by establishing herbicide tolerances for plants genetically modified to be herbicide tolerant. The movement and release of genetically modified plants in regulated by USDA under the Federal Plant Pest Act and the Plant Quarantine Act. Permits or notifications must be filed with USDA to obtain approval for field testing of new varieties of genetically modified plants under development. A determination that the genetically modified plant is not a plant pest (e.g., that the plants do not pose a risk to the environment or production agriculture) must be obtained before market introduction. The FDA has the regulatory authority to ensure the safety and wholesomeness of food and feed products, including those derived from genetically modified plants.

The FDA has adopted a decision tree approach to ensure safety of products derived from new varieties of food and feed developed by both traditional and newer genetic modification methods, including biotechnology. Under the Food Drug and Cosmetic Act the FDA has the authority to take regulatory action against a new variety of food if the genetic modifications would render the food "ordinarily injurious" to human health. Therefore, the FDA uses the same postmarket food adulterations approach and regulations for these products as are used for food and feed products derived from traditionally bred plant varieties. It is recommended that developers of genetically modified food consult with the FDA before commercialization, and this has been done with all genetically modified food products that are in the marketplace. The FDA has completed consultations on at least 29 different genetically modified crop plants to data.

The Federal Insecticide, Fungicide and Rodenticide Act (FIFRA) provides the EPA with the authority to regulate plants with bioengineered pesticidal traits. The EPA has approved at least seven different genetically modified plants with introduced pesticidal traits. These products are also reviewed by the FDA and USDA. In regard to regulation of genetically modified plants in other world areas. Health Canada regulates food safety, and Agri-Food Canada regulates feed and environmental safety as well as registration of specific plant varieties for certain crops. In Japan, the Ministry of Health and Welfare regulates food safety, whereas the Ministry of Agriculture, Food, and Fisheries regulates feed and environmental safety. In the European Union, environmental assessments are conducted before placing a product on the market, as outlined in the 90/220 EEC regulations.

The recently authorized Novel Foods Regulation governs food safety in the European Union Kingdom, Denmark, and the Netherlands. A Novel Feed Regulation is under consideration for overseeing feed safety under the current 90/220 EEC Process. Many other countries either have, or are in the process of developing, regulations for plant biotechnology products. The number of genetically modified plants that have successfully completed regulatory review in various countries include 23 in Canada, 5 in European Union, 15 in Japan, 3 in Mexico, 2 in Argentina, 1 each in Australia and Brazil.

Safety assessment Strategies for New Crop Varieties Developed through Biotechnology

The safety issues that have been raised for genetically engineered plant products are similar to those for new varieties of plants derived from conventional breeding. For example, progenies derived from conventional breeding as well as biotechnology may have progenies with altered agronomic properties when compared with the parents. Varieties with altered agronomic properties are usually readily identified in field trials. It has been suggested that biotechnology may inadvertently cause the production of new toxicants through insertional mutagenesis events that activate dormant biosynthetic pathways. This scenario seems very unlikely—particularly for those crops where there has been considerable experience with traditional breeding. It is more likely that an insertional event would result in the increase or decrease in the expression of already recognized toxicants, which has very infrequently been observed in conventional breeding, as discussed above.

In conventional breeding, the genes bearing the desired trait have often not been identified. The same applies to the protein expression products of these introduced genes. Introduction of genes through biotechnology procedures in much more precise, for the genes have been defined before their introduction. However, unlike conventional breeding, introduced genes can be obtained from almost any source, not just sexually compatible relatives of the food crop. Therefore, as part of the safety evaluation, regulatory agencies have required a molecular characterization of any gene introduced into food crops. In addition, the protein expression product of the cloned gene must be characterized as to its function, specificity of action, and safety. There are no particular concerns about the safety of the genetic material itself. Genetic material form living organisms is made from the same four nucleotide building blocks; the only difference between genes is the nucleotide composition. The human gut at any one time has been estimated to contain hundreds of milligrams of DNA from ingested food and mucosal cells sloughed into the gastrointestinal tract. These DNA molecules are efficiently degraded by nucleotidases during digestion. The contribution of genetic material from genes cloned into food is trivial compared with the other sources of DNA in the gut. It was concluded that the introduced genes (DNA) in food products pose no more health risk to consumers than the rest of the DNA ingested from food sources.

Molecular characterization

For new plant varieties developed through biotechnology, the source of the gene introduced into the plant must be identified. The transformation system used to insert the gene into the plant genome must be defined as well as the number of copies of inserted genes and the integrity and stability of the genetic insert. For genes coding for pesticidal proteins, the EPA requires the following information: (1) description of the vectors, (2) identity of organisms used for the cloning of the vectors, (3) description of the methodologies used to clone the vectors, (4) vector description (Size in kilobases), (5) restriction endonuclease sites, (6) location and function of all relevant gene segments, (7) the final delivery system, (8) description of gene segments transferred to the plant, (9) description of whether the inserted genes are expressed constitutively or inducibly, (10) localization and expression of the pesticidal substance in plant parts, and (11) estimation of the gene copies inserted, and so on.

Substantial equivalence

A general consensus has existed among regulatory agencies in major world areas regarding the use of "substantial equivalence" as an approach to assess the safety and acceptability genetically modified crops (WHO 1995; OECD 1993; FAO/WHO 1996). This concept has been adapted in part from procedures that plant breeders have followed to monitor the acceptability of new varieties of food crops developed through conventional breeding.

According to OECD, "the concept of substantial equivalence embodies the idea that existing organisms used as food or food sources can serve as a basis for comparison when assessing the safety of human consumption of a food or food component that has been modified or is new. If a new food or food component is found to be substantially equivalent to an existing food or food component, it can be treated in the same manner with respect to safety, keeping in mind the that establishment of substantial equivalence is not a safety or nutritional assessment in itself, but an approach to compare a potential new food with its conventional counterpart". The use of substantial equivalence is considered a practical approach to assess the safety of new varieties of food. Safety is defined as a reasonable certainty that no harm will result from intended uses under the anticipated conditions of consumption. The following are the three possible outcomes relative to assessing the substantial equivalence of a genetically modified food or food component to a conventional counterpart.

1. The genetically modified food or food ingredient is substantially equivalent to the conventional counterpart.
2. The genetically modified food or food ingredient is substantially equivalent to the conventional counterpart with the exception of certain well-defined differences; or
3. The genetically modified food or food ingredient is not substantially equivalent to the conventional counterpart.

Examples of new varieties of food crops that could be considered substantially equivalent to conventional counterparts are virus-resistant plants produced by introduction of viral coat protein (viral protein is already present in plant tissue of conventional infected plants). Alternatively, if the food is processed removing any added traits and the composition has not changed (e.g., processed canola oil), the oil would be considered substantially equivalent to its conventional counterpart.

The second outcome listed above will apply to most genetically modified food crops. For example, many genetically modified crops will be substantially equivalent to conventional counterparts with the exception of a new trait that imparts a desired characteristic such as pest resistance. When this is the case, further safety assessment should focus on the new trait or well-defined differences. For some genetically modified crops or food components derived therefrom, it may not be possible to demonstrate substantial equivalence to a conventional counterpart, either because differences are not sufficiently well defined or because there is no appropriate counterpart with which to make a comparison. The absence of substantial equivalence does not imply that the genetically modified crop is any less safe. The safety assessment should focus on the nature of the changes (FAO/WHO 1996).

Key Parameters for Assessment of Substantial Equivalence

An assessment of substantial equivalence should focus on a comparison of agronomic characteristics combined with compositional analysis that includes key nutrients as well as toxicants and antinutrients that have potential health significance.

1. Agronomic traits are a good starting point for evaluating substantial equivalence of genetically modified plants to their conventional counterpart. These traits will normally be examined as an integral part of the development program leading to a new food plant variety. Agronomic traits are those characteristics measured by plant breeders for a given crop. For example, in the case of potatoes these may include yield, tuber size and distribution, dry matter content, and disease resistance. Agronomic properties may vary depending on local environmental conditions. Therefore, the genetically modified variety should be grown in the same geographical regions in which it will be grown commercially.
2. Key nutrients are those components in a particular food product that may have a substantial health impact in the overall diet. These may be major constituents (fats, proteins, carbohydrates) or minor components (essential minerals, vitamins). Critical nutrients to be assessed may be determined, in part, by knowledge of the function and expression product of the inserted gene (e.g., if an inserted gene expresses an enzyme that is involved in amino acid biosynthesis, the amino acid profile should be determined). Introduction of an invertase into potatoes could influence the

carbohydrate metabolism; therefore, starch should be investigated. Alternatively, introduction of a storage protein high in methionine into the soybean could affect the amino acid content; therefore, the amino acid profile should be measured. Examples of the analysis of key nutrients in different genetically modified plants have been published.

Given differences among consumption patterns and practices in various cultures and societies, the key nutrients to be examined may differ in various countries. The critical nutrients to be addressed should be determined using consumption data for the target region. For example, in Denmark, potatoes provide an important source of vitamin C in the diet (35%), not because of their high vitamin C content (20 mg/100 gm) but because of high consumption of potatoes (140 gm/day). In the United States, potatoes are not as important a source of vitamin C owing to lower potato consumption and the availability of other dietary sources of this vitamin.

3. Critical toxicants and antinutrients are those compounds known to be inherently present in a crop variety whose potency could have an impact on health if their levels were increased significantly (e.g., solanine glycoalkaloids in potatoes, trypsin inhibitors in soybeans). Knowledge of the biologic function of the protein expression product of the inserted gene could influence the decision about which toxicants or antinutrients to examine.

The analysis of important nutrients and toxicants or antinutrients should use validated or standard methods where available (e.g., methods from the Association of Official Analytical Chemists [AOAC] or other recognized bodies). When the genetically modified line is compared with the parental line or conventional varieties, the varieties should be brown under similar environmental and agronomic conditions. If there are no statistically significant differences in measured parameters between the genetically modified line and the parent or conventional lines, then the genetically modified variety can be considered substantially equivalent to the parent or conventional variety. If statistically significant differences are observed in measured parameters, then a comparison can be made with values available in literature or other sources. If the parameter for the genetically modified line is within the normal range for conventional varieties, further evaluation is not warranted. If the parameter for the genetically modified line is outside the normal range, further evaluation may be needed.

A toxicological or nutritional evaluation may be needed to asses whether the difference between conventional and genetically modified lines for a given parameter is biologically meaningful and requires further investigation. In the future, genetically modified lines for sexually compatible crops will be crossed using traditional breeding techniques. If substantial equivalence has been demonstrated for (1) corn with a gene producing insect protection and (2) corn with a gene for herbicide resistance, then crossing (1) and (2) will produce a new variety that is likely to be substantially equivalent to the parents. If there are no expected interactions between the two traits, then no additional evaluation will be needed. Potential genetic interactions will need to be considered on a case-by-case basis. For example, if two independent modifications to the same metabolic pathway are combined by traditional breeding, further analysis of the products of the metabolic pathway may be warranted.

Safety Assessment for the Introduced Gene Expression Product

Many genetically modified food crops have been shown to be substantially equivalent to conventional crops with the exception to the introduced trait(s) that may impart one or more characteristic such as pest resistance, selectivity to preferred herbicides, modification of the ripening process, and so forth. For these examples, the safety assessment should focus on the introduced traits, the protein expression product of the cloned gene. The developer should do the following:

1. Define the biological function, specificity, and more of action of the protein. If the protein is an enzyme, assess the potential effects of the enzyme on metabolic pathways and levels of endogenous metabolites based on its mode of action and specificity.
2. Compare the amino acid sequence of the protein to known sequences in protein databases to determine if the protein has sequence homology to food proteins, toxins, or allergens.
3. Assess the inherent digestibility of the protein in vitro with simulated gastric and intestinal protease preparations.
4. Determine the level of expression of the protein in the food. This effort should focus on the raw agricultural product or a specific processed food component (e.g., oil), as appropriate.

Criteria for Concluding the Introduced Protein is "As-Safe-As" Proteins Already Present in Foods

The following criteria are important in assessing the safety of a protein introduced into food or food components:

1. The pı tein has a history of safe consumption in other food crops.
2. The protein is functionally and structurally related to proteins with a history of safe consumption in food.
3. The biological function and specificity or mode of action of the protein raises no safety concerns.
4. The amino acid sequence of the protein is not similar to known protein allergens.
5. The protein is not derived from a food source with a history of allergy.
6. The amino acid sequence of the protein is not similar to known protein toxins or antinutrients.
7. The protein is susceptible to degradation by digestive enzyme.

For certain insect control proteins such as *Bacillus thuringiensis* (B.t.) family of proteins, regulatory agencies such as the EPA require administration of the protein as a single oral high dose to mice. The scientific rationale for an acute test in mice is that known protein toxins generally manifest their toxicity via acute mechanisms. The B.t. proteins have a long history of safe use (EPA 1988); as new forms of B.t. proteins are introduced, the EPA has stated that an acute dosing test in mice will provide assurance that the new B.t. protein varieties are safe. Such studies have been conducted for the Cry3A protein expressed in potatoes, the Cry1Ab protein expressed in corn, and the Cry1Ac protein expressed in cotton and the tomato.

Enzymes that provide selectivity against herbicides or function as selective markers have also been tested in mice. Acute testing may be recommended if the protein is derived form plants, or microorganisms that have no history of consumption. Acute toxicity testing of proteins is not normally needed if the protein (1) is not present in the food (E.g., removed during processing, as in vegetable oils), (2) has a history of consumption in food, and (3) is closely related functionally and structurally to proteins with a history of consumption in food. If safety testing of proteins is undertaken, it may be possible to isolate and purify the protein from the plant. However, this may not be practical if the protein is expressed in small amounts in plant tissue.

Alternatively, the protein could be produced in an appropriate fermentation system generating a protein that is biochemically equivalent to the protein produced in the plant. Proteins that meet the safety criteria listed and are nontoxic if administered to mice are considered "as-safe-as" other proteins naturally present in foods. No further safety or nutritional assessment of the new variety of food crop is necessary.

Criteria for Concluding that Further Safety Evaluation is Required for the Introduced Protein

Some introduced proteins may require further safety and nutritional evaluation if they meet any of the following criteria.

1. The protein is functionally and structurally related to proteins that are known toxins or antinutrients.
2. The protein is derived from a food with a history of allergy.
3. The protein is structurally similar to known protein allergens.
4. The protein is not degraded by digestive enzymes.
5. The biological function of the protein has not been characterized and there are no structurally related proteins identified in protein databases.

If the introduced proteins meet one or more of the criteria listed above, additional testing may be warranted on the basis of a case-by-case assessment. Toxicology or nutritional end points that may need to be developed are summarized in the next section.

Safety Testing Recommendations

Testing for toxicity and antinutritional effects

Certain antinutritional proteins such as lectins or protease inhibitors are naturally present in plants and provide protection against insect pests. There has been an interest in inserting these proteins in food crops to provide an alternative to insecticides for control of insect pests, although the safety implications of such transformations have raised concerns. Some plant lectins exert antinutrients effects when fed to animals by binding to the brush-border epithelium of gut cells and thus disrupting nutrient absorption. If a lectin were to be introduced into a food crop to enhance protection against insect pests, it should be fed to animals to assess whether it acts as an antinutrient. Assessing its potential for binding to brush-border epithelium would be part of the safety evaluation. If the introduced lectin had antinutrient properties, a "no-effect-level" for antinutrient effects would have to be determined if an adequate safety margin existed.

Human nutritional studies may also be appropriate. Because some lectins are known food allergens, assessment of potential allergenicity will also be necessary, which will be discussed in the next section. A similar testing scheme for protease inhibitors can be envisioned. Any general safety questions regarding the introduced protein could be addressed by animal feeding studies designed to assess toxicity and nutritional endpoints. For proteins containing common amino acids, no additional testing for mutagenic, carcinogenic, or teratogenic potential is indicated because there is no evidence that proteins are genotoxic or that feeding proteins such as food enzymes have ever directly produced carcinogenic or teratogenic effects in laboratory animals.

Testing for allergenicity

Further testing of an introduced protein for potential allergy would be indicated if (1) the protein were derived from a food source with a history of allergy or (2) the amino acid sequence of the protein matched (at least eight contiguous identical amino acids) that of a known protein allergen. The immunogenic potential of the introduced protein should be tested in one of the various solid-phase immunoassays such as the radio allergosorbent test (RAST) or RAST inhibition assay or the enzyme-linked immunosorbent assay (ELISA). Solid-phase immunoassays use IgE fractions of sera from individuals confirmed allergic to the food from which the gene coding for the introduced protein was derived. Where in vitro tests are negative or equivocal, then in vivo skin prick tests could be carried out. If no positive response was detected in the prick test, then double-blind placebo controlled food challenges with patients known to be allergic to the food could be carried out under controlled clinical conditions.

The in vivo studies would only be warranted if the gene were derived from one of the following eight food groups that account for more than 90% of all food allergies: eggs, milk, fish, crustacea, peanuts, soybeans, wheat and tree nuts. The testing scheme to detect allergens summarized above works effectively as demonstrated in the case of the Brazil nut 2S storage protein. This protein was introduced into soybeans to increase the sulfur amino acid content and thereby improve the bean's nutritional value for use in animal feeds. Because there are a small number of individuals allergic to Brazil nuts, it was decided to test sera from these individuals to see if their sera contained IgE that would cross-react with the Brazil nut 2S storage protein. Sera from eight out of nine Brazil-nut-allergic individuals reacted with this storage protein. Development of this soybean line containing the introduced Brazil nut storage protein as terminated. If the developer had wished to proceed with commercialization of this genetically modified soybean, all food derived from this soybean would have had to be labeled as containing Brazil nut protein. For many genetically modified plants, the protein expression product will be derived from a gene source that has no history of inducing allergy. If the amino acid sequence of the introduced protein dos not show sequence homology to known allergens, the physicochemical properties of the protein should still be assessed to determine if it shares the profile for allergens. These properties include (1) stability to food processing conditions (high temperature and pH changes that denature food proteins), and (2) resistance to gastric acidity and digestive proteases.

The level of expression of the introduced protein in food should also be determined because allergenic proteins often comprise a significant proportion of the local protein in that food. Regarding digestibility of proteins, it is well established that protein allergens tend to be resistant to digestion and generally constitute a significant percentage of the total protein (1-18%) in the food. Allergens are often stable to heat processing. These properties increases the potential that sufficient quantities of the protein will survive both food processing and the hostile environment of the gut to elicit immunologic reactions in the intestinal mucosa. However, there are undoubtedly examples of proteins in food that represent a small percentage of total protein, are not digestible, and do not elicit food allergy. According to allergenicity experts, there is currently no validated in vivo or in vitro model to predict allergenicity potential of proteins. There is therefore a need to develop a predictive model for allergenicity.

Safety Considerations for Marker Genes

As stated earlier, marker genes are added to the genetic material inserted into plants to help identify successfully transformed plant cells. The use of marker genes and their proteins expression products in plant biotechnology has raised the following safety concerns: (1) potential horizontal transfer of the marker gene from plant cells to gut microflora that might reduce the efficacy of therapeutic antibiotics, (2) potential toxicity of the protein expression products of the marker gene, or (3) compromised antibiotic efficacy due to expression of the antibiotic marker gene product in the food. After detailed scientific review, it was concluded that the potential for horizontal gene transfer from plants to gut microflora is vanishingly small (FAO/WHO 1996; WHO 1993). Introduced genes are stably incorporated into the genome of plants, and there are no known mechanisms for direct transfer of genes from plants to micro organisms. Moreover, the DNA released from the plant during digestion is rapidly degraded. This degradation occurs well before the plant during digestion is rapidly degraded. This degradation occurs well before the plant material reaches the lower intestine, cecum, and colon, where gut microflora are found. The potential for the protein expression product of the antibiotic marker gene to compromise antibiotic efficacy is limited by (1) the digestibility of the expressed protein, (2) low expression of the protein in food, or (3) lack of available cofactors such as adenosine triphosphate (ATP) in the gastrointestinal tract that are required for antibiotic inactivation (FAO/WHO 1996). The safety assessment of the protein expression product of marker genes should focus on the approach outlined earlier for protein expression products of introduced genes (new traits).

Assessment of Safety for Nonsubstantially Equivalent Genetically Modified Plants

To date, there have been few, if any, examples of genetically modified plants that are not considered substantially equivalent to conventional counterparts (With the exception of well-defined differences in some cases). As discussed earlier, with future developments in biotechnology, products may be developed that have no conventional counterpart for which substantial equivalence can be assessed. These products could arise by transfer of genomic regions that have only partially been characterized (FAO/WHO 1996). Genomics is the study of all the genes of an organism and their organization into chromosomes, and it includes researching the links between gene structure an function, or expression, that are the foundations for bioengineering new plant products.

Desirable agronomic traits may result from the linkage of several genes functioning together to produce effects such as resistance to drought or alkaline soil conditions. Such a development would allow plants to survive and grow under harsh environmental conditions, and millions of acres of land that cannot be currently used to grow food crops would be available for cultivation of genetically modified food crops. In the future, there may be novel foods derived via biotechnology that have intended benefits, but further safety and nutritional evaluation will be needed before these foods can be marketed. In other cases, unintended changes in a new food crop may occur that could require further evaluation to understand what was changed and its significance to the safety and nutritional value of the food crop.

Toxicological and nutritional studies may be indicated in some cases to resolve safety and nutritional questions of food crops that are not substantially equivalent to their conventional counterparts. If animal feeding studies are considered necessary, their objectives must be clear and the experimental design carefully planned to avoid nutritional problems that may confound data interpretation. There may be a need to modify existing protocols for testing whole food or food components—in particular by providing adequate nutrition with respect to diet. Nutritional imbalance may mask toxic effects. Incorporation of high levels of a food into the diet may also result in nutritional deficiencies that could cause adverse effects unrelated to the food being tested.

The usual concept of safety margins may not be applicable because is of often not possible to food a whole food at a high enough level to obtain anything close to a hundred fold margin of safety. The testing of specific components or extracts of a novel food may present on option for addressing safety and nutritional testing of the whole food. For certain foods, animals are not good models for predicting safety to man because animals exhibit species-specific sensitivity to the food. For example, dogs experience transient paralysis from consumption of macadamia nuts or develop fatal cardiac arrhythmias owing to sensitivity to theobromine in chocolate; male rats of one particular strain develop cardiomyopathy when fed diets high in vegetables oils.

Feeding animals high levels of various foods in the diet has produced anemia (rats and dogs fed onions), enlarged cecums (rats fed potatoes), mucosal lesions in the stomach (rats fed tomatoes or chilli powder), pulmonary emphysema (rats fed beans), and reduced breeding performance and survival (rats fed wheat flour). If nutritional quality is the important issue, it may be appropriate to go directly to nutritional studies in human volunteers rather than to use animal models that may have limited relevance to man. The industry currently uses "sip and spit" tests to evaluate the organoleptic properties of new varieties of vegetable crops as part of the quality assessment. Once the initial safety assessment has been completed for a nonsubstantially equivalent food product, sensory or nutritional evaluation or both, may be recommended to ensure the quality and acceptability of the product for consumers.

Conclusion

Biotechnology provides plant breeders the opportunity to develop new varieties of food crops more efficiently and with greater potential benefit than has been possible with conventional breeding practices. To provide assurance that this technology will generate food "as safe as" that produced by traditional breeding programs, safety assessment strategies have been developed for products of plant biotechnology that have generally been accepted by international regulatory groups (FAO/WHO 1996). This strategy includes comparison of the agronomic properties and important nutrient-toxicant composition of genetically modified crops with their conventional counterparts. If the comparisons show no meaningful differences in the aforementioned parameters, the genetically modified food crop is said to be "substantially equivalent" to its conventional counterpart, and the safety and nutritional assessment is completed.

Many genetically modified foods will be substantially equivalent to conventional varieties with the exception of one or more introduced traits. The safety assessment will then focus on the introduced trait to determine if the protein expression product meets the criteria for being "as-safe-as" proteins already in the food supply. If protein expression products are derived from plants with a history of allergenicity; have a structural or functional similarity to known allergens, toxins, or antinutrients; or have functions that may alter the nutritional quality of the plant, then additional safety and nutritional studies may be needed. In the absence of validated assay to predict potential allergenicity, comparison of amino acid sequence homology of the protein with known allergens and assessment of its physicochemical properties are the best tools currently available to predict potential allergenicity. As biotechnology develops, future products may be developed that have no conventional counterpart for which substantial equivalence can be assessed. These varieties could be designed to have improved agronomic properties or provide important health and nutritional benefits.

The safety and nutritional assessment of these crops should be carried out on a case-by-case basis using the strategies outlined above. During the last few years, over 34 different genetically modified plant products have successfully completed regulatory review in countries around the world. The experiences gained in bringing these products to market have enabled developers and regulatory agencies to define the key safety questions and the science to answer the question. This progress will help provide guidance to the development of improved and safe plant biotechnology products for the future.

11

PANTS IN ANIMAL FEEDING

The purpose of raising most crops is to produce food—but what is food? Food is any substance that provides an organism with energy and nutrients. Plants that have become food sources contain energy-rich compounds and nutrients that are digestible and are not loaded with toxic compounds. Humans, like other monogastric animals (such as pigs, chickens, fish), readily digest starch and protein, but cannot digest the large carbohydrate molecules such as cellulose that make up plant cell walls. However, ruminant animals such as cows and sheep can digest these molecules. Mammals also differ in how they handle toxic compounds. Acorns are rich in starch and protein, but contain such high levels of toxic tannins that humans can't use acorns unless they remove the tannins. Pigs, in contrast, eat unleached acorns quite happily. Historically, experiments with small mammals (especially rats) identified the essential nutrients needed for normal human development and this research formed the basis of dietary recommendations.

Nutritionists aimed these guidelines at preventing clinical nutrient deficiencies. When epidemiological studies made clear that the prevalence of certain chronic diseases such as atherosclerosis and cancer was related to diet, scientists refined these recommendations. For example, in addition to recommending a certain number of calories per day, U.S. government guidelines now say that less than 30% of those calories should come from fat, and less than 10% from saturated fatty acids. Finally, human foods also contain many chemicals such as phytoestrogens and anti-oxidants that are not nutrients, but nevertheless can have positive effects on health.

Nutritional Requirements

Plants are autotrophic ("self-feeding") organisms with simple nutritional requirements. They need water, carbon dioxide (CO_2), and a variety of inorganic chemicals derived from soil minerals as well as oxygen (O_2), The most important feature of plants, the biosphere's major autotrophs, is that they use light energy from the sun to provide fuel for their own energy needs through the process of photosynthesis. By using sugar (the product of photosynthesis), water, and inorganic nutrients such as nitrate and phosphate from the soil, plants can synthesize all the complex organic molecules, such as amino acids and vitamins, that they need to grow and reproduce. Although plants and animals use many of the same organic molecules for their basic metabolic processes, only plants can synthesize these molecules using sugar and inorganic nutrients alone.

Heterotrophic ("other-feeding") organisms have more complex nutritional requirements. Rather than being able to use the sun's energy

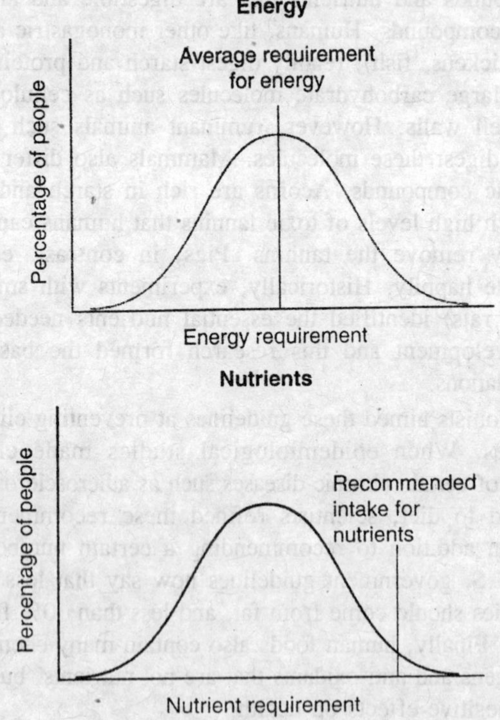

Fig. 11.1. The rationale for requirements for energy and nutrients.

directly, they must ingest energy-rich organic molecules (mostly fats and carbohydrates) to sustain life. All heterotrophs, whether they eat plants (herbivores), animals (carnivores) or decompose dead organic matter (decomposers), ultimately depend on plants to provide them with energy-rich molecules. The exact nutritional requirements of most animals are not known because they have not been studied. Humans require the following organic nutrients:

1. Ten of the 20 amino acids that make up proteins—namely leucine, isoleucine, lysine, methionine, phenylalanine, threonine, tryptophan, valine, and (for infants) arginine and histidine—must be provided by proteins in our food.
2. The fatty acid linoleic acid must be provided by fats in the diet.
3. The vitamins A, B_1 B_2, B_3, B_6, B_{12}, C, D, E, K, pantothenic acid, biotin, and folic acid must be provided by various foods.

When nutritionists estimate the daily needs for food for people, they consider energy needs and nutrient needs differently. Energy requirements vary, depending on such factors as body weight and physical activity, so nutritionists recommend an average consumption. But body tissues cannot grow and renew if the supply of even one nutrient is inadequate Nutritionists thereforeset the RDA (recommended daily allowance) of nutrients well above the average requirement.

Carbohydrates and Fats

The energy and nutritive values of plants as food depend on their chemical compositions, and foods vary in their ability to supply these dietary components. Carbohydrates are the source of most food energy for people. Not surprisingly, carbohydrates are abundant in staple food plants such as wheat, rice, maize, cassava, millets, and potatoes. Simple carbohydrates consist of one or several sugar molecules, whereas complex carbohydrates (also called *polysaccharides*) are made up of hundreds or even thousands of units of sugar molecules. Simple sugars, such as glucose or arabinose, or small oligosaccharides, such as sucrose (cane or beet sugar) or lactose (milk sugar), are readily soluble in water and are easily digested and taken up from the intestinal tract into the bloodstream.

As a result, they provide a source of quick energy. Although most adult Caucasians can digest lactose, many Asians and Africans cannot because they lack the enzyme lactase, which splits lactose into its two component sugars. Such people are lactose intolerant. When they eat milk or milk products, the lactose is fermented by their intestinal bacteria, causing severe cramps and discomfort. Babies generally have

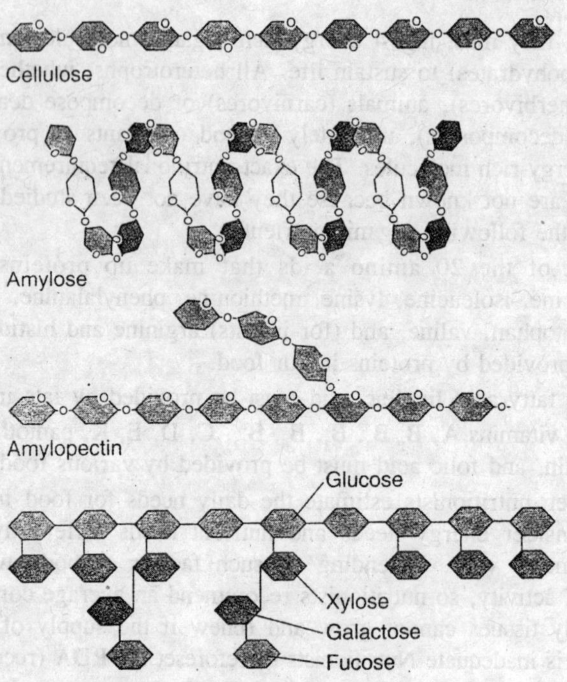

Fig. 11.2. Structure of polysaccharides.

high levels of lactase, so they have no problem digesting milk sugar. Polysaccharides, or complex carbohydrates, such as starch and cellulose, are macromolecules that consist of linear or branched chains of sugar units. They can consist of a single type of sugar, as in cellulose and starch, or different types of sugars, as with hemicelluloses, the major polymers found in soluble and insoluble fiber.

Starch and cellulose, although both polymers of glucose, have different chemical and physical properties and play different roles in the plant. Cellulose forms fibers and is found in the cell walls of plants where it provides strength. Wood is primarily cellulose. Starch, consisting of amylase and amylopectin, in contrast, is deposited by plants as food stored in specialized organs such as roots, tubers and seeds, for the plants to use whenever they need a source of energy. Complex carbohydrates are not soluble in water and must be broken down by digestive enzymes into simple sugars before the human body can use them.

Breaking down starch requires four different enzymes, which are present in the human digestive system. However, people cannot digest cellulose or hemicellulose, the other major polysaccharides present in plants. So the sugar units in these macromolecules pass unused through human bodies and provide no food energy. Nevertheless, cellulose and hemicellulose do benefit human health. They constitute the insoluble and soluble fiber that promotes peristalsis of the bowels and prevents constipation.

A high-fiber diet may lower the risk of colon cancer, but this needs further research. Fiber certainly increases the well-being of many people despite its contribution to flatulence. Some microorganisms have the enzymes necessary to digest cellulose and hemicellulose and can therefore use the sugar units in these materials as food. Ruminant animals have such microbes living in a portion of their digestive tracts—the rumen—and get energy from the release of sugars from these polysaccharides, either directly or indirectly after the microbes convert sugars to volatile fatty acids and methane.

Lipids, or fats, oils and certain other molecules are present in all organisms, being essential structural components of all cellular membranes. Their most important chemical property is their insolubility in water, which allows them to form physical barriers that control the exit and entry of many of the chemicals in cells and tissues. To enter a cell, glucose requires the presence of a specific glucose carrier protein in the plasma membrane. Lipids occur in large amounts in certain animal tissues such as brain, eggs, and some oil-rich seeds such as canola, soybeans, and peanuts. Chemically, there are many different types of lipids. Three important classes of lipids are triacylglycerols (fats and oils), phospholipids, and cholesterol.

1. Triacylglycerols have three fatty acids connected to glycerol, a simple three-carbon molecule. Triacylglycerols are a major storage form of energy in animals and plants.
2. Phospholipids are present in cellular membranes. A membrane consists of a double layer of phospholipid molecules with embedded proteins that permit the passage of small molecules through the membrane.
3. Cholesterol is structurally very different from triacylglycerols and phospholipids. Cholesterol and other molecules like it, such as vitamin D and the steroid hormones, are made out of four linked rings of carbon atoms. Humans synthesize cholesterol to make bile acids and hormones such as testosterone and estrogen. Plants have similar molecules (such as phytoestrogens).

Nutritionally, the most important lipids are the triacylglycerols (fats and oils). The size of the fatty acids (number of carbon atoms in the chain) and the number of double or unsaturated bonds determine the chemical and physical properties of the fat molecules as well as their metabolic fate in the human body. The more saturated the fatty acids and the longer the chains, the more solid the fat will be at body temperature. Fats with long, saturated fatty acids are commonly found in animal products (meat, milk, and cheese) and in some plant products (cocoa butter). Fats that are liquid at room temperature are called *oils* and are found in seeds and fish. Oils can be converted into fats by a process called hydrogenation, in which hydrogen atoms attach to the double bonds that link adjacent carbons. Hydrogenation produces *trans* fatty acids, so called because the new hydrogen atoms attach on different sides of the carbon chain. Fats in the diet also help the body absorb certain vitamins that are not readily soluble in water, such as vitamin A. Triacylglycerols from which one or two fatty acids have been removed are called *diglycerides* and *monoglycerides*, respectively. Strictly speaking, they are not fats, and they are the ingredients of "fat-free" margarine.

Lipids are an important source of energy for humans. As with carbohydrates, they must first be broken down or digested by enzymes before they can be absorbed in the bloodstream and used by the body. In addition, lipids have a nutritive role. Humans lack the capacity to make omega-3 fatty acids such as α-linolenic acid; this acid has three unsaturated bonds, but the number 3 in "omega-3 fatty acid" refers to the third carbon of the fatty acid starting from the methyl group (CH_3). Omega- 3 fatty acids must be supplied in the diet and are abundant in plant oils, nuts, and fish oils. Humans also need other omega-3 fatty acids, including eicosapentaenoic acid (EPA) and docoshexaenoic acid (DHA), but can make these from a-linolenic acid. EPA and DHA are very important for normal brain development and vision. However, the recent popularity of these fish oils stems from possible protective roles in arthritis, hypertension and heart disease. There is no RDA for omega-3 fatty acids, but nutritionists believe that 200 mg per day benefits cardiovascular health.

Diets Linked to Major Diseases

Human diets have changed dramatically over the course of history. The food sources used by hunter-gatherers were extremely varied. Most collected many different plants and supplemented a largely vegetarian diet with meat or fish when available. These mobile societies often followed their food sources as seeds, fruits, or roots of different plants became available throughout the seasons. The major cause of illness or death in these societies was infection caused by a lack of sanitation.

The human diet changed with the transition to agriculture. A still largely vegetarian diet that reflected the predominance of a few cultivated plants was supplemented with meat from wild or domestic animals. Many people in the developing world still eat such a diet. Although infectious diseases and under nutrition are still too prevalent in developing countries, the diseases that cause most deaths in the developed countries—heart disease, cancer, stroke, and diabetes—are only now emerging there.

The transition to modern affluent societies a hundred years ago led to a radical change in the diet. In such societies, plants, especially vegetables and whole-grain products, began to playa minor role and people greatly increased their consumption of meat, animal products, sugar, and alcohol. However, people's nutrient and energy requirements have remained the same or even have decreased because they lead more sedentary lives in affluent societies. As a result, obesity and high blood pressure are very common, and heart disease, cancer, stroke, and diabetes have become the leading causes of death. Epidemiologists have found a positive correlation between high fat intake, on the one hand, and certain cancers (especially colon and prostate cancer) and with coronary heart disease, on the other. Having become aware of the important links between diet and health, nutritionists now recommend that we eat more plants (five portions of vegetables or fruits per day) and fewer animal products and meat. In this respect it is important to start with a healthy diet early in life, because changing one's diet is extraordinarily difficult.

The relationship between fat intake and coronary heart disease is incredibly complex and many studies yield contradictory results. The popular press has given much attention to the Mediterranean diet (high fat intake, mostly as olive oil) and the Japanese diet (low fat intake, mostly from fish) in contrast to the U.S. diet (high fat intake, mostly from dairy products, hydrogenated oils, and meat). The first two diets are associated with a much lower incidence of cardiovascular disease. After reviewing all the studies, W C. Willett from the Harvard School of Public Health concluded, "Intake of partially hydrogenated vegetable fats and saturated fats, particularly those from dairy sources, should be minimized and that consumption of a substantial proportion of energy as monounsaturated fat would not be harmful and might even be beneficial".

Protein

Protein is a major constituent of all organisms, and a given cell may contain more than 5,000 different proteins. As noted in the previous chapter, proteins are long strings of 20 different amino acids and a given protein has 100 to 1,00001 them, with their specific order and string length determining the protein's properties. The strings are folded in specific three-dimensional structures, unique for each protein. Proteins play important roles in cells: (1) as enzymes that catalyze biochemical reactions, (2) as regulators of gene activity, (3) as transporters of small molecules through membranes, (4) as transducers of signals from the environment and other parts of the plant, (5) as structural components of the cell, (6) as toxic agents to defend the plant against pests and pathogens, and (7) as reserves of nitrogen, sulfur, and carbon for later use.

Most food proteins are structural proteins such as those found in muscle cells of meat and fish, or playa nutritive role, as in milk, egg

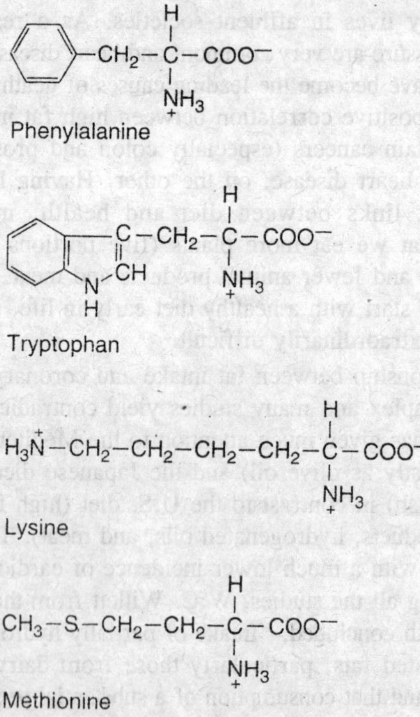

Phenylalanine

Tryptophan

Lysine

Methionine

Fig. 11.3. Structures of four essential amino acids.

white, or seeds. As autotrophs, plants can synthesize all 20 amino acids from the sugars made in photosynthesis and the nitrate and sulfate taken up from the soil. The amino acids are then linked together in a specific order using the information in the genes. Mammals, which are heterotrophs, cannot make all the amino acids and need to get some of them in their food. The amino acids that they cannot make are called the *essential amino acids*. For adults, there are eight essential amino acids.

The nutritional value of protein-rich foods depends not only on their digestibility (not all proteins can be digested equally well), but also, and especially, on the relative abundance of the eight essential amino acids. Nutritionists use these criteria to evaluate proteins for their dietary value to humans or other mammals. Human milk protein and egg protein are defined to have protein scores of 100, which means that they contain the essential amino acids in the exact proportions required by the human body. Animal proteins generally have a higher protein score than the major dietary plant proteins, because the latter tend to be low in the essential amino acids tryptophan, methionine, and lysine. A marked deficiency in even one essential amino acid drastically lowers the protein score of a protein. Maize flour, which is low in tryptophan and lysine, has a protein score of 49, and navy beans, which are low in methionine, have a protein score of 44.

Table 11.1. Protein scores of selected foods

Food	*Protein score*
Hen's egg	100
Beef	80
Cow's milk	79
Chicken	72
Fish	70
Rice	69
Soybeans	67
Wheat	62
Maize	49
Beans	44
Potatoes	34

Foods with protein scores lower than 70 are generally considered unsatisfactory for human growth and maintenance. This does not mean, however, that maize protein and bean protein have no nutritional value for humans. Foods that are deficient in different amino acids can complement each other when eaten together. A meal of maize tortillas with beans has a much higher protein score than either food alone.

As with calories, the amount of protein you require each day depends on many variables. Rapidly growing young people, physically active people, and pregnant and lactating women need more dietary protein. If your daily protein intake is very high (over 100 g per day), you metabolize excess amino acids and excrete the nitrogen (as urea), because your body has no way to store amino acids. In contrast, carbohydrates have limited storage as the polysaccharide glycogen in liver and muscle, and lipids are easily stored in fatty deposits. In low-calorie diets, body proteins are broken down for energy, because the body always satisfies its energy needs first. Physical activity, as in sports, greatly increases the body's need for energy, but when muscular activity doubles your body's caloric requirements, it increases your need for protein by only 5%.

If a woman does not eat enough protein during pregnancy, the intellectual development of the child may be slowed. This finding has focused attention on the importance of proper nutrition during pregnancy and lactation. Fetal growth (especially during the final six months of development) and milk production require that the woman's normal diet be supplemented with additional calories, proteins and other nutrients (essential fatty acids, vitamins and minerals). Women who are pregnant or lactating need to increase their daily protein intake by 20 or 30 g, respectively. Women who are undernourished therefore pass this condition on to their babies.

Improvment of Protein Value

An adequate supply of essential amino acids is important both for human nutrition and for the efficient production of livestock that supply meat, milk, and wool. For intensively fed livestock such as pigs and poultry in developed countries, the compounding (mixing) of feeds to correctly balance essential amino acids is a highly developed scientific operation. It usually involves mixing components from a number of plant sources (cereal and legume grain) with some animal sources such as fishmeal or the unusable waste from slaughterhouses. In developing countries, in contrast, people often have to rely on a single staple such as maize that does not provide an adequately balanced diet. Research has focused on improving the nutrient quality of such diets. Maize and other cereals are deficient in lysine and tryptophan, and one of the first efforts to create seeds with a better amino acid balance (higher protein score) resulted in the identification of a mutant maize line, called Opaque 2, that has a higher lysine content in its seeds (32% more lysine than the control lines). Research in the 1960s clearly demonstrated the nutritional advantage of this type of maize, but farmers did not widely adopt Opaque 2 maize, because it had poor agronomic properties (yield per hectare was 15% less). This discovery led to a 30-year major breeding effort to introduce this trait into elite maize lines, culminating in the production of quality protein maize lines. The lysine content of these lines is not quite as high as in the original Opaque 2 lines (20% more instead of 32% more), but the agronomic properties of the new lines are excellent and the nutritional benefit is still substantial. Farmers in Africa and South America, where maize is an important staple for humans, have widely adopted quality protein maize lines.

In the case of ruminant livestock such as cattle and sheep, the nutritional situation is more complicated because of the nature of ruminant digestive systems. The rumen of these animals functions as a large microbial fermentation vat in which ingested materials (cellulose, lignin, protein, fat) are digested, hydrolyzed, and converted to microbial mass (protein, cell walls) and fermentation products (methane, fatty acids). Thus ruminant animals are ideally suited to convert materials that humans cannot use (grass and other forage) into useful products (meat, milk, wool). Some human populations (such as nomadic herdspeople) depend on this conversion. Increasing the efficiency of this conversion process is a major goal of animal husbandry research. After the microbial mass is formed in the rumen, it continues its journey in the animal's gastrointestinal tract and is in turn hydrolyzed to its constituent sugars and amino acids, which are taken up into the animal's bloodstream through the wall of the intestine. The amino acids are used to synthesize new proteins for the animal's tissues. However, microbial protein has a significantly lower content of the sulfur-containing essential amino acid methionine, and there is a net loss of this amino acid during the conversion of plant protein to microbial protein in the rumen. As a result, when feed supplies are scanty, the lack of methionine limits production of milk, muscle (meat), and wool.

In times of feed shortage, the diet of ruminants in pastures may be improved with high-protein legume grains, but these are also a poor source of methionine. Efforts are now underway to create methionine-rich legumes using GM technology. Researchers have identified a protein in sunflower seeds that has an unusually high content of sulfur-containing amino acids. A further property of this protein is that it resists breakdown by microbial flora in the animal's rumen. It therefore passes undegraded into the animal's intestine, where the digestive enzymes can hydrolyze it into its constituent amino acids, including methionine, that the animal then uses to synthesize protein. Australian researchers introduced the gene cod. ing for this protein into lupin (a legume) and targeted it for expression in the seeds. The result was a 100% increase in the methionine content of the seed protein, and when this grain was fed to sheep their weight gain increased by 7% and wool production by 8% compared to sheep fed unmodified seeds. The success of this approach has prompted researchers to introduce a leaf-specific form of the gene for this high-methionine sunflower seed protein into the leaves of pasture plants, with the ultimate goal of improving the balance of essential amino acids available to grazing ruminants.

VITAMINS SYNTHESIZED BY PLANTS

Vitamins are important organic molecules that plants and bacteria can synthesize, but, as with the essential amino acids, the human body cannot (except for vitamin D). Therefore, people must obtain vitamins in their diet. Vitamins are relatively small molecules, comparable in size to amino acids or sugars. Whereas daily protein requirements are measured in grams, vitamin requirements are generally measured in milligrams (mg). But even in trace amounts they play important roles in your body.

Several vitamins are coenzymes, molecules essential for making an enzyme function. For example, you need vitamin C for the enzymatic formation of collagen, a substance important in wound healing and in the stability of the joints. You need vitamin A for synthesizing an eye pigment. Vitamin E is an important antioxidant, preventing tissue damage by chemicals. Vitamins are present in foods and are taken up in the body dissolved either in water (vitamins B and C) or in fats (vitamins A, D, E, K). Vitamin D can also be synthesized by the human body when the skin is exposed to sunlight.

Although all living organisms contain nearly all the vitamins, some are particularly rich in certain vitamins and so are especially valuable food sources. Vitamin C is abundant in citrus fruits and is found in fresh vegetables. The B vitamins are most abundant in meats, wheat germ, and yeast. Cod liver oil is a good source of the lipid-soluble vitamins A and D. Vitamin A is also found in yellow vegetables (squash, carrots, sweet potatoes), but vitamin D is not abundant in plants. Vitamin E is especially abundant in green, leafy vegetables and unprocessed plant oils. The minimum levels of vitamins are set high enough to prevent specific deficiency diseases. For example,

1. Vitamin A deficiency causes blindness and increases childhood mortality.
2. Vitamin B_1 deficiency (thiamine) causes beriberi, characterized by weak muscles and paralysis.
3. Vitamin B_3 deficiency (niacin) causes pellagra, characterized by skin lesions, diarrhea, and mental apathy.
4. Vitamin D deficiency causes rickets, characterized by weak and misshapen bones.

Unfortunately, the cereal grain staples that make up two thirds of the human diet do not offer abundant vitamins. Moreover, grain milling removes the vitamin-rich outer layers. For example, beriberi became much more common in Asia when mills started polishing rice, a process that removes the thiamine-rich outer layers of the grain. In North America, beriberi spread when bread made from white wheat flour became popular during the late 1800s. Milled grain is also deficient in folic acid, a vitamin the fetus requires for developing the neural tube and the spinal cord. Maize contains little tryptophan, an amino acid that can act as a precursor for synthesizing vitamin B_3. Moreover, the high levels of the amino acid leucine in maize proteins seem to block conversion of tryptophan into vitamin B_3. These two characteristics combine, so people who eat maize as a staple often suffer from pellagra.

There are four ways to solve vitamin deficiencies. First, if the right foods are available, people can eat a more varied diet, and thus eat foods rich in the vitamins they lack. Second, people who can afford to can buy vitamin supplements, or governments can distribute free supplements. Third, manufacturers can add vitamins to foods when they are processed. This process, termed *fortification*, has often been very successful. For example, in many developed countries both white and brown wheat flours are fortified with vitamins such as thiamine, riboflavin, niacin, and pyridoxine when the grain is milled. This process actually makes these flours, and products made from them, a better source of vitamins than the original whole grain. Another familiar example is the addition of vitamin D to milk. This fortification has greatly reduced the incidence of rickets. Fortification provides a large population with essential nutrients such as vitamins in a convenient way and is a cheap and effective way of improving a population's nutritional status. Fourth, agronomists can genetically modify food plants to contain higher levels of certain vitamins. A recent success story in this regard is the production of Golden Rice, so called because it is yellow, reflecting its high level of provitamin A.

Many people believe organically grown fruits and vegetables are richer in vitamins and minerals than conventionally grown produce. However, no data substantiate those beliefs. Nevertheless, organically grown (Europeans use the term "biological") products are a fast-growing sector of the food market. Because these products are more expensive than conventionally grown crops, the organic food markets cater to wealthier consumers.

Minerals and Water

At least 18 different minerals are essential for human life and must be taken in the diet. You need some of them, such as calcium and phosphorus, in large amounts, You need others, such as iron or magnesium, in smaller amounts. Still others—such as copper, cobalt, and molybdenum—you need in trace amounts. Because many of these minerals are quite common in the liquids people drink and the foods they eat, nutritionists may not pay enough attention to them. The minerals of special concern are those known to be deficient in certain diets: calcium, phosphate, iron, and iodine.

Calcium, phosphate, and magnesium are called the "bone builders" because people need them in large amounts for bone formation. To incorporate these minerals into bones, the body requires vitamin D. Milk and milk products contain abundant calcium and phosphate, which are also in grain products, meat, and a variety of vegetables. A 1985 nutritional survey of the United States revealed that over 40% of the people surveyed took in less calcium than the recommended allowance. Urban households on low incomes tended to have the most calcium-deficient diets. A lack of calcium causes osteoporosis or loss of bone density, a serious disease of the elderly.

Sodium, potassium, and chloride are present in all your cells and fluids as electrically charged ions or electrolytes. These electrolytes maintain blood pressure, play an important role in the acid¾base balance of the fluids (also called the pH), and are vital for transmission of nerve impulses and muscle contraction. An important source of sodium and chloride is table salt (NaCl); potassium is found in all plant foods. Deficiency of these minerals occurs in cases of chronic diarrhea, where fluid losses are excessive. Such loss can lead to congestive heart failure and death.

Iron is the most important micromineral, and iron deficiency leads to anemia, a disease characterized by a general weakening of the oxygen in the bloodstream. Anemia is more prevalent among women than men, because women must replace the blood lost during menstruation, pregnancy, and childbirth. Therefore, the RDA for iron is higher for women (18 mg) than men (10 mg). Normal diets provide between 10 and 15 mg of iron per day, which is not enough for women who lose a substantial amount of blood during menstruation.

The iron requirements of pregnant women are very high, and anemia is especially prevalent among them. Iron deficiency anemia affects 400 million women of childbearing age (15-49 years old) mostly in developing countries. One third of all mothers' deaths at childbirth result from iron deficiency anemia; the babies are often stillborn or underweight. A newborn child contains about 400 mg of iron; placenta growth and blood loss during delivery use up about another 300 mg. The total iron requirement for a pregnancy is therefore about 700 mg, or 2.5 mg per day, in addition to the basic metabolic requirement for iron. Because the body normally absorbs only 10% of the dietary iron, the diet should contain an extra 25 mg per day, for a total of 40 to 45 mg per day. Iron supplements can usually satisfy this high need for iron. But, as with vitamins, scientists can genetically engineer food crops to contain more iron. They have made GM rice that can supply 80% of the iron RDA, compared to 10% for the same amount of non-GM rice consumed (150 g). This rice is not yet available, and further testing is needed to confirm that it is both nutritionally effective and safe.

Although the human body needs only small amounts of iodine per day, the World Health Organization (WHO) estimates that many millions of people suffer from goiter, a disease caused by iodine deficiency. Goiter is characterized by an enlarged thyroid gland, a sluggish metabolism, a tendency to obesity, and enlarged face and neck. The iodine content of many foods is variable, and in many areas of the world it is so low that a normal diet does not provide the body with enough iodine. Iodine deficiency can be most easily prevented by fortifying table salt with iodide.

People take water so much for granted that they often do not realize how crucial it is for life. The loss of 10% of body water is very serious, and a loss of 20% usually results in death. You must take in water continuously to prevent your body from dehydrating and to maintain the proper balance of salts in body fluids. Water is the medium in which all biochemical transformation of the other nutrients takes place. Water helps carry nutrients from your gastrointestinal tract into your bloodstream, because the nutrients are dissolved in water. Water is also important in excretion; waste products are dissolved in it and excreted as urine. Water helps regulate your body temperature by absorbing the heat released by the respiratory activity of all your tissues; much of the heat is used to transform liquid water into water vapour during the process of perspiration. An average person (60 to 70 kg) needs to take in 1,800 to 2,500 g of water each day. About half of this is excreted as urine, and the other half leaves the body as perspiration or in expired air.

Food Plants

To defend themselves against pathogens and pests, plants synthesize many unusual chemicals—more than 50,000 different ones. The effects of only a few of them—usually potent poisons—are known, but researchers have never studied most of these chemicals. Plant biologists refer to them as *secondary metabolites* or *phytochemicals*. Classified by their biological activities, they could be considered pesticides, fungicides, bacteriocides, and so forth. The overwhelming majority occur in nonfood plants, but hundreds also occur in food plants. Quite a large number are carcinogenic, and nearly the entire carcinogenic "load" in our food comes from these types of chemicals, not from synthetic pesticides or other human-made chemicals.

These plant defense chemicals can affect our cells and metabolic processes, but often have no effect on plant cells. In some cases they need to be partially metabolized by the microorganisms in the intestinal tract to become bioactive; in other cases they are stored in separate cellular compartments of plant cells (for example, in vacuoles), where they are harmless. Only when chewing disrupts the plant tissues are the chemicals released. Some cause serious health problems. For example, the fava bean (*Vida faba*) contains chemicals (vicin and convicin) that cause favism, an acute anemia that affects individuals in Mediterranean countries with a genetic deficiency in one particular enzyme (glucose-6-phosphate dehydrogenase).

Cyanogenic glucosides are abundant in some varieties of cassava and need to be removed by extensive washing of cassava pulp because otherwise they would cause cyanide poisoning. Residues of the toxic compounds always remain, and symptoms of poisoning are endemic in areas where people rely on these cassava varieties. Potatoes and other plants, in the same family contain toxic alkaloids. When phytochemicals have obvious health effects, people usually know about them, but they may not recognize subtle effects.

Epidemiological studies and nutritional surveys have led researchers to identify dietary components that may have significant health benefits. Researchers have identified both antioxidants and phytoestrogens as potentially beneficial bioactive food components. For example, soy-rich diets (the so-called Asian diet) correlate with a low incidence of cardiovascular disease, osteoporosis, and estrogen-related cancers such as breast and en. dometrial cancer. Soy-supplemented diets relieve symptoms of menopause such as hot flashes. As a result of these findings, soy-based diets have become more popular in the United States. Soybeans contain several phytoestrogens (plant estrogens), some of which become biologically active only after microorganisms in the gut metabolize them. Phytoestrogens can block the activity of normal estrogen by binding to estrogen receptors and may elevate the level of serum proteins that bind this sex hormone. Through this dual action they may lower the risk of certain cancers correlated with elevated plasma estrogen levels. High plasma estrogen levels occur in people who eat diets heavy in meat and dairy products, and oncologists think these higher levels may induce precancerous cells to become cancerous. However, studies of diets supplemented with specific phytoestrogens have not yet identified the active ingredients in the soy-rich diet. It is possible that soy products also contain other active ingredients, which interact with the phytoestrogens to provide the beneficial effect.

Human cells cannot live without oxygen, but its role in many oxidative processes causes the appearance of molecules called *reactive* oxygen species, highly active derivatives of oxygen that damage body molecules (DNA, RNA, protein, and membrane lipids). This damage accompanies pathological processes such as cancer and inflammation and is also an integral part of normal cellular aging. The cell's antioxidant defenses normally neutralize these highly reactive molecules, but oxidative stress occurs when those defenses are compromised. The interaction of reactive oxygen species with cellular molecules creates destructive free radicals. People can eliminate these if they eat food rich in antioxidants.

All the highly coloured molecules (pigments) in human foods—the anthocyanins found in cherries and blueberries, the lipid-soluble pigments of carrots (alpha- and betacarotene), tomatoes (lycopene), and deep green, leafy vegetables (lutein), act as antioxidation. People who eat diets rich in fruits and vegetables reduce free radicals damage to their DNA (a risk factor for cancer development) and decrease their chance of getting certain cancers. However, supplementing diets with a variety of dietary antioxidants has not been shown to consistently reduce DNA damage or cancer risk. In addition to all these phytochemicals, our food sources contain proteins that have biological activity (not just nutritional value after breakdown into amino acids) and some of the large proteins are broken down into smaller pep tides with biological activity.

The practice of fortifying foods with specific vitamins and minerals, and the discovery that foods contain chemicals with unexpected health benefits has given rise to the concept of functional foods. Functional foods are formulated with specific levels of certain ingredients that promote health beyond supplying essential nutrition. For example, the Netherlands Nutrition and Food Research Institute tested margarine fortified with high levels of plant sterols, to find out how this would affect serum cholesterol levels. The study found a 7-10% reduction in low-density lipoprotein (LDL) cholesterol and no effect on high-density lipoprotein (HDL) cholesterol, also known as "good" cholesterol. Because plant sterols are not nutrients we need, such margarine would be considered a functional food. Unfortunately, many functional foods and other nutraceuticals (a combination of nutrition and pharmaceutical) for which health claims are unsubstantiated, are appearing in the marketplace. In the United States, government agencies leave this area, like herbal medicine, largely unregulated.

Consequences

Energy stores in the human body—glycogen and lipid deposits—are designed to tide a person over the inevitable situations where the demand for food exceeds dietary intake. You use these stores between meals, for example, to keep the glucose level in your blood constant (glucose is the major energy source for your brain). A myriad of complex feedback loops, hormones, and so on, provides the mechanism for regulating and using these stores.

But when people must live on diets that provide only 25 g of plant protein and 1,100 kcal per day, the stores are soon exhausted. Only about one to two days' worth of energy stored in glycogen, and at most a month in stored fats, are available for use. After this, the only source of calories left is proteins, which can be broken down into amino acids and these in turn metabolized for energy. But as mentioned earlier, humans do not store proteins—all are used in some vital way. So breaking down an essential protein means it is then not available to carry out its role in the body.

Among the most accessible proteins for breakdown are those in blood, including antibodies, which are proteins the immune system makes to fight infections. As a result, people who are chronically undernourished are highly susceptible to infections. The poor sanitation of many regions where people are poorly nourished adds to the danger, creating a breeding ground for infectious agents. Children are especially vulnerable. Many infectious diseases that kill children affect those whose immune systems are severely compromised by inadequate nutrition. What is more, poverty puts treatments (such as antibiotics) or prevention (vaccines) out of reach. A second impact of undernutrition, again most dramatically seen in the young, is growth retardation. During the first few years of life, when a child undergoes rapid growth, the body must have the necessary calories, proteins, and other nutrients with which to build more tissues.

Most importantly, nutrient deprivation harms brain development. The human brain grows most rapidly during later stages of prenatal development and the first year of life, and is most sensitive to undernutrition at these times. Before birth, the fetus obtains its food from the mother. As she must eat more to gain the necessary nutrition for both herself and the fetus, dietary deprivations at this time can harm both. Although the mother's nutritional status is sacrificed in favour of the fetus, both animal studies and human observations show that severe caloric restriction causes lower birth weight and later mental retardation. Undernutrition during the first year of life, even if later remedied, often results in a physically smaller brain, as reflected by a reduced head circumference. Children who suffered from malnutrition when they were younger often score lower on intelligence and adaptive behaviour tests than their counterparts who have been adequately nourished and live in a similar environment. These studies suggest that poor nutrition of infants (especially during the first year) may permanently restrict their mental abilities.

Vegetation and Diet

A small number of plants—principally cereal grains, legumes, and root crops—supply most of the energy and nutrients in the human diet. In general, seed staples have a favourable protein content (8 to 15% of the dry weight), whereas roots and tubers have a much lower protein content (1 to 3%). Scientists often express the ratio between protein and carbohydrates as grams of protein per 100 kcal. Because an average adult should have a daily food intake of at least 50 g protein and 2,500 kcal, or about 2 g protein per 100 kcal, staples that contain 2 or more grams protein per 100 kcal are therefore also good sources of protein, It is no accident that early human societies adopted staples that meet or exceed the protein-calorie ratio standard. Currently, more than half the human diet (55%) comes from the grain staples wheat, rice, and maize. Another 13% is provided by the protein-rich seeds of legumes (peas, chickpeas, beans, lentils, soybeans, cowpeas, and so on). Animal products supply 20% of human beings' dietary protein.

The lower protein-energy ratio of plant foods in comparison with animal foods and the low protein scores of plant proteins when compared with animal proteins also help partly explain the correlation between nutritional deficiencies and dependence on plant protein. Plant proteins are usually low in certain essential amino acids, and heavy reliance on a single staple can result in an inadequate intake of these, People can correct this deficiency by supplementing their diet with plant proteins from different sources (for example, by mixing cereal grains and legumes) or by including small amounts of animal proteins in their diet.

If animal proteins have a higher nutritional value for human beings than plant proteins because they have a higher protein score, why not convert more plant protein into animal protein by raising more livestock and producing more milk and cheese? In the past, production of animal protein depended largely on food sources that human beings could not eat. Herbivores such as cattle, sheep, and goats can live entirely on grass and other plants that human beings cannot digest. Animal protein produced in these ways is said to come "cheaply," because food sources of little value to human beings are being converted into foods valuable to humans.

Eating meat nearly every day is so culturally ingrained in developed countries that it is easy to forget that many vegetarians never eat any meat and vegans don't eat any animal products at all. Some do so out of choice, and others are vegetarians because of poverty. Can plants and plant products be an exclusive food source for humans? And can people remain in good health if they only eat plants? The answer to this question is "Yes, indeed"—if people carefully consider their intake of vitamin B_{12}' which neither plants nor animals can make. Only bacteria and other microorganisms such as yeast make this cobalt-containing vitamin. By carefully selecting a variety of plant proteins, vegetarians can obtain a diet with a high protein score by eating complementary foods, usually cereals and legumes. It is easier to ensure a proper amino acid balance by eating a small amount of animal protein to supplement a larger amount of plant protein. The large amounts of animal protein now consumed by most people in technologically advanced countries are nutritionally superfluous and may carry particular health disadvantages, such as types of cancer and heart disease.

Diets rich in plants may have particular health benefits. As already noted, phytoestrogens from soybeans and the antioxidants present in highly coloured fruits and vegetables can prevent certain types of cancers. The nutritional recommendations from the USDA include eating five portions of fruits and vegetables per day.

Genetic Modification of Plants

Because monogastric (poultry, pigs, and fish) and ruminant (cattle, sheep, and goats) animals have very different digestive systems, the goals of the animal feed industry for genetically modifying animal feed are diverse. For ruminant animals the goals are to lower the lignin content of the pasture plants (to improve their digestibility) and to increase soluble, readily digestible components. For monogastric animals such as pigs, chickens, and fish that are fed primarily on seeds, the goals are to eliminate antinutrients. Antinutrients are chemicals that lower animals' absorption of nutrients from the feed.

The cell walls of many cereal grains are rich in arabinoxylans and glucans, which are polymers of xylose and glucose, respectively. These polymers are antinutritional in monogastric animals, particularly poultry, because they increase the viscosity of the gut cantent. The feed industry commonly pretreats cereal grains with the enzymes xylanase and glucanase from microbial cultures that break down these polymers to the corresponding monomers, which are not antinutritional Researchers modified a fungal xylanase gene so that it would be expressed in seeds and introduced the gene into barley. The enzyme was highly expressed in the developing seeds and remained active during storage at 37°C for at least six months, making these seeds a great potential source of feed.

Similarly, researchers engineered a bacterial gene for a heat stable, β-glucanase for high-level production of the enzyme in barley endosperm. They predict that this added enzyme will help break down the β-glucans in barley cell walls and thus increase the energy value of barley in poultry diets. Barley is well known for having low metabolic energy for poultry. The heat stability of the β-glucanase means that the enzyme can survive the pasteurization process needed during feed processing to prevent chick infection by salmonella.

Phytate (myo-inositol hexaphosphate) is the main storage form of phosphorus in cereal and legume feed grains. When the seeds germinate, they produce the enzyme phytase, which liberates the phosphate from inositol. However, neither the mature, ungerminated grain nor the digestive system of monogastric animals has a significant level of phytase activity, so animals use the phosphorus in grain very poorly. Phytate is also a known antinutritional factor because it binds zinc, iron, and other essential minerals. To overcome these two problems with phytate, manufacturers routinely add fungal phytate preparations to stock feed mixes. This greatly improves phosphorus use and increases availability of essential mineral elements. Researchers recently made genetically engineered pigs that secrete phytase in their saliva. These pigs could digest the phytate in their feed. This will obviate the need for separate treatment of feed mixes with the enzyme preparations isolated from fungi and will improve growth rates in animals.

Another research approach is to look for mutant seeds that store phosphorus as inorganic phosphate rather than as phytate. Researchers have identified such a line of low phytate maize, and animals and humans that eat this maize have an increased ability to absorb iron.

Food Safety

Is our food safe to eat? The popular press bombards us with messages about food safety. The food supply in developed countries is much safer than ever in the past, but warnings occur about pathogenic bacteria in poultry and eggs; hormones in beef, carcinogenic pesticide and fungicide residues on fruits and vegetables; mycotoxins (toxins produced by fungi) on maize, soybeans, and peanuts; the unknown effects of radiation to kill bacteria; and antibiotic residues in meat. Although some pesticides and fungicides are carcinogenic, the residues that remain on fruits and vegetables are low enough so that they do not threaten the population. The main threat from pesticides and fungicides is to the workers who apply them, especially in developing countries, and—to a lesser extent—to adjacent ecosystems. Scientists estimate that over 95% of our carcinogen load in food comes from the natural chemicals in food plants. These chemicals are part of the defense mechanism plants use to ward off insects, bacteria, and fungi.

Mycotoxins should be of greater concern, especially if seeds are stored for prolonged periods of time under conditions that permit the growth of fungi. If the seeds are also insect infested, as often in developing countries, then synergism arises between the insects and the fungi, because the inroads made by the insects promote the growth of fungi. One benefit of genetically engineered *Bt* maize is that the reduction in insect damage results in much less growth of the fungal pathogen *Fusarium*, which produces the potent mycotoxin fumonisin.

Food-borne illnesses associated with microbial pathogens are a serious health threat in developed as well as developing countries. The problem is compounded by the emergence of antibiotic-resistant pathogens as a result of feeding antibiotics to poultry, pigs, and cattle to promote their growth. The Centers for Disease Control and Prevention report that in the United States alone, food-borne pathogens and contaminants account for 76 million cases of illness and 5,000 deaths annually. This is probably a substantial underestimate, because many cases of food poisoning are not reported. On a global scale, there are annually as many as 1.5 billion episodes of diarrhea, mostly in developing countries. The World Health Organization estimates that biologically contaminated food causes 70% of these. Providing clean food and clean drinking water is a major challenge for developing countries. These preventable problems put an unnecessary stress on the already overburdened health systems of these countries.

Irradiating food is undoubtedly the most effective way to prevent a substantial part of those deaths and illnesses. Although all major international public health organizations have endorsed the practice as safe and effective, it is not yet generally used because some consumer groups have organized opposition to it.

In the United States, consumers seem to have confidence in the government agencies that regulate food safety. In Europe, partly as a result of several food safety breakdowns in the late 1990s (such as the outbreak of mad cow disease in Great Britain and the contamination of animal feed with dioxin in Belgium), consumers have less faith in their regulatory agencies. The problem needs an international solution in part because international trade in food is continually increasing. For example, according to the U.S. Food and Drug Administration 38% of the fruits and 12% of the vegetables consumed in the United States are imported.

Genetically Modified Foods

Medicines produced by genetic modification techniques have been widely used since 1982, when regulatory authorities first approved the use of insulin isolated from genetically modified bacteria. Food processing aids, such as chymosin (rennet) used in cheese making, which was usually produced from calf stomach, can now be obtained by genetic engineering. Chymosin has been approved since 1990. Food, food ingredients, and feed produced from genetically modified plants are subject to a stringent regulatory process in the countries where they are produced or used. In the United States, for instance, three agencies—the Food and Drug Administration (FDA), the U.S. Department of Agriculture (USDA), and the Environmental Protection Agency (EPA)—have legal and regulatory power to ensure safety of feed and food and to set tolerances for pesticides. The products of genetically modified plants must be shown to be as safe as the products of plants derived by conventional plant breeding. These agencies can remove a food from the market if it poses a risk to public health.

Some consumers have voiced concern about the safety of food from genetically modified plants. They worry that scientists may introduce new allergens or toxins into the plants when they insert a new gene. Clearly, the same problems may arise in conventionally modified crop varieties, but these are not subject to regulation. Occasionally, new traditionally bred varieties have been withdrawn from the market by their producers because they contained unacceptable levels of toxins (glycoalkaloids in potatoes for example). Some people also worry that if the selectable marker gene used to develop the genetically modified plant is a gene conferring resistance to antibiotics, this gene could spread to pathogenic microbes and thus increase resistance to clinically important antibiotics. In fact, many pathogenic bacteria are already antibiotic resistant, but this situation arose from misuse of antibiotics by physicians and their gross overuse in the animal feed industry. No antibiotic resistance has been traced to the genetic engineering of crops.

These and other concerns about GM crops are addressed in depth during the evaluation of any new genetically modified plant for regulatory approval by the same three agencies (see above). The assessments are stringent, science-based and comprehensive. For example, regulatory boards pay special attention to the allergenicity of novel proteins. Some people develop allergic reactions when they consume certain foods—peanuts, soybeans or wheat, for example. These seeds contain allergens, proteins that may cause the body to synthesize a special class of antibody, which triggers severe inflammation. Scientists have traced such allergic responses to certain regions (short sequences of amino acids called *epitopes*) of the food proteins on the surface of the protein. People who know they are allergic to peanuts know not to eat them—but what happens if genes encoding peanut proteins show up in other plants? For example, researchers isolated a gene for a high-methionine protein from Brazil nuts and introduced it into soybean with the aim of improving the methionine status of soybean meal for stock feed. The Brazil nut protein was successfully expressed in soybeans. However, it was well known that some people are highly allergic to Brazil nuts, although the actual allergen had not been identified. When scientists ran allergenicity tests on the methionine-rich protein, it proved to be a major culprit, so they did not develop this particular strategy for improving protein quality further. Given the widespread use of soy proteins in prepared foods, this modification was considered too great a risk to human health by the company that did the work. The result was that the project was discontinued long before the new soybeans reached commercial production.

The approval process for a single GM crop can take up to five years and involves numerous tests that potential distributors of the crop carry out. For instance, for maize plants carrying the gene for the insect toxin from *Bacillus thuringiensis* bacteria, these tests ranged from toxicity studies in animals, to digestibility of the toxin in humans, to toxicity in non-targeted insects. Government scientists and regulators then examine the results of such tests, often comprising thousands of pages. Each country has its own regulations and regulatory agencies, but confidence in such agencies varies. It is important that the procedures regulatory agencies use be transparent to the public and that all regulatory decisions be based on sound science. The perception of risk by the public is a complex phenomenon that social scientists have long studied and biotechnology companies must address in the future.

Despite assurances from government agencies, the public in some countries is concerned about the safety of GM foods. Such uneasiness is not limited to GM foods. Historically, the public has actively opposed some food-related health measures, such as pasteurization of milk in Great Britain (as noted earlier), and fluoridation of water in the United States. Government agencies use science (epidemiology or toxicology) to determine the risk associated with chemicals that occur naturally in foods or that are added during food processing. The public often takes a different view. This discrepancy caused Peter Sandman from Rutgers University in the United States to propose the following equation:

$$\text{Risk} = \text{Hazard} + \text{Outrage}$$

In this equation "Hazard" is the scientifically determined risk factor, or the probability that eating the food will have an adverse outcome. "Outrage" covers the non-quantitative, non-biological properties of the food. The public sometimes pays little attention to hazard but is greatly influenced by outrage. Scientists and governmental advisory groups, in contrast, pay little attention to outrage and concentrate entirely on hazard. "Outrage" factors that Sandman identified include

- Whether the risk is voluntary
- Whether the risks and benefits are equitably distributed in society
- Whether the risk is from natural or synthetic sources
- Whether the risk is subject to individual control
- Whether the risk is familiar or not

For example, the public is not concerned about natural carcinogens in food, but very concerned about synthetic pesticide residues that might be carcinogenic. Physicians and cancer researchers estimate that of the one third of cancers caused by dietary factors, less than 1% are related to synthetic pesticides. Also, the benefits of GM foods have so far not been equitably distributed—most of the benefits have gone to companies and farmers, not to the public. Controversies about risks can generally not be solved by better communication of those risks to the public, although poor communication can exacerbate the problems.

One result of the controversies has been a move to label foods that are produced by GM technologies. A number of European Union countries have made such labeling mandatory. In the United States and Canada (as of 2002), labeling is not mandatory but is used rather, in reverse, to advertise that foods are "GMO-free" (GMO means "genetically modified organism"). This allows companies making those foods to develop a niche market, much as the organic food industry has. The U.S. government has so far rejected mandatory labeling because it takes the position that what matters most is the composition of the crop or the food prepared from it, not the method of producing it. Thus the nutritional content is reported on all types of packaged foods, but not how the crops were bred, grown, or processed.

12

Weeds Interfering Crop Production

Weeds reduce crop yields primarily by competing with crop plants for light, water, and nutrients. In regions where farmers cannot afford herbicides and farm machinery, weeds are possibly the most important factor limiting increased food production. Hand weeding in such places consumes countless hours and is usually done by women and children. In the United States, farmers spend an estimated $10 billion annually to control weeds. The economic importance of weeds is emphasized further by the fact that herbicides comprise as large a share of the worldwide agrochemical market as all other pesticides combined. As with insect and disease management, development of effective weed management strategies starts with understanding the pest. Thus we begin this chapter by defining what a weed is, discussing the attributes of plants that make them weedy, and describing how weeds interfere with crop plants.

Weeds

Everyone has some notion of what a weed is, yet defining weeds with a short statement that is both complete and accurate is surprisingly difficult. A widely used definition is "a plant growing where it is not wanted." The appeal of this definition is its inclusiveness. Common lamb's-quarters growing in a soybean field reduces soybean yield, and therefore a farmer considers it a weed. Similarly, volunteer maize plants are not desirable in a soybean field because they too reduce soybean yield. Volunteer plants are, therefore, rightfully defined as weeds even though one does not normally think of a major crop species as a weed. To take this a step farther, common lamb's-quarters is considered a weed of worldwide importance; young leaves of this plant, however, can make a tasty salad. Downy brome is a troublesome weed to wheat producers, but valuable forage for livestock producers.

Morning glory growing up a trellis is an attractive flowering plant, but twining around maize plants it, is a damaging weed. What emerges from these examples is that whether a plant is a weed depends on your perspective of that plant and where it is growing. Thus the anthropocentric (human-centered) definition of a weed, a plant growing where it is not wanted, is appropriate. Any plant can be a weed if a human decides it is a weed. Simply defining weeds as plants growing where they are not wanted, however, fails to reveal anything about the biological characteristics common to most weeds. Furthermore, although by this definition any plant can be a weed, few plant species actually are *troublesome* weeds.

Of over 250,000 plant species in the world, only a few hundred are troublesome weeds. What biological characteristics of these relatively few plant species make them weeds? Although no single factor determines how weedy a plant is, most weeds share two common characteristics: competitiveness and persistence. On a given area of land, a finite amount of resources (light, water, nutrients, CO_2) is available to support plant growth. Modern agricultural practices try to capture all available resources for crop growth, but frequently weeds can secure some resources for their growth. How extensively weed species draw these resources away from the crop is a measure of their competitiveness.

Weed persistence is their ability to survive year after year on a given area of land, despite farmers' attempts to control them. Rapid seedling establishment, high growth rates, prolific root systems, and large leaf areas are traits contributing to the competitiveness of weeds. Many weed species are not especially competitive on an individual plant basis; instead, they exert their competitiveness through sheer numbers. Such species produce several thousand seeds per plant, and some may produce a million or more per plant! High reproductive output contributes not only to weed competitiveness by numbers, but also contributes to weed persistence. Even if nearly all weeds in a field are killed, the few survivors will produce ample seed to ensure infestations in following years.

Seed dormancy, however, is probably the single most important factor contributing to the persistence of weeds. Seeds lie dormant in the soil, where they form a soil seedbank. Seed dormancy is a mechanism that has evolved to ensure the survival of the species. Cultivated soils typically contain thousands of seeds per square meter of surface area, waiting for the right opportunity to germinate. Many perennial weed species reproduce vegetatively from their roots or from specialized structures such as stolons, tubers, and rhizomes. Canada thistle, for example, has a deep, spreading root system capable of vegetative reproduction. This plant is extremely persistent; even after repeated destruction of the aboveground portion of the plant, the root system will send up new shoots. In fact, patches of Canada thistle and other perennial weeds are often spread by tillage, which breaks up and disperses the vegetative propagules.

Another key trait contributing to the persistence of weeds is developmental plasticity. This is the ability of a plant to alter its growth and development in response to the environment, to optimize reproductive output. For example, many weeds respond to moisture stress and other stresses by initiating flowering even at a very young age. Furthermore, these weeds can then go on to produce mature seeds just a few days after flowering. How did weeds come to possess the traits just described? In other words, where did weeds come from? You can better understand what it is about plants that makes them weeds when you consider the ecological and evolutionary histories of weeds. Before doing so, however, take a look at the taxonomic classification of weeds. Taxonomists have grouped all plants into about 450 different families. About two thirds of the world's most troublesome weeds, however, appear in only 12 of these families. Nearly all major crops, including the grass crops such as wheat, rice, and maize; legumes such as soybean, common bean, and peanut; and other crops also come from these 12 plant families.

Weeds and crop plants are not as different as you might have thought! That weeds and crops are not entirely different makes sense from an ecological perspective, because to be a weed, a plant must adapt to the same environment to which the crop is adapted. Ecologically, weeds are often classified as "ruderals." Ruderals are early successors of disturbed environments. Consider a riverbank that is often disturbed by floodwaters. One strategy for occupying such an environment is to have a rapid life cycle, so that the plant can produce seed before the next flood occurs. In addition, it is important to produce numerous, dormant seeds: numerous because most seeds will be swept away and never land on a favourable environment, and dormant because all the seed should not germinate at the same time, as the environment is so unpredictable. These are the same traits that adapt a plant to human cropping systems, which also undergo frequent disturbance. Plants that are both ruderal and competitive are the worst weeds. Another ecological concept, the niche theory, is also helpful in understanding why weeds succeed. A niche is simply a species' place within a community. An element of the niche theory is that species are specialized to occupy specific niches, and therefore no one species can use all available niches. This is why most natural ecosystems are characterized by numerous species. Cropping systems based on a single crop species, however, provide many unoccupied niches (for example, the space between crop rows, and the area above the crop canopy). In nature, niches seldom go unoccupied. Consequently, monoculture cropping systems are destined to have weeds!

Weeds Reduce Crop Yield

The primary way in which weeds exert their negative effects on crops is by reducing crop yield. They do so through plant-plant interactions that include competition, allelopathy, and parasitism. Competition is the most important type of plant-plant interaction for the vast majority of weeds and therefore is a key topic in weed science. The severity of yield loss caused by weed competition depends on numerous factors:

- The crop grown
- What weed species are present
- Density and spatial distribution of the crop
- Densities and spatial distributions of each weed species
- The duration of competition (when the weeds emerge relative to the crop)
- Availability of resources (light, water, nutrients, and CO_2) for which plants are competing
- Various climatic factors (such as temperature)
- Various edaphic (soil-related) factors (such as soil texture)
- Presence of insects and diseases (which may be general or specific to the crop or certain weeds)

Because of this myriad of interacting factors, it is difficult to predict yield loss simply based on weed density. This is one reason why economic treatment thresholds are not widely used in weed management. In most cropping systems, preplanting operations such as tillage or herbicide applications remove all existing vegetation. During the first stages of crop growth, competition does not occur because resources are sufficient for both crop and weed seedlings, and because the seedlings are not spatially overlapping. As both crop and weed plants grow larger, competition arises. If weeds are not killed now, they will reduce crop yield. If, however, weeds are killed during the first few weeks the crop is establishing, subsequently emerging weeds usually will not be strong competitors with the much larger crop plants. The window of time beginning when weed competition starts and ending when the crop can outcompete any newly emerging weeds is the critical weed-free period.

The significance of the critical weed-free period is that it defines a relatively narrow window of time during which weeds must be controlled to prevent yield loss. In particular, for competitive crops such as maize and soybean, the critical weed-free period is usually no more than a 4-week period that begins a week or two after planting. Less competitive crops, however, such as sugarbeet and onion, have critical weed-free periods of 10 weeks or more. Allelopathy, a type of plant-plant interaction distinct from competition, is the production of a chemical (allelochemical) by one plant that hinders the growth of another.

Many weeds are suspected of producing allelochemicals; because of the difficulty in studying allelopathy, however, its role in agricultural systems is poorly understood. In practice, the effects of competition and allelopathy often are not differentiated, and in both cases the net result is decreased crop yield. Parasitic weeds are physically attached to crop plants, siphoning off resources rather than competing for them. In addition to competition, allelopathy, and parasitism, weeds can interfere with crop plants through indirect interactions.

Certain weeds may increase crop disease severity by acting as a host for one or more stages of a pathogen's life cycle. For example, wheat producers consider barberry a weed because it is an alternate host for the pathogen that causes wheat stem rust. Besides reducing crop yield, weeds may also interfere with crop harvesting and thereby reduce the effective yield. And weeds present in the crop at harvest can reduce crop quality:

1. Weeds in cotton fields at harvest can stain cotton fibers.
2. Juice released from nightshade berries during harvest stain soybeans.
3. Aerial bulblets of wild garlic harvested with wheat impart an off-flavour to the processed flour.

Negative economic effects of weeds also occur through livestock poisoning, by affecting human health and well-being and through negative aesthetic effects. Homeowners seem to prefer weed-free lawns! Similarly, golfers demand weed-free greens, and golf course owners invest heavily in weed-control chemicals. The aesthetic impact of weeds also plays a role in crop production. In highly managed and productive cropping systems, for example, farmers may apply excessive herbicide treatments because they don't want to see any weeds in their fields. In some cases, the land is owned by one person and farmed by another, and the landlord insists that the fields be weed free.

Weed Management

In the preceding sections, we considered weeds from anthropocentric, biological, and ecological perspectives, and discussed their negative impacts. Now, how should weeds be managed? There are three general approaches to managing weeds: prevention, control, and eradication. Because weeds are persistent, once they become established in a field it is nearly impossible to eradicate them. For this reason, the use of preventive strategies is an important—although often overlooked—component of weed management.

The goal of preventive weed management strategies is to keep weed seeds and vegetative propagules from being introduced into a field. Weed seeds are often harvested along with the crop. Consequently, farmers who plant their crops using seed from a previous year's harvest often inadvertently sow numerous weeds. Thus adequately cleaning crop seed or buying weed-free seed is an example of a weed prevention strategy, and the first component of an overall weed management approach. By considering how weed seed might be introduced into a field, the farmer can implement other effective weed prevention strategies. As examples, new weed species may be brought into a field via farm machinery, dispersal from weeds growing in field borders, and irrigation water. Thus, cleaning implements between fields, mowing weeds in field borders to prevent seed production, and installing screens in irrigation systems are weed prevention strategies.

The enormous cost and effort necessary to eradicate a particular weed species is justified only in rare circumstances. Such circumstances include an especially noxious weed that has been introduced only recently into a new area. Once a new species becomes more firmly established (has increased in numbers, increased its range of infestation, and has contributed numerous seeds to the soil seedbank), eradication is impractical if not impossible. One of the few examples of taking an eradication approach to weed management is the attempted eradication of witchweed from the United States. Successful eradication of a weed species requires a long-term, concerted effort consisting of several preventive and control strategies.

Biological and Chemical Weed Control

Despite preventive strategies, some weeds are always present, and as just stated, eradication is rarely a practical goal. Thus the primary weed management approach is to use weed control steps to reduce the negative impact of established weed populations. Weeds are controlled through cultural, mechanical, biological, and chemical means.

Cultural weed control includes all practices that promote the growth of the crop over the growth of weeds. Planting high-quality seed at the proper depth and when soil moisture and temperature are adequate will-give the crop a head start over the weeds. Over the past few years, many soybean producers in the midwestern United States and elsewhere have adopted narrow-row planting. In addition to improving soybean yield, narrow rows benefit weed management. The narrower the rows, the earlier in the growing season leaves of plants in adjacent rows will overlap, shading the soil and hindering growth of weeds. Rotating crops from year to year is also a very effective cultural weed management strategy. Annual weeds compete well with and persist in annual crops but are less persistent in perennial crops. Consequently, rotating a perennial forage crop with annual crops such as maize and soybean reduces the severity of annual weed infestations.

Cultural weed management strategies also include growing two crops together, or "intercropping." For example, oats are often seeded with perennial forages such as alfalfa. The rapidly establishing oat crop competes with and suppresses weeds while the perennial crop is becoming established. Intercropping is a common strategy in developing countries for maximizing food production per unit area and decreasing the amount of hand weeding required. The benefit of intercropping, in terms of weed management, is that two crops are better able than one at occupying niches that would otherwise be available to weeds. Organic farmers often include in their crop rotations the planting of a highly competitive plant species that is grown but not harvested. Such "mulch crops" are used not only to improve soil quality and add nitrogen via nitrogen fixation but also to manage weeds. The weed management benefits of this strategy are, in some cases, partly caused by allelopathy. As noted earlier, allelopathy is a way in which some weeds hinder crop growth. Allelopathy can also be used to hinder weed growth. Rye, for example, is a commonly used mulch crop that is allelopathic to weeds.

Mechanical weed control, the second of the four basic weed control strategies, includes the oldest forms of weed control: hand pulling and hand hoeing weeds. In countries where herbicides and farm machinery are not affordable, hand weeding is still the predominant form of weed control. Even in more developed countries, hand weeding is still commonly used in high-value, small-acreage crops. The disadvantage of hand weeding, of course, is that it is very labour intensive: it takes one person over 200 hours to hand-weed a one hectare field! In contrast, farmers using large machinery can till 10 or more hectares per hour. Primary tillage operations, which break apart and loosen the soil, as well as secondary tillage operations, which are used for seedbed preparation, are effective at destroying weeds. In addition, after planting the crop farmers use selective cultivation, such as tillage between crop rows, to destroy weeds. An important disadvantage of soil tillage is that it greatly increases the soil's susceptibility to wind and water erosion. No-till cropping systems reduce soil erosion, but also eliminate one of the most effective weed management options.

Mowing, another example of mechanical weed control, is an effective way of preventing seed production, but it does not kill most weeds. Perennial weeds, in particular, are rarely killed by mowing. By its broadest definition, mechanical weed control also includes flooding (commonly practiced to control weeds in rice), nonliving mulches (such as plastic sheeting or dead organic material, commonly used in high-value crops), and the use of fire (occasionally used to control weeds in noncrop areas and for initial clearing of land).

Biological weed control is the third of the four basic weed control strategies. Although this can be a successful strategy, to date biocontrol of weeds has been used primarily in pasture and noncrop systems. One reason for the limited use of weed biocontrol in cropping systems is that weed suppression by biocontrol agents usually occurs slowly and incompletely, and therefore the agents do not suddenly halt weed competition. A second reason is that most biocontrol agents have a very narrow host range, and therefore multiple biocontrol agents are needed to control the multiple weed species in anyone field. Despite these limitations, efforts to develop new weed biocontrol strategies continue, and a few biocontrol strategies are currently used in cropping systems.

Mycoherbicides, which are microbial organisms formulated to be applied in a similar manner as are herbicides, have been developed to control specific weeds. For example, Collego® is a formulation of a fungus used to control northern jointvetch in rice and soybean. And biocontrol agents are not limited to microbes and insects: For example, certain breeds of geese selectively eat grasses and nutsedge in broadleaf crops, so a few farmers use them as weed biocontrol agents.

The fourth and final weed control strategy is *chemical weed control*, or the use of herbicides. Herbicides offer many advantages over other forms of weed control:

1. They control multiple weed species.
2. Some can move into roots and thereby kill perennial species.
3. Some possess soil residual activity and thus continue to protect the crop by controlling newly emerging weeds long after herbicide application.
4. They usually cause little or no injury to the crop plant, and thus can kill weeds grow ing right next to crop plants (whereas mechanical control of such weeds would injure the crop plants).
5. They can be applied to large areas of land in a short period of time.
6. They control weeds quickly, bringing competition with the crop to an abrupt halt.
7. They allow farmers to control weeds without tillage, thereby reducing soil erosion.

Of course, anything that sounds too good to be true probably is, and herbicides are not without their disadvantages:

1. They may be acutely and chronically toxic to humans and other nonplant species.
2. They may move with surface waters or leach through the soil and thereby contaminate water supplies.
3. Some herbicides persist in the soil long enough to damage a susceptible crop grown in the same field the following season.
4. Improper herbicide application or unusual weather conditions may result in crop injury or poor weed control.
5. Drifting of herbicides during application may result in off-site, nontarget plant injury (such as the neighbour's prized roses).
6. Herbicides may negatively impact the soil ecosystem.
7. Repeated use of a herbicide may result in a population of weeds that is resistant to that herbicide.

A further disadvantage of herbicides is that because they are so effective, farmers tend to overrely on them and to abandon other good farming practices. For example, since the advent of herbicides many farmers have abandoned the practice of rotating annual grain crops with perennial forage crops. A decrease in diversity of cropping systems leads to increases in disease and insect problems and to a decrease in soil quality. Despite the disadvantages of herbicides, however, they have been used extensively since the middle of the twentieth century, and will continue to be used well into the foreseeable future. Worldwide, farmers spend some $15 billion annually on herbicides. In the United States, more than 95% of all maize and soybean hectares receive at least one herbicide application annually, resulting in the use of 100 thousand tons of herbicide active ingredient!

Herbicides

Herbicides are chemical molecules that can move into plants and disrupt a vital process. They exhibit phytotoxicity at low dosages: The use rate of many herbicides is in the range of 1-2 kilograms of active ingredient per hectare, and some of the more potent herbicides are used at rates as low as a few grams per hectare. The first steps in herbicide phytotoxicity include uptake and translocation of the herbicide by the plant. Soil-applied herbicides may be taken up by plant shoots as they grow through the soil, or by roots. Other herbicides are applied to plant foliage and can penetrate cuticles to gain access into plants. Once inside a plant, most herbicides move with the xylem stream and are distributed to transpiring tissue. Some herbicides are translocated through both the xylem and phloem. These are called *systemic herbicides*, because they are distributed throughout the plant system. Systemics are effective at controlling perennial species because these herbicides are translocated to, and cause death of, the root system. A few herbicides, called *contact herbicides*, are not translocated in plants. Effectiveness of contact herbicides depends on thorough coverage of the foliar tissue. Regardless of whether a herbicide is xylem-, phloem-, or non-mobile, it must, at some point, move into individual cells and disrupt a vital process.

What vital processes do herbicides disrupt? Most herbicides bind to, and thereby block the activity of, a specific enzyme. The protein with which a herbicide interferes is referred to as the herbicide's *site of action*. By binding to their sites of action, herbicides inhibit or disrupt specific physiological processes and thereby cause plant death. A herbicide's *mode of action*, which is a more general term than site of action, refers to the process by which herbicides kill plants. The modes of action of many herbicides involve plant-specific processes—in particular, processes that take place in chloroplasts. Consequently, many herbicides have low toxicity to humans. For example, it stands to reason that a herbicide that targets photosynthesis would not likely be highly toxic to humans. Similarly, herbicides that block synthesis of essential amino acids (those not synthesized by humans) should, a priori, be nontoxic to humans. The fact that many herbicides target processes not present in humans explains why, in general, herbicides are less toxic to humans than are insecticides, which usually do target processes shared by humans. This does not mean, however, that herbicides are nontoxic to humans; herbicides may have acute effects in humans unrelated to their herbicidal effects. Also, long-term exposure to some herbicides may result in carcinogenic or other toxic effects.

Herbicides are often grouped based on similarities in chemical structures. Herbicides belonging to the same chemical family have the same site of action and often share many other herbicidal properties. Different herbicides within the same chemical family can vary in crop selectivity, species of weeds controlled, and soil persistence and mobility. Small modifications to a chemical molecule can significantly alter its herbicidal properties. Compare the structures of 2,4-D and 2,4-DB. 2,4-D is an active herbicide molecule, whereas 2,4-DB, per se, is not. However, most dicot plants metabolize 2,4-DB to 2,4-D, so when 2,4-DB is applied to these plants, the net result is that it is herbicidal. Legume plants do not rapidly metabolize 2,4-DB to 2,4-D; consequently, 2,4-DB can be used as a selective herbicide on alfalfa and other legumes. 2,4-D, in contrast, kills most legumes. This example illustrates how tinkering with the structure of a herbicide molecule can result in slightly different herbicides with new uses. When a company involved in herbicide discovery identifies a promising new herbicide chemistry, it evaluates thousands of molecules that are variations of that chemistry, looking for those with the best herbicidal properties (broad-spectrum weed control, crop safety, and low use rates) and the best toxicological and environmental profiles.

The 2,4-DB example just given also illustrates the basis for the crop selectivity of many herbicides. That is, differential metabolism of herbicides between crop and weed species usually accounts for crop selectivity. Most herbicides do not remain unaltered in plants. Rather, many biochemical reactions metabolize herbicide molecules, usually resulting in the degradation of the herbicide into nontoxic molecules.

The ability of crop species to metabolize certain herbicides into nontoxic metabolites faster than weed species metabolize them allows the selective use of these herbicides. The 2,4-DB example is one of a few exceptions to the norm, because a herbicide metabolite (2,4-D in this case) is more phytotoxic than the parent herbicide. Usually herbicide metabolites are less phytotoxic than the parent herbicide. Plant metabolism of herbicides may also result in the chemical combining of the herbicide molecules with sugars and other organic molecules, rendering them nontoxic. For example, the enzyme glutathione-S-transferase conjugates atrazine and many other herbicide molecules with glutathione, a tripeptide molecule.

Some herbicides are nontoxic to the crop because the crop has a different site of action from that of the targeted weed species. This is true for the aryloxyphenoxy-propionate and cyclohexanedione herbicides on dicot crops. The site of action of these herbicides, acetyl-CoA carboxylase (ACCase), catalyzes the first step in fatty acid biosynthesis, the formation of malonyl-CoA. Dicots have two forms of ACCase—one is plastid-encoded, the other is nuclear-encoded—whereas grasses have only the nuclear-encoded ACCase. The aryloxyphenoxy-propionate and cyclohexanedione herbicides inhibit only the nuclear-encoded ACCase. Consequently, farmers can use these herbicides on many dicot crops to control grass weeds.

Weeds Attempt to Control Themselves

During human attempts over the years to control them, weed populations have not remained static. Strategies that farmers use to kill weeds have resulted in numerous examples of Darwin's theory of natural selection in action. Because weed species evolve in response to weed management practices, some weeds are becoming harder to control. A common consequence of repeated use of a particular weed control strategy or a change in cropping systems is a change in the predominant weed species that are present. These so-called *weed species shifts* occur because certain weed species are more tolerant than others of particular control strategies.

Shifting from conventional tillage to no-till cropping systems usually increases perennial and biennial weed infestations. Also, small-seeded weed species, such as pigweeds and lamb's-quarters, become more common under no-till because the small seeds depend less on burial by tillage for germination. Heavy use of a particular herbicide also causes weed species shifts. For example, repeated use of 2,4-D, which controls only dicot weeds, may result in an increased prevalence of grass weeds. This shift occurred in cereal crops in the 1950s and 1960s, when 2,4-D was about the only herbicide option available.

Weed species shifts are changes in the types of species; evolution in response to weed control may also occur within weed species. Most weed species show substantial genetic variation. Variants within a species that possess genetic attributes allowing them to survive a particular weed control strategy increase in frequency on repeated selection by that control strategy Weed scientists often use the term "biotype" to distinguish such variants from the species as a whole. One way in which weed species have evolved to avoid control is to look more like the crop. Because these crop mimics are more difficult to distinguish from the crop, they are likely to be overlooked by farmers during hand weeding. The common practice of handweeding rice fields has resulted in the selection of a biotype of barnyardgrass that looks more like rice than does normal barnyardgrass.

Sometimes crop mimicry results because a weed hybridizes with the crop species. For example, cultivated rice occasionally hybridizes with wild rice, resulting in wild rice strains that mimic cultivated rice. A purple-leafed strain of cultivated rice was developed so farmers could easily identify the crop during hand weeding. This solution was short-lived, however, as hybridization between this strain and wild rice soon resulted in purple-leafed biotypes of wild rice! Another example of crop mimicry is the evolution of weed seeds that are more similar in size and shape to crop seed. This occurs because the result is that the weed seeds are less likely to be removed during harvesting and cleaning of the crop seed.

Consequently, the weeds are planted with the crop the following year. Examples include the evolution in lentil fields of a biotype of common vetch that produces flat rather than round seeds; and the evolution in flax fields of a biotype of flaseflax that flowers at the same time as the crop, produces fruits that do not shatter (and hence the seeds are harvested with the crop), and produces seed nearly identical in size and shape to the crop seed. Because herbicides are so effective at controlling weeds, they exert a strong selection pressure. Consequently, weed populations have evolved that are resistant to herbicides. There is no evidence that herbicides have ever actually created a mutation that conferred herbicide resistance. Rather, the evolution of herbicide-resistant weed biotypes is analogous to the evolution of crop mimics.

Herbicide-resistant individuals preexist in weed populations because of natural variation, and herbicides simply select for those individuals.

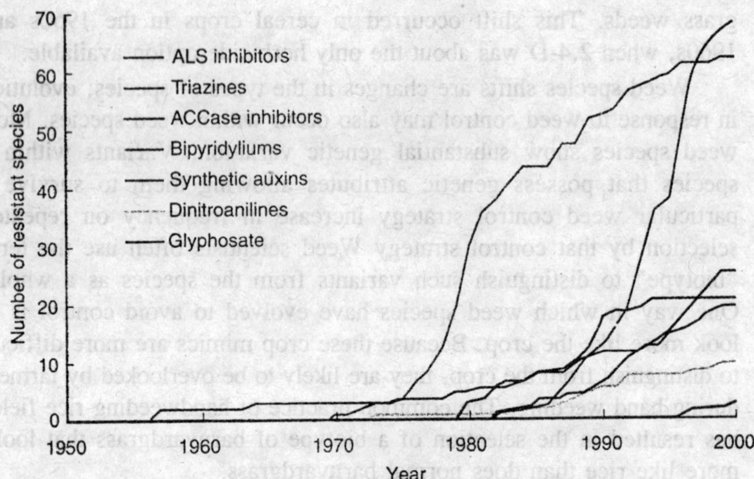

Fig. 12.1. Evolution of herbicide-resistance weeds.

After repeated herbicide selection, alleles of genes conferring herbicide resistance increase in frequency within weed populations, eventually resulting in resistant populations. The first example of a problematic herbicide-resistant weed population was reported in 1970 with the identification in Washington State of common groundsel resistant to triazine herbicides. Continued use of these herbicides has led to several additional triazine-resistant weed populations.

Widespread use beginning in the 1980s of herbicides that target acetolactate synthase (ALS) and of herbicides that target ACCase quickly led to many more herbicide-resistant weeds. By 2001, over 150 weed species had evolved resistance to at least one herbicide, and for nearly every major herbicide there is at least one weed that has evolved resistance to it. In some cases, biotypes of weeds have evolved resistance to multiple herbicides with distinct modes of action. The mechanism of herbicide resistance is usually a change in the herbicide target site that reduces its affinity for the herbicide. In other words, the herbicide is no longer able to bind and inhibit the target site. Often the change is very subtle—a single amino acid—and thus requires no more than a single base mutation within the gene that encodes the target site. Although subtle, the change is specific: The change must be sufficient to abolish herbicide binding, yet it cannot severely alter the normal function of the target site.

The basis for triazine resistance in weed populations typically is a substitution of glycine for serine at amino acid position 264 in the

D1 protein of photosystem II in the chloroplast, which is the herbicide target site. This change renders the D1 protein insensitive to triazine herbicides, yet the protein is still able to carry out its normal function of transferring electrons through photosystem II. A cost is associated with this change, however, in that the triazine-resistant D1 protein is not as efficient as the wild type protein in electron transfer, which in turn reduces the photosynthetic efficiency of the plant. Consequently, triazine-resistant plants are slightly less fit than wild type plants in the absence of triazine herbicides. This explains why triazine-resistant individuals are initially at a very low frequency in natural weed populations.

Site of Action

The development of herbicide-resistant weed populations has been most problematic for herbicides that target the enzyme acetolactate synthase (ALS). This is partly because several, single amino acid substitutions can occur in ALS that render the enzyme insensitive to ALS-inhibiting herbicides but have little to no effect on the enzyme's normal catalytic activity. Populations of weeds resistant to ALS-inhibiting herbicides have arisen in as few as four years after repeated use of these herbicides. Other herbicides are much less likely to select for resistant weeds. Glyphosate, for example, is the world's most widely used herbicide, yet only a few glyphosate-resistant populations have evolved.

The likelihood for resistance-development to a particular herbicide strongly depends on that herbicide's site of action, and on whether point mutations in that site of action will abolish herbicide binding but not normal activity of the site of action. For this reason, sustainable use of herbicides (management of herbicide resistance) requires knowledge about the physiology and biochemistry of how herbicides work and about the genetics of herbicide resistance. Site of action-mediated herbicide resistance is caused by a single gene change and, if it involves a nuclear gene, may be dominant or recessive. Resistant ALS genes, for example, are dominant (or semidominant), and this further explains why resistance to ALS inhibiting herbicides evolves so rapidly. For a recessive resistance gene, the resistant allele is only selected if it exists in a weed in a homozygous condition.

Geneticists estimate that alleles of genes that confer herbicide resistance exist in natural populations typically at a frequency of 10^{-6}; the likelihood that a weed will be homozygous for such a low frequency allele is exceedingly small, especially for outcrossing weed species.

Site of action-mediated resistance to dinitroaniline herbicides is recessive. Not surprisingly, resistance to dinitroanilines has occurred only rarely. Once herbicide resistance develops in a weed population, it may spread to other populations. For dominant, nuclear-encoded resistance, such as resistance to ALS-inhibiting herbicides, pollen movement readily spreads the trait. As noted earlier, resistance to triazine herbicides usually arises via a change in the D1 protein.

The gene encoding this protein is in the plastid genome. Because plastids of most plant species are inherited maternally, triazine resistance normally spreads only by seed movement, not via pollen. The previous discussion of herbicide resistance has focused on site of action-mediated resistance. Although this is the most common mechanism of resistance in weed populations, other mechanisms of herbicide resistance have evolved. The second most common mechanism is metabolism of the herbicide. As noted, herbicide metabolism is often the basis for crop selectivity of herbicides.

Sometimes weeds evolve the ability to rapidly metabolize herbicide molecules, rendering them less sensitive to those herbicides. Although the magnitude of resistance is often much less than is seen with site of action-mediated resistance, metabolism-based resistance may result in resistance to multiple herbicides with different sites of action. What can farmers do to prevent the evolution of herbicide resistance and preserve the effectiveness of currently available herbicides? As previously mentioned, herbicide-resistance management requires that farmers understand how and why herbicide resistance develops.

Herbicide resistance evolves in response to herbicide selection pressure. Therefore, the best way to prevent resistance development is to avoid repeatedly applying the same selection agent. Farmers are encouraged to use herbicides with different sites of action from year to year. In Australia, the government sought to help farmers do so, by requiring that herbicide containers be labeled with an alphabetic code indicating that herbicide's site of action. Other countries are now considering similar requirements. Besides rotating herbicides, farmers are also encouraged to use nonherbicidal weed control strategies to prevent development of herbicide resistance. In practice, however, farmers are often driven by the philosophy of "It worked last year, so I am going to use it again this year" and feel driven by economics to use the same one or two herbicides year after year. Consequently, herbicide resistance will continue to be a major problem confronting farmers in their ongoing efforts to control weeds.

New Strategies for Managing Weeds

Because weeds will continue to find ways to adapt to weed control efforts, farmers will always need new strategies. Biotechnologists have developed new weed management tools, and many more are in progress. Before we describe these tools, however, it is important to note that they are just tools. As people involved in crop production continue to learn over and over, for pest management there is no easy solution. Weeds, like any other pest, must be managed with multiple strategies. Strategies offered by biotechnology simply add to a farmer's collection of weed management tools. By far, the predominant new tools for weed management arising from biotechnology have been herbicide-resistant crops (HRCs).

In fact, HRCs are currently the biotechnology products that dominate all crop production. There are at least two reasons for HRC dominance over other crops created from biotechnology. First, compared to the manipulation of other traits, such as crop yield or crop composition, HRCs can be created with a single, easily selected gene and thus is a very straightforward process. Second, there is a large economic incentive for a company to develop HRCs: By developing a crop resistant to a herbicide that the company markets, the company can capture return on its investment by selling both the crop seed and the herbicide. What exactly are HRCs, and how are they developed? HRCs are simply crop varieties that are resistant to a herbicide that is normally lethal (or highly injurious) to that crop. Researchers develop them through either a selection or gene insertion process. Selection can be accomplished simply by screening large numbers of plants with the herbicide of interest, looking for a naturally occurring variant within the crop species that is resistant. In practice, however, it is more efficient to create variation by subjecting the crop to a mutagenic agent (chemical or radiation) before performing the selection. Crops that are resistant to sulfonylurea, imidazolinone, and cyclohexanedione herbicides have been obtained by selection and are now commercially available.

HRCs obtained by selection are analogous to herbicide-resistant weeds and have the same mechanisms of resistance. In fact, the occurrence of herbicide-resistant weeds is what inspired scientists to develop HRCs by selection! Scientists use gene insertion (transgenic) approaches when they cannot get an acceptable level of resistance to a particular herbicide from selection. The inserted gene may encode a form of the herbicide target site that is insensitive to the herbicide.

Genetic engineers have obtained resistance to glyphosate by inserting into crop plants a modified gene from an *Agrobacterium* species that encodes a resistant form of the herbicide target site, 5-enolpyruvyl-shikimate-3-phosphate synthase (EPSP synthase). Researchers isolated this bacterium as resistant to glyphosate, and subsequent research revealed that the unusual form of EPSP synthase caused the resistance.

The gene was isolated and modified (to ensure adequate expression in plant cells and to ensure targeting of the encoded EPSP synthase to chloroplasts, where it functions) and then transferred into plants. Resulting crops are called Roundup Ready® because they are resistant to Roundupâ herbicide, which contains glyphosate as the active ingredient. Transgenic HRCs that metabolize particular herbicides have also been developed and commercialized. LibertyLink® crops are resistant to glufosinate (sold as Liberty® herbicide) because these crops have been engineered to metabolize this herbicide. Glufosinate, also called *phosphinothricin*, is essentially a synthetic, truncated version of a toxin produced by *Streptomyces hygroscopicus*.

Glufosinate

Phosphinothricin-L-alanyl-L-alanine

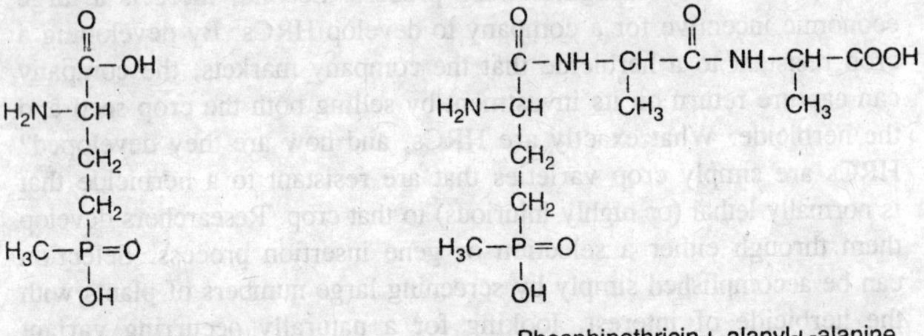

Fig. 12.2. Chemical structures of phosphinothricin-L-alanine, a natural compound produced by Streptomyces hygroscopicus, and the herbicide glufosinate.

To prevent auto toxicity, this organism also produces an enzyme, phosphinothricin-N-acetyl transferase, which acetylates phosphinothricin into a nontoxic. metabolite. Genetic engineers used a gene encoding this enzyme to make transgenic, LibertyLink® crops. Herbicides already exist that can be used selectively for growing major crops, so what new advantages do HRCs offer? Essentially, HRCs enable "better" herbicides to be used on crops. Glyphosate, for example, is a good herbicide in that it effectively controls most major weeds (including troublesome perennial species), is essentially nontoxic to humans, has low environmental persistence, has a low likelihood to select for resistant weeds and has foliar activity (and therefore farmers can use it based on need, after determining the severity of the weed infestation). Furthermore, because glyphosate is so effective, it reduces the need for tillage and thus can be used to reduce soil erosion.

Given all these advantages, it is not surprising that within five years after their introduction, Roundup Ready® varieties comprised over 50% of the U.S. soybean market. A survey of 452 farmers in 19 U.S. states released by the American Soybean Association in the autumn of 2001 showed that about half the U.S. soybean growers are indeed using conservation tillage practices in combination with Roundup Ready® soybeans and making fewer tillage passes over their fields compared to 1996. These practices not only limit soil erosion, but they reduce production costs as well.

In some areas of South America, farmers are planting these soybeans without authorization from their government or from Monsanto, the producer of these genetically modified (GM) soybeans, because of the cost savings they provide. What are the disadvantages of HRCs? Much of the opposition to HRCs centers on fears and concerns common to people's reactions toward all transgenic crops or "genetically modified organisms" (GMOs). Remember that many HRCs have been obtained by selection and not by gene transfer, and thus are not classified as GMOs. Nevertheless, potential disadvantages are associated with HRCs, especially if they are not used correctly. One of the

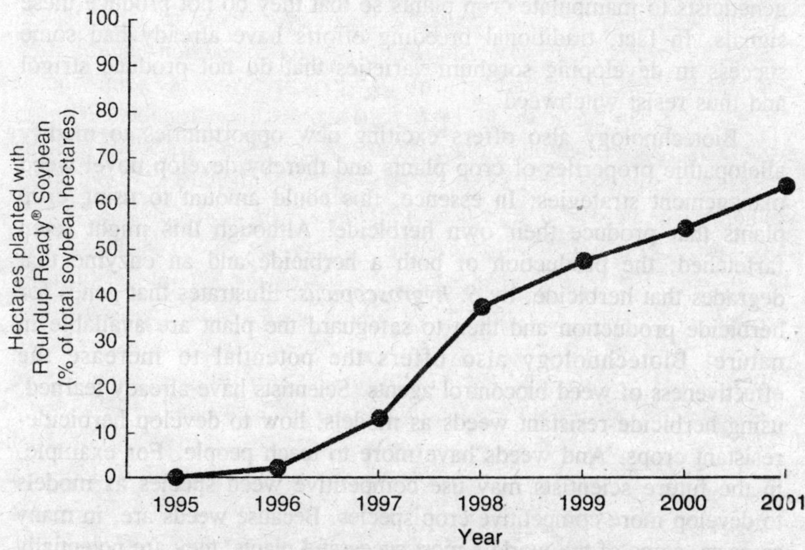

Fig. 12.3. Adoption of Roundup Ready' soybean.

major potential disadvantages of HRCs is that they may be overused. Overuse may result in farmers abandoning other weed management strategies, further reducing diversity of crop systems and creating new problems. The availability of HRCs tempts farmers to rely on a single herbicide, which, as discussed before, can lead to the selection of herbicide-resistant weeds and can bring about shifts to more tolerant weed species. Herbicide-resistant weeds may also arise by gene flow from HRCs. Although transmission of herbicide resistance between crops to weeds has been possible before the advent of HRCs, the single-gene nature of resistance in HRCs increases the likelihood for resistance transmission. Another potential problem is that HRCs themselves may become weeds that are difficult to control. For example, a farmer growing Roundup Ready® maize one year and Roundup Ready® soybean the next would have volunteer maize growing in the soybean crop that glyphosate would not control.

Resistance in volunteer crops can be compounded when different herbicide-resistant varieties of a particular crop are grown in proximity. These plants resist glyphosate, glufosinate, and imidazolinones! Biotechnology offers weed management more than just the development of HRCs. Parasitic weeds can be especially difficult to control, and some use chemical signals from the host plant for inducing both seed germination and attachment of haustoria. Biotechnology may enable geneticists to manipulate crop plants so that they do not produce these signals. In fact, traditional breeding efforts have already had some success in developing sorghum varieties that do not produce strigol and thus resist witchweed.

Biotechnology also offers exciting new opportunities to modify allelopathic properties of crop plants and thereby develop novel weed management strategies: In essence, this could amount to using crop plants that produce their own herbicide! Although this might seem farfetched, the production of both a herbicide and an enzyme that degrades that herbicide, by *S. hygroscopicus*, illustrates that genes for herbicide production and then to safeguard the plant are available in nature. Biotechnology also offers the potential to increase the effectiveness of weed biocontrol agents. Scientists have already learned, using herbicide-resistant weeds as models, how to develop herbicide-resistant crops. And weeds have more to teach people. For example, in the future scientists may use competitive weed species as models to develop more competitive crop species. Because weeds are, in many respects, some of the world's most successful plants, they are potentially a rich source of genes for crop improvement.

13

INSECT, MITE, AND NEMATODE PESTS

Despite today's best management efforts, the insects, mites, and nematodes that feed on plants reduce the amount of food and other crops that farmers produce. Therefore, people generally consider these invertebrates "pests." Crop losses occur regardless of whether crops are grown in organic, conventional high-input, or subsistence farming systems. Obviously, farmers want to minimize these losses, to get a higher return (food or money) for their capital and labour. Much controversy and confusion, however, surrounds the various approaches used today to control agricultural pests. A number of strategies have been developed to reduce crop losses from invertebrate pests. Multiple strategies can be woven together in *integrated pest management* (IPM).

A specific IPM program may be developed for each crop species and geographic production region. The primary goal of integrated pest management is to keep pest levels below the *economic injury level*, the initial point at which crop losses from the pest can justify the cost and use of a control strategy. Many pest control strategies exist; one of the most widely used is extermination with chemical pesticides. Because of the adverse health effects, environmental impact, and lack of specificity of many such chemicals, chemical control is less desirable than biological control.

Biological control seeks to suppress outbreaks of pests in agricultural crops by using living organisms that kill the pests, such as predators, parasites, and disease. Another effective and ecologically sound strategy is to breed pest-resistant plants. This approach requires that genes for desired resistance traits be transferred as genes into the susceptible crop variety from other plants or organisms. Combination classical plant breeding with genetic engineering can improve pest resistance in crops.

Genetic engineering is also being used to improve the effectiveness of biological control agents, biopesticides, and chemical pesticides. Two major differences between agroecosystems and natural untouched ecosystems are that (1) agroecosystems are usually simpler, with fewer species of plants and animals, and (2) agroecosystems are managed by humans. Considerable controversy exists over how to best manage agroecosystems for long-term sustainability, which strategies of pest control should be used, and how and when they should be applied. Thus many biological, societal, and economic factors require that pest control strategies and integrated pest management programs be dynamic, changing from one geographic crop production region to another, and in some cases from one year to the next. Pest control approaches used in high-input systems in technologically advanced countries differ from those of subsistence farmers in less developed countries.

Crop Yields

Insects, mites, and nematodes reduce crop yields worldwide each year by more than 20% on average. However, individual fields may sustain losses of 50 to 100% from one or more pests. Insect pests are the most damaging, responsible for 50 to 60% of these losses; next come nematodes, which cause 40 to 50% of the losses; least damaging are mites, responsible for 10 to 20% of the losses. The USDA estimates that 3% of the annual U.S. wheat crop is lost to insects alone: enough bread to feed two million people for more than 20 years. Pest damage to crops is greatest in the tropics, with losses averaging 20% from insect and mite pests, and an additional 20% from nematodes. Less well known is the fact that insects and mites also damage stored agricultural products, such as the primary food grains (rice, maize, and wheat) and root crops (potatoes and sweet potatoes) and legumes (beans).

In developing countries, especially in the tropics, crop storage losses from insect pests alone average 30%, whereas the crop losses during storage are much lower, averaging less than 10% in developed nations located in temperate climates. Thus developing countries in the tropics, with the greatest need for food, suffer the greatest losses from pests—losing on average about 50% of their crops to combined pre- and postharvest pests. Surprisingly, since the initial widespread use of synthetic insecticides in the 1950s the annual percentage of crops lost to insects and mites has remained relatively constant, at about 12%. However, if you look at individual crops worldwide, you see that the percentage lost to these pests was greater in 1990 than in 1965. On average, the percentage lost to all invertebrate pests is the same today as in 1990. The amount of pest-caused yield loss or decrease in yield quality farmers tolerate depends on their economic circumstances. For subsistence farmers yield loss means less food, whereas losses for commercial farmers in developed countries mean less profit, or a revenue loss that puts the farmer out of business.

Invertebrate pest outbreaks, and the resulting yield losses and high control costs are a major cause for the low profit and riskiness of farming. Worldwide, analysts valued total preharvest losses for the 1988 through 1990 period at US$90 billion for eight principal food and cash crops (barley, coffee, maize, cotton, potato, rice, soybean, and wheat). You can see that more effective methods of invertebrate pest control are needed to reduce pest-induced losses. Using genetically engineered pest-resistant crops may be the best strategy for today's farmer to achieve sustainable improvement in pest control and yields.

How do Pests Damage Crops?

Invertebrate pests can damage crops in a number of ways. Some insects such as certain beetles, grasshoppers, and caterpillars are chewers that consume plant tissues; especially desired are young leaves and immature flower buds and fruit because they are better nutritionally and contain fewer or lower levels of defense chemicals. Chewers may directly reduce yield by damaging or consuming plant parts to be harvested, or indirectly, by reducing photosynthesis and growth. Other plant-feeding insects such as plant bugs and thrips, and all plant-feeding mite and nematode species, use sharp mouth parts called *stylets* to penetrate specific tissues; they inject salivary secretions and microorganisms, then suck up the partially digested plant cell components. These stylet users are specialists at these sipping and sucking techniques that damage leaves, fruit, and roots, reducing the plant's ability to grow and produce fruit.

Aphids use another feeding strategy, inserting their stylets into the phloem and sipping the photosynthate directly. As noted earlier, some pests damage seeds when they are stored. Some pests are crop specialists, feeding only on one or two crop species, such as the famous boll weevil, which feeds and reproduces almost exclusively on cotton. Others, such as the two-spotted spider mite and many nematode species, are generalists feeding on many crops species. Many insects, such as the corn rootworm, and nematodes live and feed below ground on plant roots. Some insects and mites burrow in leaves, stems, or fruit, feeding and living inside plant tissues, making these pests difficult to control with conventional synthetic pesticides.

Plant-feeding nematodes are the most abundant invertebrates in many agroecosystems, reaching densities of 30 million per square meter of soil. They are smooth, tiny roundworms, usually less than 2 mm long, and so thin they are almost invisible to the naked eye. Many live unseen in the soil, as much as 10 cm or more below the surface, where they feed on roots and underground tubers. Their feeding can result in wounding, scaring, galling, and death of roots, reducing the ability of roots to contribute to plant growth, thus reducing plant yield.

Pests Transmit Plant Pathogens

Do pests damage crops in ways other than by feeding on them? A few leafhoppers feeding on a sugar beet plant will cause little yield loss. However, if these insects are carrying the curly top virus, a serious pathogen of sugar beets, feeding by the leafhoppers can infect the entire plant with the virus resulting in plant stunting and yield loss. Worse yet, the disease will be spread throughout the crop, causing considerable yield and economic losses. Thus in addition to crop damage caused by feeding, insect, mite, and nematode pests can cause additional yield losses by carrying and infecting crops with disease-causing pathogens.

Worldwide, over 200 plant diseases are known to be transmitted by insects, mites, and nematodes, three quarters of which are virus diseases. Many of these diseases dramatically reduce crop yields and yield quality. Pest-transmitted diseases have been especially destructive in tropical countries. For example, African cassava mosaic virus, transmitted by whitefly insects, causes crop losses of 30 to 40% in all African cassava-producing countries. Another way that pests injure crops is through the saliva they inject into the plant during feeding. Saliva commonly contains various chemicals, including enzymes that can injure plant cells or stimulate abnormal plant growth and/or abscission (loss) of flowers or fruit, resulting in greater reductions in photosynthesis and yields than from the feeding damage alone. How many pest species attack a crop?

The number of invertebrate pest species found in a field on a particular day may be as high as 50; however, in temperate climates only 5 to 10 will be doing damage, and of these only 1 to 5 would be

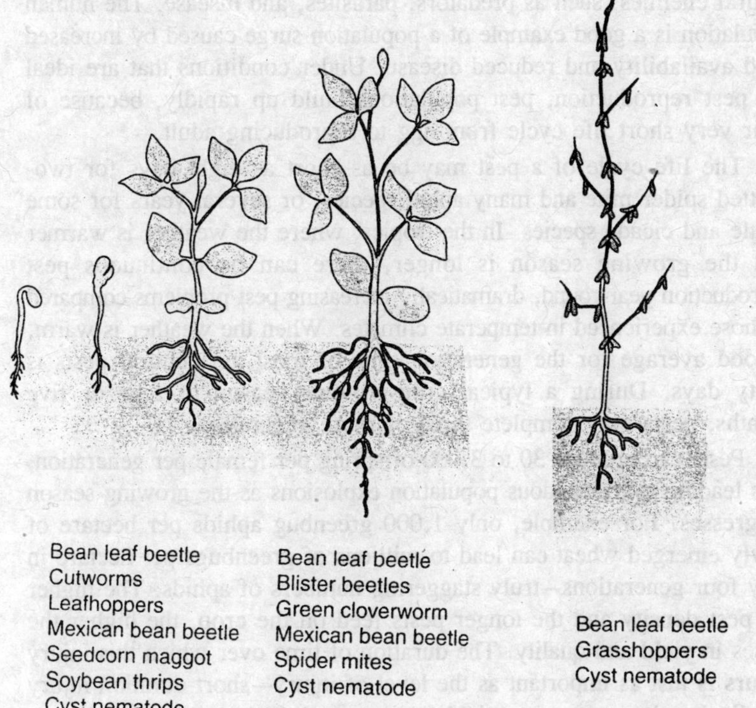

Fig. 13.1. The pests that attack soybean plants change throughout the season as the plants develop.

doing sufficient damage to be considered for a control strategy. This complex of damaging pests usually consists of one nematode, one mite, and four insect species. As the crop grows and matures, different pests appear and disappear in response to changes in nutritional quality of the crop, weather, natural enemies, and diseases. In the tropics, a particular field may be infested with twice the number of pest species found in temperate climates.

Pest Outbreaks

Why do pests suddenly appear and cause economic crop loss? Populations of a particular herbivorous insect, nematode, or mite species in both natural plant communities and agricultural crops can suddenly increase to tremendous numbers. These outbreaks are a result of the changes in biological or environmental conditions that would normally hold these pest populations in check. The most important natural population regulating factors are weather and abundance of food and natural enemies, such as predators, parasites, and disease. The human population is a good example of a population surge caused by increased food availability and reduced disease. Under conditions that are ideal for pest reproduction, pest populations build up rapidly, because of their very short life cycle from egg to reproducing adult.

The life cycle of a pest may be as short as eight days for two-spotted spider mite and many aphid species, or several years for some beetle and cicada species. In the tropics, where the weather is warmer and the growing season is longer, there can be continuous pest reproduction year-round, dramatically increasing pest problems compared to those experienced in temperate climates. When the weather is warm, a good average for the generation time of most agricultural pests is thirty days. During a typical crop-growing season of four or five months, pests may complete three or more generations.

Pests can produce 30 to 3,000 offspring per female per generation, thus leading to tremendous population explosions as the growing season progresses. For example, only 1,000 greenbug aphids per hectare of newly emerged wheat can lead to millions of greenbugs per hectare in only four generations—truly staggering numbers of aphids. The higher the pest density and the longer pests feed on the crop, the higher the losses in yield and quality. The duration of time over which this injury occurs is just as important as the level of injury—short-duration injury usually has less effect on yield loss or quality than injury of longer duration. Agricultural activities over the centuries have increased the opportunities for pest outbreaks in crops. Some of the most important activities favouring such outbreaks are

1. Growing crops as a monoculture that provides large areas of the pest's favourite food crop.
2. Improving crop suitability for the pest with irrigation, fertilizer, and herbicide.
3. Growing the crop in areas where no natural enemies of the pest exist.
4. Growing the crop in a new area causing native species of insects, mites, or nematodes to feed on the crop and become pests.
5. Unknowingly introducing pest species of insects, mites, or nematodes from one geographic area of the world to other areas where the crop is grown, and here they adapt to the new cropping environment.
6. Accidentally removing the natural enemies of pests through farming practices such as cultivation, weeding, irrigation, mono culture,

and planting of nursery crops, and the use of herbicides, plant growth regulators, broad-spectrum pesticides (which are toxic to a broad range of invertebrates).

7. Continued use of the same chemical pesticide, resulting in pests adapting to the pesticide so that it is no longer toxic; this is known as *pesticide resistance*.

Farming Practices may Contribute to Pest Damage

As you can see, many of the farming practices that enhance yields also contribute to increased pest problems. The widespread use of broad-spectrum pesticides is one of the greatest factors aggravating pest outbreaks. Farmers have found that repeated application of pesticides over 20 or more generations of the pest all too frequently results in the development of pest populations that are resistant to the pesticide. In other words, resistant pests can feed, grow, and reproduce normally in the presence of the pesticide.

Pesticides kill not only the crop pests but also the insects, mites, and nematodes that function as natural enemies of the pest. When the natural enemies are removed, the crop pest targeted with the pesticide may rapidly return to even more damaging numbers than before it was treated; this is known as *pest resurgence*. Further, loss of natural enemies may induce the outbreak of *secondary pests*; these are pests that were held in check by natural enemies and caused no economic damage until broad-spectrum pesticides were applied. A good example of this problem was the epidemic of brown planthopper in Indonesia after the government started subsidizing pesticide use up to 75% of the cost of the pesticides. Before the use of nonspecific chemical pesticides, brown planthopper populations were held in check by their natural enemies. Pest resurgence and/or outbreaks of secondary pests result in the need for more pesticidal sprays throughout the growing season. This pesticide-induced paradigm has been called the *pesticide syndrome* or *pesticide treadmill*, and once a farmer is caught on the treadmill it is difficult to get off. The best way to avoid the treadmill is to minimize use of broad-spectrum pesticides.

The high frequency of pest outbreaks in crops is most commonly caused by (1) the accidental introduction of foreign pest species, without their natural enemies, to regions of the world where they were never present before, such as the pink bollworm introduced along with cotton seed to Australia, and (2) aerial migration of large numbers of pest insects and mites that move rapidly from one field or crop to another or from native vegetation to cropland, over distances less than a kilometer to greater than 100 kilometers. Aerial migration is a common survival strategy for many pests like plant bugs, armyworms, aphids and spider mites, but not nematodes. Neither introduced nor migratory pests are controlled by natural enemies because these enemies were left behind on the previous crop or native vegetation or in another country. To remedy the lack of natural enemies in many crop ecosystems, programs have been conducted to locate effective natural enemies, and release them in the pest-infested crop. These natural enemies are also called *biological control agents* and the pest management strategy that uses them to control crop pests is known as *biological control.*

PESTICIDES

Pesticides have been and will probably continue to be the most effective method to quickly stop pest outbreaks that threaten to destroy crops. Comparisons of pesticide effectiveness for different pests invariably show the dramatic effect of pesticide use: Without pesticides, crop damage ranges from 35 to 100%, whereas with pesticides, crop losses drop to 0 to 20%, depending on the pest species, pesticide, and crop. These differences assume cultural practices, such as monocultures, that do not necessarily minimize pest damage. The detrimental effects of some pesticides on wildlife, the environment and human health have been well documented and brought to the public's attention, first in the classic book *Silent Spring*, written in 1962 by the biologist Rachel Carson. Later, special interest groups such as Greenpeace, the Union of Concerned Scientists, and the Audubon Society focused the public's attention on these pesticide safety issues. In response to public concern over pesticide risks, a new governmental agency, the U.S. Environmental Protection Agency (EPA), came into being in the 1970s as the watchdog over the environment.

The EPA is also responsible for evaluating the use, risks, benefits, and registration of pesticides and other potential environmental hazards. Because of these events, the chemical nature of pesticides has changed dramatically since the 1940s, when DDT was the insecticide of choice for every insect and mite problem. Many such high-risk pesticides have been banned from use in agriculture in the United States, Europe, and a number of other countries. They were replaced gradually, beginning in the 1950s and 1960s, with safer organophosphate-and carbamate-based pesticides. Many of these in turn were replaced in the 1970s and 1980s with yet safer pyrethroid-based pesticides. Synthetic pyrethroids are widely used today because of their safety for people, wildlife, and the environment; however, they are broadspectrum, killing not only the target pest but most of its natural enemies as well.

Synthetic pyrethroids are being replaced today with even safer, more selective microbial toxins, new natural and synthetic pesticides and transgenic pesticide-producing crops. This shift in pesticide use from one chemical class to another is caused by the newer pesticides being safer for humans and the environment, as effective or superior at controlling the targeted pest, and safer for natural enemies of the pests. Also, safer application methods have evolved, from dusts and sprays to granular, underground, and plant-produced transgenic pesticides. According to EPA statistics, the amount of insecticides and miticides used in U.S. agriculture decreased by 50% from 1979 to 1997, from 85,000 tons down to 37,000 tons, in part because many older organophosphate and carbamate insecticides have been replaced by pyrethroid insecticides. An additional 5 to 10% reduction in conventional pesticide usage is estimated globally, for 1997 to 2000, because of genetically engineered insect resistant crops partially replacing synthetic pesticide sprays. The newest, safest, and most specific pesticides include the following:

1. Crystal proteins (Cry proteins), also called *endotoxins*, or *Bt proteins*, of the bacterium *Bacillus thuringiensis*. These toxins attack the insect's digestive tract, resulting in death.
2. Synthetic hormones that mimic the effects of molting hormones. These chemicals cause immature insects to prematurely molt and die. Synthetic hormones are very specific and affect only a few related insect species.
3. Azadirachtin is gaining acceptance globally, especially on vegetable crops. Made from extracts of seeds from the neem tree of India, the EPA registered it in 1993 for commercial use against insect pests in agriculture in the United States.

When insecticides are very specific, as are Bt proteins, the farmer must mix two or more pesticides to control the two to five pest species that occur simultaneously in a field. This approach is more expensive than using a single broad-spectrum pesticide. Although pesticides can be very effective, primarily because they are toxic and kill more than 90% of the targeted pest, crop losses still occur.

Economic estimates for the benefits of insecticide used in the United States suggest a return of $3 to $5 for every dollar invested in these chemicals. These returns go to farmers but also have helped make food less expensive for consumers. However, the farmer does not pay the societal costs of pesticide usage such as loss of fish and other wildlife, damage to water supplies, health care costs for the people suffering from pesticide poisoning, and the cost of regulating and monitoring pesticides. These costs are currently borne by society at large, and are estimated at between $4 billion and $10 billion per year in the United States. New technologies and society's desire for a safer environment have inspired new and safer pesticides, application methods, and pesticidal crops for use in agriculture. Novel pesticidal technologies commonly take ten years to develop, evaluate, and commercialize. Today's transgenic pesticidal plants and new low-risk natural and synthetic pesticides will be important tools to stop pest outbreaks and increase crop yields for the coming decade.

Pest Management

The recognition of widespread pest resistance to pesticides and the unacceptable consequences of excessive pesticide use led to the development of IPM approach. It is an ecologically based philosophy and methodology that emphasizes the need for pest management strategies to minimize pest outbreaks and minimize pesticide use. It is an approach that relies primarily on natural control factors such as pathogens, predators, parasites, and weather, and integrates these with other pest-suppressing tactics such as cultural practices and pest-resistant crops.

Broad-spectrum chemical pesticides are used, but only as a last resort to manage pest outbreaks. Moreover, pesticides are used only when absolutely justified by expected crop losses and when all other means of control (sustainable biological control strategies, including cultural control) have been exhausted. Although IPM programs have as their primary objective the maximization of economic profit on the farm, proponents of IPM see many other goals. These include protecting the environment by minimizing pesticide use, increasing regional cooperation among farmers, substituting local skill and inputs for imported inputs (such as pesticides), especially in developing countries, creating new products, and prolonging pesticide life span. For more than 50 years, pest control specialists have been developing IPM programs that carefully describe strategies and knowledge for controlling agricultural crop pests. Development of sophisticated programs, and their adoption by farmers and their crop consultants, has been greatest in the industrialized nations and least in the developing nations, where farmers have less access to practical IPM information tailored for their cropping systems.

How and Why do Pest Populations Fluctuate?

The key to successful IPM is understanding what regulates the population fluctuations of a pest in a particular region and environment and implementing strategies to suppress outbreaks. No general recipe for pest control works for every pest. In a typical IPM program, pest and natural enemy densities are monitored at specific intervals during the growing season (usually weekly). Additional data gathered in the field (such as age of plants, temperature, humidity, pest densities, densities of certain pathogens that attack the pest, and so on) and other data concerning cost of the pest control intervention, price of the crop, and expected crop yield are all evaluated (this may require analysis using computerized programs). Then, if necessary, pest management action is taken based on these data and past experiences. The real question is whether the pest, at this level of population density and in this environment, is likely to cause economic injury.

The level of pest density at which action is taken is called the *action threshold* or *economic threshold* and is below the economic injury level. The EIL is the lowest level of pest density that if unchecked would result in economic loss sufficient to justify the expense of pest control intervention. Until farmers can predict that pest injury will exceed this level, they have no economic justification for using a control strategy, such as applying a pesticide. The EIL is the central guidepost for decision making in pest control today. Understanding biological factors affecting fluctuation of pest populations is crucial

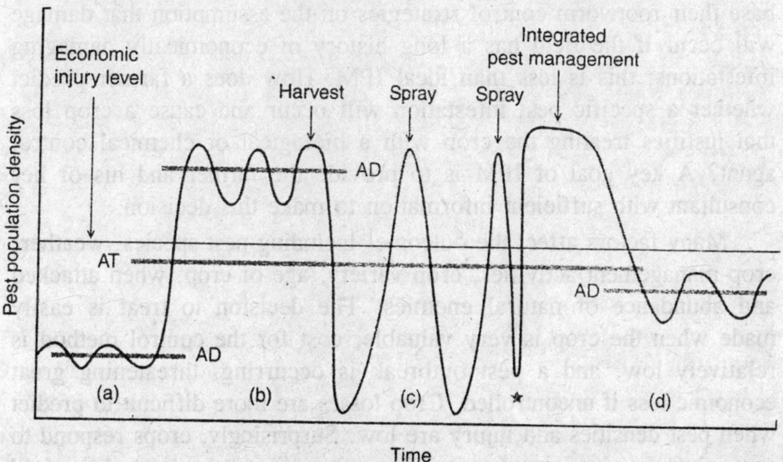

Fig. 13.2. Changing pest population density in an agricultural ecosystem.

for pest control, as shown by the work of scientists from Iowa State University who studied outbreaks of the green cloverworm in Iowa soybean fields. This pest, like many others, does not overwinter in Iowa but arrives each spring as a migrant from the southern United States. If green cloverworm arrives late in spring, its population never builds up and the soybean crop suffers no damage. If it arrives too early in the spring, a local fungus attacks it and causes the cloverworm population to collapse before it can cause harm. However, in the years when it arrives at just the right time, green cloverworm densities become high enough to cause economic damage. This example makes clear that pest control programs should be fine-tuned to take into account each particular situation and that there is no general prescription even for controlling a single pest species.

What is the Role of the Farmer in IPM?

Ideally, the farmer should determine the population density of a pest species by counting the number of individuals on crop plants from representative samples taken in the field. With flying insects, this can be done using traps baited with volatile substances normally emitted by females to attract males (known as *pheromones*). For some pest species, counting their numbers may not be feasible. For example, corn rootworm does considerable damage to maize grown in continuous culture in the upper Midwest United States. Counting rootworms in the same year as the maize is growing is too expensive, and using only the previous year's damage data is unreliable. Thus farmers must base their rootworm control strategies on the assumption that damage will occur if the field has a long history of economically damaging infestations; this is less than ideal IPM. How does a farmer predict whether a specific pest infestation will occur and cause a crop loss that justifies treating the crop with a biological or chemical control agent? A key goal of IPM is to provide the farmer and his or her consultant with sufficient information to make this decision.

Many factors affect the outcome, including pest species, weather, crop management activities, crop variety, age of crop, when attacked and abundance of natural enemies. The decision to treat is easily made when the crop is very valuable, cost for the control method is relatively low, and a pest outbreak is occurring, threatening great economic loss if uncontrolled. Crop losses are more difficult to predict when pest densities and injury are low. Surprisingly, crops respond to pest injury in a variety of ways. For example, in wet years, less yield is lost for the same density of green cloverworm in soybean than in

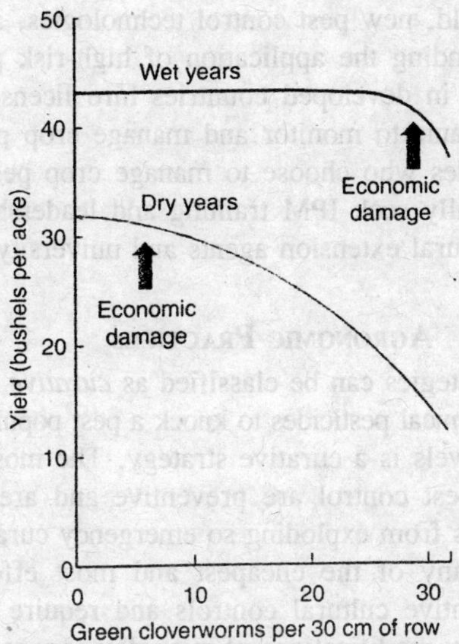

Fig. 13.3. Soybean crop response curves for defoliation by the green cloverworm.

dry years. Moreover, very light plant injury in many crops can result in no yield loss. In fact, in some crops, such as cotton, light injury early in the growing season may result in a slight increase in yield, a response known as *plant compensation to injury*.

The downside of light pest injury and compensation is that the injury may cause delayed crop maturity and harvest. The curves, called *crop response curves*, are useful to farmers in deciding when to apply a control tactic in the practice of IPM. For example, a soybean farmer can determine from these crop response curves that in a wet growing season the crop can tolerate up to 30 green cloverworms per 30 cm of row without serious economic loss, whereas in a dry year only 5 green cloverworms per 30 cm can be tolerated before control is economically justified. Many farmers may not have access to the knowledge necessary to monitor and understand the interactions of their pest-crop agro ecosystem and implement a successful IPM program. This can result in using pesticides as "insurance" against crop loss, so farmers make applications of pesticides that may not be necessary.

Global Integrated Pest Management Facility, Extension Services, World Bank, FAO, USAID, and other national and international organizations sponsor pest control educational programs to help improve and tailor IPM practices, safety, and crop productivity for all farmers worldwide. Because of the complexity of IPM programs, swiftly changing pest status in the field, new pest control technologies, and regulatory requirements surrounding the application of high-risk pesticides, the majority of farmers in developed countries hire licensed and highly trained IPM consultants to monitor and manage crop pests. Farmers in developed countries who choose to manage crop pests without an IPM consultant usually seek IPM training and leadership from local and national agricultural extension agents and university IPM training programs.

Agronomic Practices

Pest control strategies can be classified as *curative* or *preventive*. The crisis use of chemical pesticides to knock a pest population outbreak down to tolerable levels is a curative strategy. The most ecologically sound methods of pest control are preventive and are used first to keep pest populations from exploding so emergency curative strategies are not needed. Many of the cheapest and most effective control strategies are preventive cultural controls and require nothing more than smart and timely modifications of everyday agronomic practices, such as which variety and crop to grow; when to plant, irrigate, or cultivate; and whether to destroy crop residues. The farmer can select, mix, and match these agronomic practices (preventive IPM tools) to achieve the most inexpensive, effective, and sustainable pest control for each situation. *Crop rotation* is probably one of the most widely used cultural control methods in the IPM toolbox.

As noted earlier, many kinds of pests thrive only on specific crop species or groups of crop species. Populations of such pests tend to build up if the same crop is planted in the same field year after year. Rotation works well for many soil-dwelling nematode and insect pests that are present in the soil before the crop is planted. For example, growing a field of maize one year, then following it with soybean the next year is an excellent method for controlling corn rootworm in the U.S Midwest. However, for some tropical root-knot nematodes that feed on a wide range of crop species, rotation is not a practical option.

Surprisingly, almost all agronomic practices can have a positive or negative effect on which pest species and natural enemies are present, and on their population buildup. For example, use of high irrigation, high fertilization, and high plant density of cotton create an attractive and optimum food source for boll weevil, tobacco budworm, and bollworm, thus increasing potential for outbreaks and economic

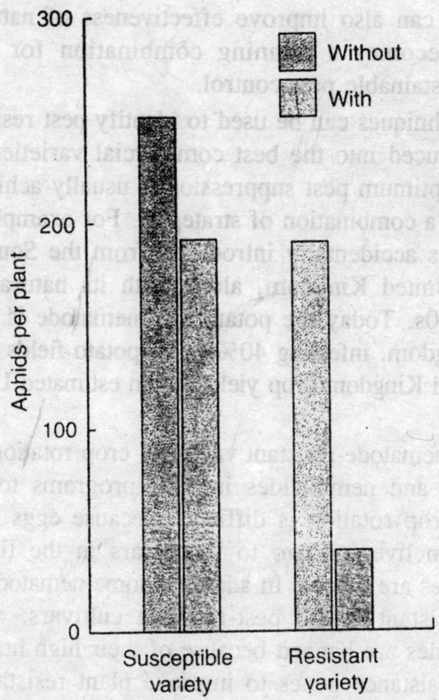

Fig. 13.4. Population size of the greenbug aphid, Schizaphis graminum, after 4 weeks on a susceptible and a resistant variety of barley, either without or with the greenbug parasite, Lysiphlebus testaceipes.

damage. Many pests emerge in the spring at a specific time, so farmers can avoid these pests by planting the crop at an appropriate interval before or after pest emergence to avoid pest buildup. For example, delayed planting of winter wheat reduces damage from Hessian fly, and delayed planting of cotton reduces boll weevil damage. In the tropics, many subsistence farmers practice *intercropping* to reduce pest damage; they grow two or more crops together in a field. Today this practice is almost nonexistent in temperate industrialized farming systems of developed nations. A review of pest abundance in intercrops found 49 of 69.pest species had reduced abundance in intercrops compared to associated mono cultures.

From an ecological view, intercropping is thought to reduce pest abundance by increasing habitat diversity, which encourages greater abundance of natural enemies and reduces pest population growth rates and crop injury. One of the most effective strategies for preventing pest outbreaks is growing *pest-resistant varieties* that sustain less pest injury. Such varieties are the foundation of an IPM program. Pest-resistant varieties can also improve effectiveness of natural enemies of pests. This becomes a winning combination for inexpensive biorational and sustainable pest control.

Molecular techniques can be used to identify pest resistance genes that can be introduced into the best commercial varieties to improve pest resistance. Optimum pest suppression is usually achieved in IPM programs through a combination of strategies. For example, the potato cyst nematode was accidentally introduced from the South American Andes into the United Kingdom, along with its natural plant host, potato, in the 1700s. Today the potato cyst nematode is a major pest in the United Kingdom, infesting 40% of all potato fields and reducing the value of United Kingdom crop yields by an estimated US$60 million per year.

Farmers use nematode-resistant varieties, crop rotation, destruction of crop residues, and nematicides in IPM programs to control this pest. However, crop rotation is difficult, because eggs of the potato cyst nematode can live for two to five years in the field and only hatch when potatoes are grown. In addition, some nematode populations are becoming resistant to the pest-resistant cultivars; and the most effective nematicides are banned because of their high human toxicity. New nematode-resistance genes to improve plant resistance and new safer nematicides, or other effective control strategies, are needed globally for nematode control.

Plants Defending

If plants are such good food sources, and if pests are so abundant and voracious, why don't you see pests devouring every green thing? Over millions of years, plants have evolved effective physical barriers and a mind-boggling array of chemical defenses to protect themselves against herbivorous animals. Plants produce more than 50,000 specialty chemical compounds, called *secondary metabolites* or *secondary plant chemicals*, that repel, kill, or reduce pest growth and reproductive potential. However, volatile plant chemicals can also function as attractants, and insects have evolved olfactory mechanisms to identify these chemicals. This is how insects find and are attracted to their food source, the crop plant. Some plant chemicals even attract insects that are predators or parasites of plant pests.

When an insect predator "smells" a particular plant chemical produced during pest feeding, the predator expects to find on that plant its own food—the pest insect. Herbivores have developed mechanisms for coping with these chemical defenses, resulting in a never-ending game of plants producing new defensive compounds and herbivores succeeding in overcoming the plant's defenses. This is known as *coevolution*. Pest adaptation to plant-produced toxins is similar to the adaptation of pests to synthetic pesticides.

A major difference between the plant's use of natural pesticides and human's use of synthetic pesticides is that people commonly use a single toxin to which the pest can quickly adapt after a few years of exposure, whereas plants produce many toxic chemicals that together make pest adaptation much slower, taking thousands of years. Plants and herbivores have become specialists. Anyone plant species synthesizes a limited number of defensive chemicals (may be 50 to 100 compounds), and each pest species has adapted over time to feed on this narrow range of plant toxins. Thus, each herbivore species can feed on only a limited number of plant species. Some pests have become so specialized that they can only survive and reproduce on the few plant species that produce the toxins to which they are adapted.

Plant defensive compounds may be present in special glands, located on the epidermis of the plant, on waxy surface layers, and deep within leaves, roots, or stems. In most plants, chemical defenses are present in all tissues but they are especially concentrated on the surface to deter pest damage as early as possible. Plants use their chemical defenses in two ways: (1) like an ever-present minefield, called *constitutive defense*; and (2) *inducible defense*, where the defense chemicals are produced only when the plant is under attack and being wounded by the pest—why waste precious energy building all plant defenses when they may not be needed? Many secondary plant metabolites are volatile, and pests use them as cues to help locate and identify the crop species they are adapted to feed on. Understanding pests and their interaction with secondary plant metabolites offers great potential to improve future pest control for most crops. Altering secondary plant chemistry of a specific crop not only can improve pest resistance but can cause a crop to become repellent or unattractive to the pest, or cause it to be more attractive to natural enemies of pests.

Breeding Pest-resistant Crop

Plant breeders can improve pest-resistant crops by finding individual plants in wild populations that show the desired resistance trait and by transferring the genes for desired expression of the trait to conventional crop varieties. Archeologists think humans have been improving crops for thousands of years by selecting plant types that resist pest injury. Only very recently has it become possible to identify the specific genes responsible for a pest-resistant trait. Two early success stories in the development of pest resistance are the Hessian fly on wheat, and the grape phylloxera on European grapes. The grape phylloxera insect, which feeds on the roots of grape vines, was accidentally introduced into European vineyards. In the United States, its native country, the phylloxera had co-evolved with native grapes as an unimportant insect herbivore. However, European grapes had no defenses against the insect and were extremely vulnerable. By 1884, most of France's wine industry was failing because of phylloxera outbreaks attacking and killing the roots of the vines. A program to develop pest-resistant European grapes was begun. The most phylloxera-resistant root stocks from grapes native to the United States were selected and sent to Europe where European grape varieties were grafted on to them. Within a few years, this practice saved the French wine industry by eliminating outbreaks of grape phylloxera.

Today, well over 100 insect-, mite- and nematode-resistant crop varieties have been bred in the United States. Possibly twice that many pest-resistant varieties, in all major crops, are grown worldwide. Over half of these resistant crop varieties are for pests of the major grain food crops: rice, maize, sorghum, and wheat. Pest-resistant varieties are developed by teams of scientists that include entomologists (insect and mite specialists), nematologists, plant pathologists, geneticists, agronomists, plant breeders, and recently biotechnologists. In developing countries, international agricultural research centers have been the primary developers of pest-resistant crops.

Using pest-resistant crops in IPM is advantageous because they (1) provide season-long pest control; (2) are usually less expensive than pesticides; (3) require no expensive, labour-intensive applications and equipment; (4) produce higher dollar returns to the growers than pesticides; (5) are compatible with biological control and pesticides; (6) are environmentally safe and usually harmless to people; and (7) provide cumulative benefits over time in that the longer and more widely a pest-resistant variety is grown, the more effective the pest suppression.

Plant Defenses

In many cases, pests evolve to overcome the defense mechanisms introduced by plant breeding. When only a single trait for pest resistance is present in the crop variety, the pest may quickly adapt to the pest-resistant trait, such as a higher level of a secondary plant metabolite, or a genetically engineered protein toxin. The source of the chemical resistance trait is not the cause of pest adaptation. Pests are challenged to adapt to any toxin they are frequently exposed to; the toxin acts as a selective agent driving the pest population to become more resistant so that the pest population survives. A limited number of pest species (fewer than 100) have developed resistance to natural plant toxins in pest-resistant crops, far fewer than have developed resistance to conventional pesticides (over 500 species). For example, after a period of 5 to 10 years of growing greenbug-resistant wheat varieties, strains of greenbug began to adapt and flourish on the resistant wheat varieties. Resistant pest populations, such as resistant greenbugs, are called *biotypes* and commonly look physically different from susceptible individuals of the same species. Resistant biotypes have developed in 14 insect species attacking a dozen crops, from alfalfa and maize, to rice and vegetables. In some cases, such as Hessian fly, as many as eight separate biotypes have developed consecutively over time that overcame the consecutive releases of eight new resistant wheat varieties. As new biotypes of a pest develop to overcome a pest-resistant crop variety, plant-resistance programs have to find new genetic sources of pest-resistance and breed them into current crop varieties to produce a new crop variety that is resistant to the new pest biotype. The development of resistant biotypes again makes the point that pests can adapt to any toxin, regardless of its source.

This stepwise human-managed "coevolution" of crop protection chemicals, whether natural or synthetic, whether applied as topical sprays or genetically produced in the resistant plant, will continue to challenge IPM. Many scientists believe that certain approaches can be used to slow the time it takes for a pest to adapt to the chemicals in pesticides and pest resistant varieties. On average, pest adaptation may occur 10 to 20 years after a new pest-resistant variety is grown commercially. However, many pest-resistant crop varieties have been grown for more than 20 years without resistant pest biotypes developing. Scientists cannot accurately predict if or when resistant pest biotypes will develop in a specific pest-crop agroecosystem. The diversity in crop genes needed to repeatedly combat development of new pest biotypes is another reason for maintaining extensive crop gene banks and natural biodiversity; they are essential to maintain sustainable crop productivity with pest-resistant varieties.

Pest-resistant varieties work by reducing target pest densities and damage compared to susceptible varieties. Even pest-resistant crops are not immune to damage. They do not kill or repel all pest individuals from the crop; their goal is to keep pests below economically damaging numbers. Studies on insect pest reductions in 10 major crops using pest-resistant varieties showed pest densities averaged 12 times lower on resistant compared to susceptible crop varieties. Growing pest-resistant varieties also may aid in controlling the spread of plant diseases transmitted by pests. For example, a nematode infestation of a crop encourages the incidence and severity of fungal diseases such as Fusarium, and verticillium wilt. Pest-resistant varieties also slow down pest population development, giving farmers more time to implement other IPM tactics and giving natural enemies and diseases of the pests more time to overtake pest populations and further suppress their density and damage.

Plant Resistance to Pests

If growing crops that are resistant to pests is so widespread and effective, why do farmers still use pesticides? The answer is that pest-resistant crops do not reduce all pests to levels that farmers can tolerate, whereas properly applied pesticides do. Also, well-adapted resistant varieties are not available for all pests on all crops. Conventional pest-resistant varieties have been estimated to reduce pesticide use by more than 30% a year for all pesticides used in the United States on maize, barley, grain sorghum, and alfalfa; however, there are still unacceptable crop losses if conventional pesticides are not used. Beginning in the late 1980s, molecular biologists found new pest-resistance genes that could be used to achieve a higher level of target pest control, greater than 80%, than achieved with conventional pest-resistant varieties. This is equal to or greater than the level of control achieved with conventional pesticides. The first genes available for genetic engineering of crops for pest resistance were the Cry genes of *Bacillus thuringiensis*, also called *Bt genes*. When this bacterium forms spores, each spore contains a large protein crystal that has a mixture of proteins encoded by different genes. Each type of protein has a slightly different toxicity for insects and other invertebrates.

The first crops with Cry protein pest-resistant varieties—cotton, maize, and potato—were released in the mid-1990s. Growers found these transformed insect-resistant varieties effective in controlling some of their worst pests, and possessing all the advantages of traditional pest-resistant crops that were discussed earlier. Farmers worldwide have rapidly adopted these transgenic varieties. In 2001, commercial varieties of insect pest-resistant *Bt* maize, *Bt* cotton and *Bt* potato were planted on 12 million acres in the United States, Argentina, Canada, China, Australia, South Africa, Mexico, Spain, France, Portugal, Romania, and the Ukraine. Farmers worldwide have used these insect-resistant varieties to control insect pests that have become resistant to conventional insecticides, such as budworms and bollworms (many species worldwide), or insects such as stalk-boring worms in maize that are difficult to kill with conventional pesticides because they feed inside plant tissues where they cannot be reached by insecticides. *Bt* crop technology also has value for farmers in developing countries. In 1999, 1.5 million small farmers in China adopted pest-resistant *Bt* cotton varieties to grow on their farms. The compelling reasons were the effective control of the insecticide-resistant bollworm (*Heliothis armigera*), the insecticide savings and the dramatically increased yields, income, and personal safety. In 1992 bollworm outbreaks, insecticide-resistance, insecticide applications and crop losses cost China approximately US$1.2 billion. The cotton industry was reduced 40% and China began to import cotton. For China and its subsistence farmers, *Bt* cotton is an incredibly beneficial and valuable IPM technology. The Chinese farmers who grew *Bt* cotton in 1999 saw yield increases averaging 25% over conventional cotton, with less than one insecticide spray on the *Bt* cotton, compared to 14 on conventional cotton. In fact, about 70% of Chinese *Bt* cotton farmers applied no insecticide sprays on their *Bt* cotton in 1999. They also observed a 25% increase in natural enemies of cotton pests on *Bt* cotton. In the United States, *Bt* cotton also has been shown to reduce insecticide use by about 2.2 sprays while increasing yields about 10% and raising farm profits about $100-$200 per hectare, compared to conventional cotton.

Benefits of *Bt* Crops

For any pest-resistant crop variety, the pest control and economic benefits to farmers vary from field to field, year to year, and region to region. These benefits can result from (1) reduced conventional insecticide costs, (2) increased yields and crop value (an undamaged crop has greater value), or (3) both. Other, less economically definable benefits to the farmer are (1) better natural biological control of target and nontarget pests; (2) reduced environmental pollution; (3) safer farm environment to live and work in; (4) less risk of crop loss from all pests; (5) less risk of lawsuits from accidental poisonings; (6) complementation of other IPM strategies by *Bt* crops for a more sustainable farming operation; (7) better relationships with neighbours, especially home/city/school owners and administrators; and (8) greater peace of mind. One of the most effective ways to reduce conventional pesticide use on crops and allow natural enemies to become more abundant and effective at controlling pesticide-induced pests is to develop pest-resistant varieties targeting the key pests and those resistant to conventional pesticides. Current transgenic *Bt* crops do just that.

Many species of caterpillar pests damage maize, some by boring into the stalks and ears, where they cannot be satisfactorily controlled with current pesticides or pest-resistant varieties. Yield losses to such pests reduce pre harvest yields by more than 10% worldwide. Caterpillar tunneling and feeding also increase incidence of bacterial and fungal infection, thus increasing the incidence of some microbial toxins (mycotoxins) in stalks and grain. This of course reduces grain quality and value. One of these mycotoxins, aflatoxin, produced by the common fungus *Aspergillus flavus* in maize kernels pre-and postharvest, is extremely toxic to humans and livestock. Reducing crop pest injury to crops such as maize and rice by using pest-resistant *Bt* varieties or other types of pest resistance has the potential to significantly reduce mycotoxins in food.

The first Cry genes used were particularly effective against certain lepidopteran insects (that is, caterpillar stage of some moth species). Since then, Cry proteins that are toxic to beetles, mosquitoes, and nematodes have been discovered. This opens up the possibility that genetic engineering with Cry genes will target many other plant pests in the future.

New Genes for Insect-Resistant Transgenic Crops

Possibly the next major breakthrough in commercially available engineered pest resistance will be maize varieties that are resistant to damage from beetle larvae known as corn *rootworms*. Farmers control soil insects and nematodes with the insecticide/nematicide terbufos, and in the United States alone some 6,000 tons are applied annually to the maize crop. Replacing this insecticide/nematicide with a safer, more effective transgenic pest-resistant crop would be desirable. The source of new pest-resistant genes is likely to be the bacterium *Bacillus cereus* that produces insecticidal proteins, known as *Vip proteins*, during its vegetative growth stage. These proteins make up another new class of pesticidal chemicals, the Vip class, for "vegetative insecticidal protein." The initial field trials with transgenic Vip maize for control of corn rootworms look promising.

Another approach to slowing down insect growth is to use genes that encode protease and amylase inhibitors, defensive compounds that are abundantly found in seeds. Insects use amylases and pro teases to digest the starch and protein in their food. Plants have evolved amylase and protease inhibitors that inhibit the digestive amylases and pro teases of insects. Often such inhibitors have considerable specificity, inhibiting the digestive enzymes of some insect species but not of others. A specific amylase inhibitor protein found in the common kidney bean completely inhibits growth of the pea weevil and the cowpea weevil. When researchers introduced this bean gene into peas and the protein was expressed in peas at the same level as is normally found in beans, the development of pea weevil larvae was completely inhibited. Similarly, genetic engineers are inserting genes for protease inhibitors to make plants resistant to specific insect pests.

Is the level of pest control always the same with pesticidal crops? Something unexpected occurred during the first years when *Bt* cotton was farmed on a commercial scale—the level of target pest control was not as consistent as expected. The level of target pest control appeared to change (1) under different environmental conditions, (2) during different stages of plant growth, and (3) with different *Bt* varieties. For example, control of bollworm is greater for preblooming *Bt* cotton than for blooming. Exactly how these factors influenced the level of target pest control is unclear; however, they are thought to apply to all natural and transgenic pest-resistant crops produced now and in the future. Stated simply, the level of pest control produced by a plant trait can be influenced by any genetic or environmental factor that affects primary or secondary metabolism of the crop. Thus genetic background of a variety, weather, soil fertility, moisture stress or insect injury, to name a few, can all influence effectiveness of plant-produced pesticides, or any other gene product of the crop plant.

TRANSGENIC PESTICIDAL CROPS

We cannot stop some pests from developing resistance to conventional pesticides, biopesticides, or new transgenic pesticidal crops. Evolution of pests that are resistant to a particular toxin is a gradual process, requiring on average six or more years. Scientists already know that genes for resistance to Cry proteins exist in wild populations of some target pests of the new *Bt* crops, and researchers can therefore predict with some confidence that resistance will develop. In reality, all pesticides, whether natural or produced by humans, are valuable limited resources, and their loss results in serious social and economic costs. Methods to delay or avoid development of pests that are resistant to control tactics are necessary if agriculture is to become more sustainable. The likelihood of developing resistant populations of a pest is very difficult to predict for a specific pest and toxin; however, based on more than 500 real-world cases, scientists can make some generalizations. Resistance is likely to develop rapidly if all the following occur at the same time:

1. The pesticide is very effective.
2. The pesticide is widely used.
3. The pest does not travel over long distances (many miles) and specializes on a few plant species.
4. The pesticide is applied frequently for a long time.
5. The pest population harbors many resistance genes, even if these occur in rare individuals.

When the Environmental Protection Agency first registered *Bt* cotton, *Bt* maize, and *Bt* potatoes in the United States for commercial production in the mid-1990s, the developer of these products, Monsanto Company, filed plans that it hoped would delay the development of pest resistance to the specific Cry protein insecticides in Monsanto transgenic *Bt* crop varieties. The Environmental Protection Agency has made such resistance management plans a requirement for registering transgenic insecticidal crops. The question of how to manage transgenic *Bt* pesticidal crops to minimize selection pressure on target pests is a perplexing and controversial question for which there is not a single best answer.

How does insecticide resistance develop in insect pests? When a specific population of insects is exposed to an insecticide, all insects die except those that are partially or fully resistant. Contrary to popular belief, pests don't suddenly mutate when exposed to a toxin, and develop an entirely new toxin-resisting trait that was never present before. Rather, the trait for resistance to a specific insecticide is already present in at least one individual in the treated population. Repeated treatments of the same population of a pest with the same pesticide, and increased matings of partially resistant individuals with other partially resistant individuals, produce a larger and larger number of highly resistant individuals in a field population. The methods outlined in resistance management plans for transgenic plants try to avoid matings of partially resistant insects with other partially resistant pests because part of their offspring would be much more resistant to the specific toxin than would the parents. These strategies are most successful if pests that are genetically heterozygous for pesticide resistance are susceptible or only partially resistant, so they are easily killed by the dose of the pesticide used.

Scientists have suggested several approaches to reduce the selective pressure from a pesticide that causes the pest population to adapt to it over time. Reducing selective pressure delays resistance. The approaches are:

1. In mixtures, two or more toxins (pesticides) are combined in the same field by mixing varieties that produce different toxins or combined (referred to as "pyramided") in the same plant.
2. In *rotations*, different toxins are used in rotation, one year one toxin-producing variety and the next year another toxin-producing variety.
3. *Mosaics* are special patchworks of toxin-producing varieties where varieties with different toxins are planted in adjacent fields.
4. *Refuges* are variety plantings with no pesticidal plants that are expected to be colonized by and produce susceptible pests that mate with any rare partially or fully resistant pest individuals produced on a pesticidal crop. The susceptible pests are expected to maintain overall population susceptibility by these matings with partially or fully resistant pests.
5. *High-dosage* is the use of pesticidal crops that produce a dose of the pesticide in their tissues that is several-fold higher than needed to kill all susceptible and partially resistant pests in the population.

These approaches can be combined. For example, refuges can be used with mixtures and planted in mosaics. Mixtures and rotations of conventional broad-spectrum spray pesticides have been used with limited success to delay development of pesticide resistance in some insect and mite pests.

Plants in nature do not rely on a single genetic trait (*monogenic resistance*) to resist a pest but rather on many resistance traits (*polygenic resistance*), so scientists would be wise to take a tip from what works in nature and develop polygenic resistance in crop varieties with at least two or more pest-resistant traits pyramided together in each variety, and planted in mosaics, mixtures, or rotations with a refuge. Pyramided traits are expected to be most effective when combined with a fully implemented integrated pest management program that uses all available pest suppression methods in a truly holistic crop management approach. In the United States, the current resistance management programs for *Bt* crops use many of these strategies.

Predators, Parasites, and Pathogens

Why is biological control so valuable to successful long-term pest management? The answer is that natural biological control of pests is the most ecological, sustainable, effective, inexpensive, and long-term pest control method available to farmers. In many cases, the farmer need do nothing to make biological control work on the farm except minimize use of conventional broad-spectrum pesticides. However, biological control commonly does not provide a high enough level of control for every pest in a farmer's field to avoid economic damage. Frequently, the pests most poorly controlled by natural biological control are introduced foreign pests and migratory pests, as discussed earlier. Biological control agents move into a farmer's field from three sources: (1) Native parasites, predators, and diseases naturally occur in the geographic region where the farmer grows the crop, known as *natural biological control* or *natural enemies*. (2) Introduced predators, parasites, or diseases are collected from foreign lands, commonly the native home of an introduced pest, and then released and established in a new geographic crop production region where these biological control agents never existed before, known as *classical biological control*. After release they survive by feeding on the introduced pest. Eventually, populations of the introduced pest and introduced natural enemy reach a new population balance at a lower density and hopefully the introduced pest no longer causes outbreaks and serious economic losses. The aim is to permanently lower the population density of the introduced foreign pest and thus eliminate or reduce the need for conventional broad-spectrum pesticides. Such a lowering can often be achieved with perennial crops, such as orchard fruits, that are part of a stable agroecosystem, or in other stable natural ecosystems, such as forests and rangelands. However, decreasing pest density is very difficult to achieve with high-value annual row crops because they do not constitute a stable ecological system. (3) Additional biological control tactics used are *augmentation* of naturally occurring biological control by the periodic release of natural enemies that have been reared in the laboratory, or the *conservation* and encouragement of natural enemies already in the field. These tactics may be more appropriate for annual row crops.

The 1997 guide to beneficial insects, mites, and nematodes of North America lists approximately 106 insect, 17 mite, and 6 nematode species that are cultured and available commercially for augmentative biological control of crop pests. However, their use in commercial agriculture has been limited because of their lower effectiveness and high cost relative to conventional broad-spectrum pesticides.

Invertebrate predators, such as lady-beetles (also known as ladybugs or ladybird beetles), capture and quickly kill and eat the pests. On the other hand, parasites are insects and nematodes that act like a "fatal disease" in that they deposit a number of their eggs or immature offspring, in or on a single pest, and the immature parasites slowly feed and grow over a long period, days to weeks, on the living pest, eventually killing it and becoming a free-living adult. Parasitic insects are commonly small wasps (Hymenoptera) that feed on nectar as adults and are some of the most abundant and effective of the natural enemies of crop pests.

A dramatic recent example of classical biological control comes from sub-Saharan Africa where a "new" pest, the cassava mealy bug, began attacking cassava, the staple for 200 million Africans. In the early 1980s, the mealy bug caused enormous crop losses. Researchers at the IITA research center in Nigeria, searching for enemies of the mealy bug in its South American homeland, found a small parasitic wasp that kills cassava mealy bugs by laying its eggs in the mealy bug's body. Periodic distribution of the wasp in the 28 African countries of the "cassava belt" is now the responsibility of the center. The dollar value of the program has been calculated at $150 returned to the farmers and society for each $1 invested, making it a very successful pest control program.

Microorganisms that cause diseases of invertebrate pests can also be used as biopesticides. These microbial control agents include pathogenic bacteria, fungi, and viruses, and are applied much like sprayable synthetic pesticides. These disease agents are commonly cultured in the laboratory of pharmaceutical companies using the same fermentation equipment that is used for culturing the bacteria and fungi used to produce medical antibiotics. Sprayable *Bt* and *Saccharopolyspora spinosa* are produced in this way, and are the most widely used biopesticides. The oldest microbiological agent in use are the spores of *Bacillus thuringiensis* (*Bt*). It has been sprayed commercially since the 1950s in the United States for control of certain species of lepidopteran caterpillars (butterflies and moths). Different formulations of *Bt,* with different strains producing different Cry proteins, are widely used in vegetable crops in Hawaii, Florida, and California, and especially by home gardeners and organic farmers everywhere. Overuse of *Bt* sprays has led to emergence of pests such as the diamondbacked moth, which can resist certain Cry toxins. Sales and usage of traditional biopesticides such as *Bt* and baculovirus account for less than 2% of the total global insecticide/miticide/nematicide market. The reason for their low usage is mainly their relatively low effectiveness at controlling the target pest, narrow spectrum of pest species controlled, and higher expense compared with conventional broad-spectrum pesticides.

Entomopathogenic fungi (fungi that cause diseases of insects) illustrate the potential for using other microorganisms in biological control. At least 750 different species of such fungi have been identified, and attempts to use them as biological control agents of insects, mites, and nematodes began in the late 19th century. Variable test results have plagued the commercial development and use of such biopesticides.

Nematodes and viruses that attack pests also have considerable potential as biological control agents, but much more research is needed to produce cost-effective, usable formulations that can be marketed to farmers. Because biological control is species specific rather than general, the commercial market for anyone pest is much smaller than for multiple pests. Such small markets offer little incentive to industry for developing highly specialized control agents. For some pests, such as cutworms and root weevils, nematodes are the most widely used biological control agents because they can be easily mass-reared, applied with spray equipment, and provide rapid and satisfactory control.

INDEX